T0140860

Beiträge zur organischen Synthese

Band 100

Beiträge zur organischen Synthese

Band 100

Herausgegeben von
Prof. Dr. Stefan Bräse

Karlsruher Institut für Technologie (KIT)
Institut für Organische Chemie
Fritz-Haber-Weg 6, D-76131 Karlsruhe

Institut für Biologische und Chemische Systeme – Funktionelle Molekulare Systeme
Hermann-von-Helmholtz-Platz 1
D-76344 Eggenstein-Leopoldshafen

Céline Leonhardt

Novel Organic Emitters for Thermally Activated Delayed Fluorescence

Logos Verlag Berlin

λογος

Bibliographic information published by the Deutsche Nationalbibliothek

The Deutsche Nationalbibliothek lists this publication in the Deutsche Nationalbibliografie; detailed bibliographic data are available in the Internet at http://dnb.d-nb.de
.

ISBN 978-3-8325-5661-7
ISSN 1862-5681

Logos Verlag Berlin GmbH
Georg-Knorr-Str. 4, Geb. 10, 12681 Berlin

Tel.: +49 (0)30 / 42 85 10 90
Fax: +49 (0)30 / 42 85 10 92

https://www.logos-verlag.de

It’s all about where your mind’s at.

– *Kelly Slater*

German Title of the Thesis

Neue organische Emitter mit thermisch aktivierter verzögerter Fluoreszenz

Table of Content

Kurzzusammenfassung

Seit Adachi *et al.* im Jahre 2012 das Potenzial von thermisch aktivierter verzögerter Fluoreszenz für organische Leuchtdioden endeckten, eröffneten sich neue Möglichkeiten für effiziente organische Leuchtdioden (OLEDs).[1] TADF-Emitter für OLEDs könnten phosphoreszente Emitter der zweiten OLED Generation, die zwar effizient sind, aber auf umweltschädigenden Materialien basieren, vom Markt ablösen. OLEDs die auf TADF basieren sind nicht nur effizient, sondern können zudem mit niedrigem Kostenaufwand hergestellt werden. Eines der wichtigsten Vorteile ist die Nutzung rein organischer Moleküle. Dadurch können die Moleküle mit dem Baukasten der organischen Synthese synthetisiert und so modifiziert werden, dass die bestmöglichen optoelektronischen Eigenschaften erzielt werden. Dabei muss darauf geachtet werden, dass ein starker Charge-Transfer-Effekt im Molekül entsteht. Am besten gelingt dies, wenn Elektronen schiebende Gruppen, Donoren, mit Elektronen ziehenden Gruppen, Akzeptoren, verbunden werden. Der Charge-Transfer-Effekt und somit das Ausmaß an TADF-Eigenschaft kann durch die Stärke der entsprechenden Donoren und/oder Akzeptoren, sowie den Dieder Winkel zwischen Donor und Akzeptor, gesteuert werden. Im ersten Projekt wurden zwei neue Akzeptorgruppen für blaue TADF-Emitter, das Mono- und Di[1,2,4]-triazolo[1,3,5]-triazin (MTT & DTT) entwickelt. Es wurden verschiedene Donoren eingebaut und der Diederwinkel durch verschiedene funktionelle Gruppen eingestellt. Die Emitter wurden anschließend strukturell und photophysikalisch untersucht. Zwei Emitter mit guter TADF-Eigenschaft stachen heraus. Das **DMAC-DTT** emittierte mit einer Wellenlänge von 488 nm (3wt% mCP) und einer Quantenausbeute von 31% und das **DMAC-MTT** mit einer Wellenlänge von 530 nm (10wt% mCP) und 62% Quantenausbeute. Das zweite Projekt baute auf der CzBN Serie von Zhang *et al.* auf.[2] Der Nitril-Akzeptor wurde zu Tetrazol und Oxadiazol Akzeptoren umgesetzt. Durch verschiedene Derivate wurde die Toleranz gegenüber funktionellen Gruppen und sterischen Einflüssen untersucht. Außerdem wurde herausgefunden, dass die Installation einer zusätzlichen Phenylgruppe zwischen Donor und Akzeptor und damit der Einfluss des Dieder Winkles, auschlaggebend ist, ob die Moleküle TADF zeigten. Die Tetrazol Derivate zeigten kein bis wenig TADF, wohingegen die Oxadiazol Derivate TADF zeigten. Emissionswellenlängen von 445 bis 513 nm, abhängig vom Substitutionsmuster, wurden erzielt. Zum Schluss wurde das eingeführte Design auf ein D-A-A-D Gerüst angewendet. Allerdings stellte sich heraus, dass diese Moleküle kein TADF zeigten.

Abstract

Thermally activated delayed fluorescence (TADF) has drawn immense attention since Adachi and co-workers exploited its phenomenon in 2012.[1] Organic light emitting diodes (OLEDs) have become one of the most promising techniques to replace efficient, though environmentally concerning, phosphorescent dyes. Simultaneously, TADF OLEDs ensure a high efficiency and low-cost production. One of the main advantages is that purely organic molecules can be exploited. This offers a flexible fine-tuning of the optoelectronic properties by simply applying the synthetic organic chemistry toolbox. Here, a few design principles need to be considered. The chemical structure must allow a sufficient charge transfer. To make this feasible, electron-donating groups (donors) are connected to electron-accepting groups (acceptors), so the charge transfer can be controlled by the strength of each group and by the dihedral angle between the donor and acceptor. Two different design approaches have been engaged in this thesis. In the first project two novel acceptor cores were designed to develop two new scaffolds for blue TADF emitters. The mono- and di[1,2,4]-triazolo[1,3,5]-triazine (MTT & DTT) acceptors were decorated with **DMAC**, **DPAC,** and **PXZ** donors and, for some, further enhanced by tuning the dihedral angle with methyl or trifluoromethyl groups to study them structurally and photophysically. The aim was the realization of efficient TADF with tunable emission colors. The **DMAC-DTT** and **DMAC-MTT** were the most efficient ones, showing a blue emission of 488 nm (3wt% mCP) with 31% PLQY and 530 nm (10wt% mCP) with 62% PLQY, respectively, in the MTT and DTT series.

In the second project the acceptor of the known CzBN series by Zhang was modified.[2] The nitrile acceptor was transformed into a tetrazole or oxadiazole and derivatized to study the tolerance of steric challenges and functional groups. Further, it was learned that extending the phenylene spacer between the donor and acceptor shuts off the TADF pathway. The tetrazole derivatives performed poorly, respecting TADF, whereas the oxadiazole derivatives showed good TADF. The emission colors covered a range of 445 to 513 nm depending on the substitution pattern on the acceptor. Lastly, the developed structure was adopted to design a D-A-A-D frame with oxadiazole as the acceptor. However, these structural conformers shut off the TADF light-emitting pathway no matter what donor-spacer-bridge combinations were tried.

1 Introduction

The energy used for lighting devices, such as lamps, smartphones, televisions, and billboards, accounts for approximately 15% of global electrical power consumption or 5% of anthropogenic greenhouse gas emissions.[3] What is more, the need for these devices is expected to increase, e.g. the number of global smartphone users of 3.6 billion in 2021 is estimated to be 4.5 billion in 2024.[4] To meet the growing demand for lighting sources, new energy solutions must be found that are both more efficient and consider the associated environmental impacts.
This work aims to contribute to the search for new efficient and sustainable emitters for OLEDs. First, the historical background that gave today's state-of-the-art technologies will be outlined.

Thomas Alva Edison invented the incandescent light bulb in 1879. Incandescent describes light that is conditioned by the rise of temperature. He demonstrated a new way of generating light by exploiting electricity. This invention was a milestone and the starting point for the lighting technologies that are known today. A light bulb generates light by electrically heating a wire filament until it glows. However, this method is inefficient since only 5% of the used energy is utilized to turn it into visible light. 95% is lost in the form of heat.[5-6]
A major success in enhancing the efficiency of lighting technologies was the introduction of light-emitting diodes (LEDs). As opposed to incandescent light bulbs, LEDs are composed of inorganic crystalline semiconductors to which a current is applied, resulting in the generation of light via electroluminescence (the recombination of electrons and holes). The luminous efficiency reaches 150 lm/W and is almost thirteen times more efficient than a light bulb (12 lm/W).[7] Next to its enhanced efficiency, another advantage is that they produce white light, which is favorable because it is close to natural light. Despite

their high efficiency, they show disadvantages, too. The inorganic semiconductors are based on expensive and harmful materials like Gallium Nitride (GaN), which causes high production costs and severe environmental issues.[8] LEDs show a high intensity of short-wave emission with high-energy blue light, which harms eyesight. Moreover, they lack light quality and have issues with sensitivity to high voltage.[9] Considering the advantages and disadvantages, the LED is still superior to its predecessors. Nevertheless, finding better technologies that address the disadvantages of light bulbs and LEDs is urging.

The invention of the organic light-emitting diode (OLED) might has become the solution. OLEDs can be built purely organic in highly efficient devices with lower production costs and less environmental impact. The key phenomenon, which is responsible for its light generation, is electroluminescence. Luminescent organic semiconductors, processed in an ultra-thin film, convert electric power into light. The detailed composition and working principle of an OLED will be described later in Chapter 1.2.

LEDs and OLEDs are often applied in lighting panels, liquid crystal displays (LCD) and OLED displays. The superior features of an OLED display are bright and high contrast colors, a "true black"[10], fast response time, and the design of flat, flexible, and even transparent displays.[11] In an OLED display each pixel creates light. A change in the electric field causes an alteration of the excitation and hence the emitted color. For the "color" black, the pixel is shut off and does not consume power, thus the term "true black". In an LCD with LEDs, the light is generated with a backlight, and the colors are created through a filter.[12] Because of that backlight, the color black in an LCD always shows some glow. In other words, a "true black" can never be achieved in an LCD.[13] Furthermore, the backlight is why an LCD display cannot be foldable or transparent, unlike an OLED display.

1.1. Photoluminescence

The Discovery of Fluorescence and Phosphorescence

Eilhard Wiedemann first introduced the term luminescence in 1888 for all phenomena of light that are not incandescence.[14] Today luminescence is defined as a spontaneous emission of radiation from an electronically excited species.[15] The different types are classified by the way an emission process occurs.

Photoluminescence is the emission that arises from the direct photoexcitation of these emitting species. Well-studied photoluminescent processes are fluorescence as well as delayed fluorescence and phosphorescence.[15]

As we know today, fluorescence was first applied in 1565 by Nicolás Monardes as a method to counterfeit objects. It was discovered from an infusion (emitting a blue color) of wood from Mexico, which the Aztecs initially used to treat kidney and urinary diseases.[16-19] In 1852, George Gabriel Stokes identified an important phenomenon, which is known today as the Stokes Law or Stokes Shift. The wavelengths of dispersed light are always longer than those of the absorbed original light.[20] His terminology of dispersive reflection was later formed into the term fluorescence.[21]

Phosphorescence, a just as important photoluminescent process as fluorescence, was described ten years earlier by Edmond Becquerel. He reported the emission of light of calcium sulfide when exposed to sun light. Some considered that he was the one who discovered the phenomenon of the later called Stokes Shift. However, as it is known now, Becquerel described phosphorescence.[22] Later on, a milestone in exploring photoluminescence was reached by Becquerel in 1858, who was the first to measure the decay time of the phosphorescence of multiple compounds with his phosphoroscope. These experiments were the first time-resolved photoluminescence measurements.[23]

Fluorescence

Fluorescence is a three-stage process, containing stage 1: Excitation via absorption; stage 2: Excited-state lifetime; and stage 3: Fluorescence emission (Scheme 1). A photon from an external source (laser, lamp, etc.) creates an excited electronic singlet state S_1 or higher singlet states S_n. These excited singlet states only exist for a short time (~1-10 ns) until the electron drops by *de*-excitation back to the singlet ground state S_0. This means that an electron reaches an energetically higher state with excitation from the highest occupied molecular orbital (HOMO) into the lowest unoccupied molecular orbital (LUMO) whilst simultaneously an excited state is created.

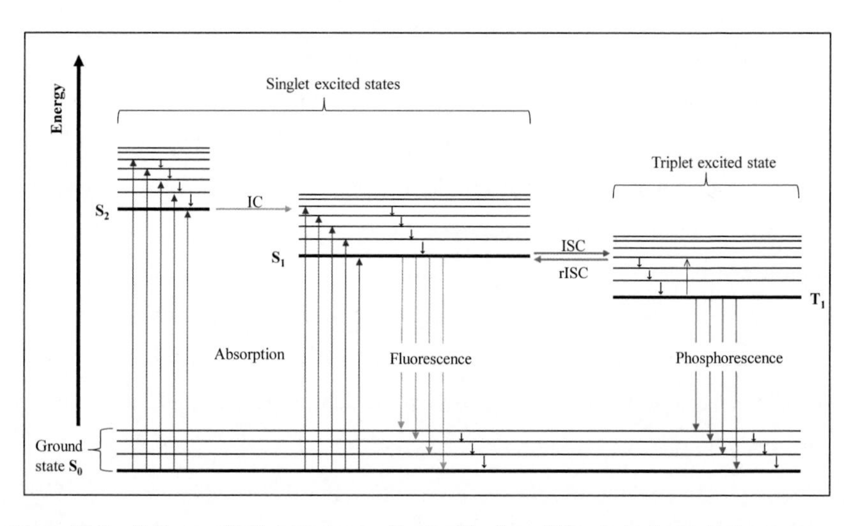

Scheme 1 Jablonski diagram with IC - internal conversion (transition from a higher electronic state to a lower one with the same multiplicity) and vibrational relaxation (small black arrows). The figure is recreated following the literature.[24-25]

During these excited state lifetimes, the molecule still interacts with its environment. Consequently, the energy of S_1 is dissipated, and not all molecules return to the ground state via the fluorescence channel (S_1 to S_0). Collisional quenching,

Fluorescence Resonance Energy Transfer (FRET), and intersystem crossing (ISC) result in the depopulation of S_1. With excitation, higher vibronic singlet states are populated. The spontaneous fluorescence decay pathway occurs from the lowest-lying singlet state (S_1), which means that the higher populated vibronic states must lose energy to end up in S_1.
This energy dissipation leads to the lower energy of the emitted photon and, therefore a longer wavelength than the absorbed light (Stokes Shift).[26]
The luminescence only lasts as long as the molecule gets excited externally. When the excitation source is switched off, the luminescence stops within 10^{-8} s.[27-28]

Phosphorescence

Phosphorescence is a radiative transition between two states of different electron spin multiplicities, following the *de*-excitation pathway from the lowest-lying triplet state T_1 to S_0 (Scheme 1). The initial excitation process is the same as in fluorescence. However, not only excited singlet states are populated, but also excited triplet states T_n. T_1 lay energetically lower than S_1, which makes it more favorable. Via intersystem crossing, the electron can be moved from S_1 to T_1. ISC is associated with a spin reversal, also referring to spin forbidden. This forbidden transition, however, becomes possible depending on the molecule's architecture, e.g., when it contains heavy atoms with a high spin-orbit coupling (SOC).[29] The excited electron, now in T_1, can decay to the singlet ground state S_0, accompanied by the emission of light. This process, as was the case for ISC, is spin forbidden. Still, there is a slight possibility, which is why phosphorescence occurs as a rather weak emission (because the electron spin must be reversed again). Furthermore, this causes a slow emission release (because the energy is trapped for a while), making phosphorescence an afterglow, long-lasting process in milliseconds up to hours, depending on the circumstances.[30]

Delayed Fluorescence

Harvesting triplet excitons can create a delayed fluorescence via reverse intersystem crossing (rISC), e.g., through thermal activation (Scheme 1). One of these delayed fluorescence mechanisms is thermally activated delayed fluorescence (TADF). The fundamental processes of TADF will be discussed and illustrated in detail in chapter 1.2.4. Another mechanism is triplet-triplet-annihilation (TTA), which will also be described in the following sections.

While TADF has been known for over 90 years, it has only been exploited for technological applications for ten years. In 1929, Perrin proposed that a so-called "dark state", by which he meant a triplet excited state (as it was suggested to be named in the mid-1940s [31]), can be activated thermally to a singlet state, resulting in a delayed fluorescent decay.[32]

In 1961, Parker and Hatchard discovered the delayed emission, namely E-type delayed fluorescence, in the red-emitting organic molecule eosin.[33] Wilkinson and Horrocks introduced the, now used, terminology TADF back in 1968. In 1996, Berberan-Santos and Garcia showed the first useful application for TADF. They discovered that the quantum yield of a C_{70} fullerene, which showed only weak fluorescence, can be enhanced by TADF.[34]

The recent success of E-type delayed fluorescence is probably owed to Adachi and coworkers, who, in 2009, introduced Sn^{4+} porphyrin complexes as TADF emitters into the world of organic light-emitting diodes (OLEDs).[35] Three years later, in 2012, they installed purely organic TADF emitters in OLED devices.[1, 36-37] In 2013, the progress of purely organic emitters was continued by Monkman and coworkers. They reported a series of TADF emitters based on the dibenzothiophene-S-S-dioxide acceptor and decorated it with different donors, such as carbazole derivatives. Further, they elucidated the general TADF mechanism in their emitters.[38] The potential of organic TADF emitters was discovered, and the research has skyrocketed from 40 journal articles and reviews until 2009 to more than 3200 published until today (according to a SciFinder^n search on 22/12/2022, keyword "Thermally activated delayed fluorescence").

1.2. Electroluminescence – Organic Light-Emitting Diodes (OLEDs)

Bernanose *et al.* observed the electroluminescence of organic materials in 1953 by applying a high alternating voltage to a thin film of cellophane coated with acridine derivatives (orange emission). In 1963, Pope *et al.* investigated luminescence in single anthracene crystals placed between asymmetric silver paste electrodes.[39] Helfrich and Schneider, in 1965, also worked on single anthracene crystals and explored electroluminescence emerging from electron-hole recombination. They set up a device built between a hole-injecting electrode and an electron-injecting electrode. A blue light emission appeared when the two electrodes were contacted with negative and positive anthracene ions at 50 V.[40] A milestone was set when Tang and VanSlyke built the first efficient OLED device in 1987 and paved the way for today's OLED research and production. They constructed a double-layered organic thin film incorporated between two electrodes. One layer consisted of 8-hydroxyquinoline aluminum (**Alq$_3$**) as the emitting material, which was processed micro-crystalline to enhance efficient electron transport. The structure of **Alq$_3$** is shown in Figure 1. The other layer was composed of aromatic diamine, responsible for the holes' transport. This two-layer construction with separate electron and hole transport layers made it possible for electron-hole recombination, which occurs in the middle of the organic layer. The result is an operating voltage reduction and improved device efficiency.[41] With this first OLED device, a green fluorescent emission, an external quantum efficiency (EQE) of 1%, a luminous efficiency of 1.5 lm/W, and a brightness of >1000 cd/m^2 at a driving voltage below 10 V was achieved.[42]

1.2.1. Working Principle and Architecture

To explain the principles of an OLED device, a simplified architecture consisting of a single-layer OLED placed between two electrodes is shown in Scheme 2. The work function of the electrodes matching the energy levels of the HOMO and the LUMO of the emissive molecule is important. This means that the anode has to level the HOMO energy, and the cathode has to level the LUMO energy to implement a charge carrier injection from the electrodes into the organic emissive material.[43] In Scheme 2, the four steps to generate light are illustrated: 1) An external voltage is applied to inject electrons from the cathode to the LUMO and simultaneously extract holes from the anode to the HOMO. 2) The charges are transported toward the electrode with opposite polarity due to the external electric field. 3) If electrons and holes overcome the Coulomb barrier and come close to each other, they recombine under the formation of excitons. An exciton can be described as an originating excited state in an organic molecule. 4) The exciton, lying in the molecule's excited state, decays radiatively, letting the OLED device generate light.[44]

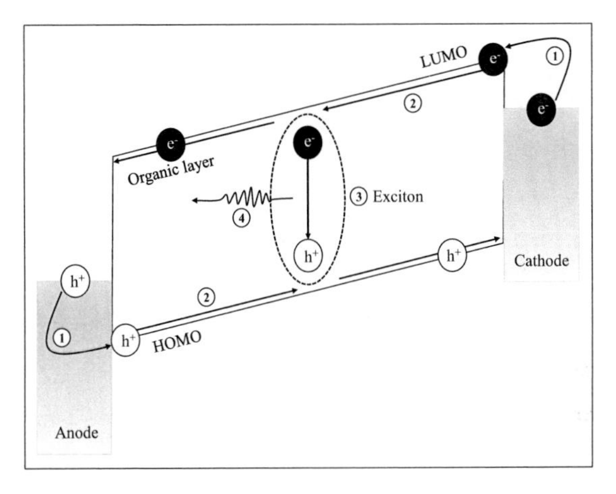

Scheme 2 Simplified single-layer OLED stack. Steps - 1) Charge carrier injection- electron at the cathode, the hole at the anode. 2) Charge carrier transport. 3) Recombination of electrons and holes. 4) Radiative decay of the exciton. The figure is recreated following the literature.[32]

Practically applied, an OLED architecture is more complex than previously described. Such a multi-layered device is shown in Scheme 3. Typically, the anode consists of thin indium tin oxide (ITO) deposited on a glass substrate, so the generated light can escape the device. The hole transport layer (HTL) is layered on the anode. On top is the emission layer (EML). Usually, the EML is not just one layer carrying the organic emitting material but multiple different ones consisting of a guest/host system. Next is the electron transport layer (ETL), finished off with a metallic cathode. An encapsulation around all layers is necessary to avoid contact with oxygen and humidity.
In practice, the inlet of an OLED is much more complex, and every layer can be adjusted to maximize efficiency. Often the HTL and ETL act simultaneously as a blocking layer for the opposite charge carriers, and parts of the EML are doped not to be conductant to hinder exciton quenching pathways.[45]

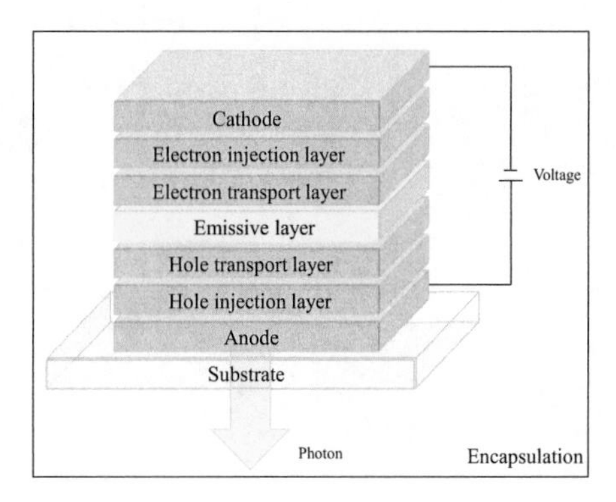

Scheme 3 Multilayer stack architecture of an OLED device. The figure is recreated following the literature.[32]

Overall, OLEDs must achieve three goals: A high internal quantum efficiency (1), running at a low operating voltage (2) while having a high light-outcoupling efficiency (3). One and two are of great importance for the design and synthesis work since they are related to the chemistry of the materials.[46]

Over the last years, four generations of OLEDs have been developed, distinguished by their method of emitting light. The first generation takes fluorescent emitter materials, the second generation uses phosphorescent emitters, the third generation exploits TADF emitters, and the fourth generation is based on hyperfluorescence. During the following chapters, the first, second, and third generations will be discussed and compared with each other. The fourth generation will be picked up and discussed in the outlook section.

1.2.2. First Generation – Fluorescence

The distribution of singlet and triplet excitons in an electrically excited molecule is unequal. It depends on the spin statistics and is composed of the three projections of the triplet state and one of the singlet state, together accounting for the total spin. Therefore, the distribution of excitons populating singlet excited and triplet excited states is 25% and 75%, respectively.[47] Fluorescent dyes are composed of organic molecules with only weak SOC, leaving the 75% triplets unprocessed (Scheme 4). An OLED with a fluorescent emissive layer can only use 25% of the generated excitons. The other 75% dissipate their energy through non-radiative decay. The result is a theoretical internal quantum efficiency (IQE) of 25%. However, the light outcoupling efficiency of a typical OLED of ~20% must be considered leaving a fluorescent OLED with a maximum EQE of 5%.[32] The spin limitation is often accidentally exceeded by triplet fusion, a decay channel part of TTA.[48] With that, an IQE of 62.5% can be achieved.[49] The prominent emitter **Alq$_3$** for this class was introduced in section 1.2.

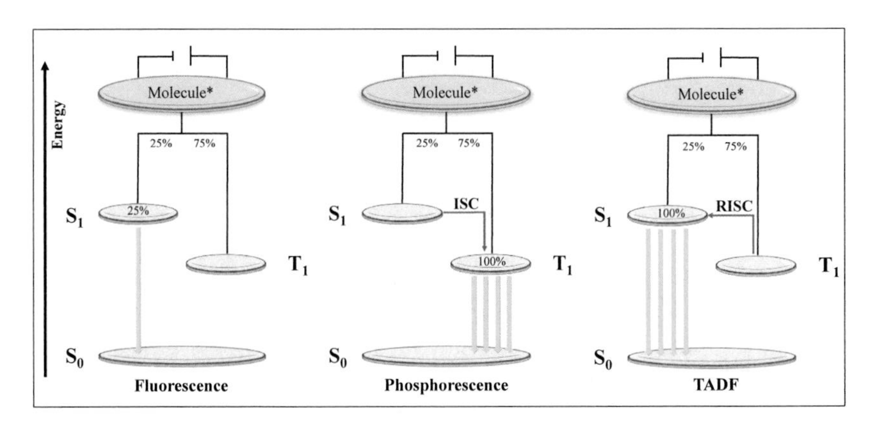

Scheme 4 Fluorescence, phosphorescence and TADF pathways after excitation of a molecule. The figure is recreated following the literature.[1]

1.2.3. Second Generation – Phosphorescence

An important success in OLED research was the discovery of methods to overcome the limitation inflicted by charge recombination spin statistics.
One method to battle these limitations is the use of phosphorescent dyes. Phosphorescent OLEDs can achieve a theoretical IQE of 100% because they can use the triplet excitons and further harvest singlet excitons via ISC (Scheme 4).[50-52] Heavy metal complexes having a strong SOC are used to loosen the spin prohibition, enhancing ISC and the transition from T_1 to S_0.[53] Indeed, these materials have shown extraordinary OLED performances with EQEs over 30%.[54-55] Iridium metal complexes with organic ligands are mostly used, such as **Irppy$_3$** (Figure 1).[56-57] But other metals like Platinum[58-59], Ruthenium[60] and Osmium[61-62] have also been established.[63]

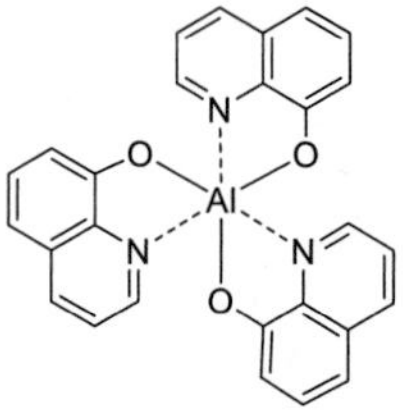

1) **Alq$_3$** 2) **Irppy$_3$**

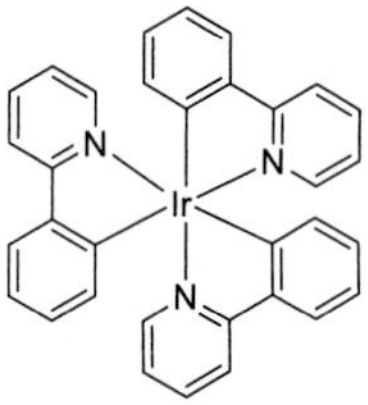

Figure 1 Fluorescent dye **Alq$_3$** (left) and **Irppy$_3$**, an Iridium-based phosphorescent dye (right).

Nowadays, phosphorescent dyes are widely used in the display industry. They are state-of-the-art technology for red and green OLEDs, although metal-organic triplets face difficulties regarding their relatively long triplet lifetime. The limitation factor is the operation at a high voltage because, at high currents in combination with a long triplet lifetime, the triplet excitons have a high probability of undergoing TTA, which quenches the emission.[64] More challenges arise for blue phosphorescent OLEDs with their inability to emit good quality blue light.

Not only do these limitations spread the urge to look for other materials, but also that most phosphorescent materials bring economic and environmental issues simply due to their heavy metal composition. These metals are mostly toxic, scarce, and expensive.[65-66] Further, their mining and processing severely impact the environment.[67]

1.2.4. Third Generation – Thermally Activated Delayed Fluorescence

TADF materials promise a noble metal-free, purely organic alternative to phosphors with their ability to theoretically harvest all generated excitons in an OLED device (100% IQE). This is especially good news for blue OLED materials.
The singlet excitons are harvested through the fluorescent channel S_1 to S_0 (prompt fluorescent component, range of ~10 ns). The triplet excitons are lifted with a thermal activation to higher energetic vibrational triplet levels and then converted to singlet excitons via rISC (Scheme 4). From here, the light emission also occurs through the fluorescent channel S_1 to S_0, however, delayed to the prompt fluorescence (delayed fluorescent component, ns to ms[68]).
A competitive process to TADF is TTA. When two excited triplet states from neighboring molecules encounter each other, called TTA, they form one singlet excited state S_1 and one ground state S_0.[69] This ideally leaves one-half of the non-emissive triplet excitons to be used for singlet emission. The excited triplet states must have at least half of the excitation energy of the singlet state. The newly generated exciton in S_1 can then decay, accompanied by light emission in the form of delayed fluorescence. In chapter 1.2.2., the described upper limit for TTA accounts for 62.5% (= 0.25 + 0.5 x 0.75).[49] The above describes one of two pathways to how TTA operates. The other pathway results in the deactivation of one of the two triplet excitons and not the generation of a new excited singlet state followed by delayed fluorescence. Here, with the energy of the two triplets, one of the excitons is uplifted to a higher-lying triplet state. What follows is the relax-

ation of this higher triplet exciton via IC back to T_1 without enhancing the potential radiation. In fact, this loss channel is a challenge for OLEDs operating at high currents and is especially critical for TADF OLEDs because it reduces the number of excited triplet states.[70-74]

Crucial parameters for efficient TADF are the rate of reverse intersystem crossing, k_{rISC}, and the energy gap between S_1 and T_1, ΔE_{ST}. The importance of k_{rISC} will be discussed later in chapter 1.2.4.1.

ΔE_{ST} must be minimized so the excitons can overcome this barrier through thermal vibration aided by ambient temperature (ΔE_{ST} < 0.37 eV).[75] The energy difference between a singlet and triplet state depends on the exchange interaction of the unpaired electrons. This requirement can be met by charge-transfer states (CT) with far-distanced unpaired electrons. A separation of the frontier orbitals, or more specifically, a separation of HOMO and LUMO, is doing exactly that, reducing ΔE_{ST}.[76]

Separated frontier orbitals, however, work contradictory with the fact that the TADF emission must have high luminescent efficiency with a small transition dipole moment. To achieve high luminescent efficiency for a strong emission, an overlap between HOMO and LUMO is required.[77]

To fulfill all mentioned requirements, a TADF emitter is usually constructed by electron-donating groups, the donor, and electron-accepting groups, the acceptor. Such a donor-acceptor system leads to CT states with separated HOMO and LUMO. A functional bridging system is often required to connect the donor and acceptor to create a small overlap of HOMO and LUMO, hence a high luminescent efficiency.

So far, a rather simplified mechanism of TADF has been explained. The complete underlying principles, after all, are not fully understood yet. Monkman and coworkers have suggested a more complex mechanism[32], as illustrated in Scheme 5.

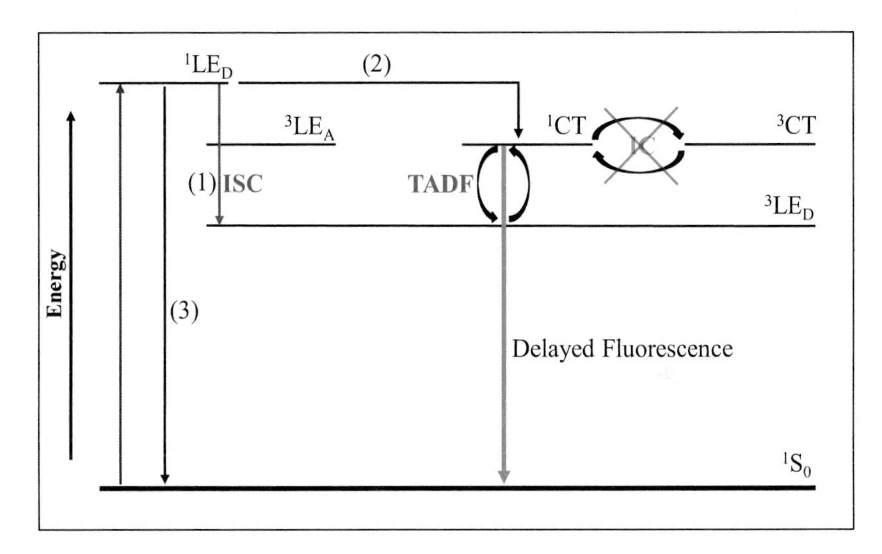

Scheme 5 Proposed mechanism for TADF by Monkman. $^1S_0$ – singlet ground state, ^1LE_D – Locally excited singlet state of donor, ^3LE_A – Locally excited triplet state of acceptor, ^1CT – singlet charge transfer state, ^3CT – triplet charge transfer state, ^3LE_D – Locally excited triplet state. The figure is recreated following the literature.[38]

Monkman described the TADF mechanism in an OLED as follows.[32, 78] The excitation of a donor-acceptor molecule (D-A) creates a locally excited singlet state of the donor, ^1LE_D. From here three pathways are possible: First, (1) the transition of ^1LE_D to ^3LE_D via ISC, where the singlet donor excited state becomes a triplet donor excited state. The second pathway (2) is the charge transfer process, CT. Here, the electron gets pushed from the donor to the acceptor to form the ^1CT. The third possibility (3) is the decay from ^1LE_D back to the ground state $^1S_0$, which is non-radiative.

A competition is created between the three ways. Which pathways occur depends on the fastest rate of the mechanism. It is found that the rates for (1) and (2) are fairly similar in many molecules. Their rates must be higher than the non-radiative pathway (3) to avoid or minimize this loss mechanism.

For TADF, the ^1CT transition is the most important process, meaning as little as possible of the transition from ^1LE_D to ^3LE_D is wanted. It was believed that the ^1CT

and ^3CT states are isoenergetic, or molecules are designed so they would be isoenergetic, so a cycle is created (red marked black cycle in Scheme 5). Any state that found itself in the ^3CT could, by a thermally activated process, go back to the ^1CT (through internal conversion, IC, and reverse internal conversion, rIC) and vice versa. That would give a good emission of light, which is outcoupled of the OLED device. Nonetheless, this process is forbidden, which makes the previous assumption incorrect. The actual process is slightly different.
The molecules that are designed for TADF are fashioned in such a way that the donor is twisted or perpendicular to the acceptor. This leads to a strong coupling between the ^1CT and the ^3LE_D. This creates a cycle between ^1CT and ^3LE_D (black cycle in Scheme 5). Here, thermal activation is still required to successfully harvest triplet states from the triplet donor states and the CT singlet states by an electron spin flip to the local donor ^3LE_D. This is the crucial TADF step. The emission occurs from the ^1CT state via delayed fluorescence to the $^1S_0$ ground state.

1.2.4.1. Theory of TADF

For a long time, solely the importance of ΔE_{ST} was considered. It was found that k_{rISC} plays an equally important role in efficient TADF. To understand the role of k_{rISC}, a closer look into its theory is necessary. The delayed fluorescence mechanism occurs when the rate of prompt fluorescence k_F and nonradiative decay k_{NR} is much lower than the rate of ISC k_{ISC} (1).

$$k_F + k_{NR} \ll k_{ISC} \tag{1}$$

If this correlation is fulfilled, the singlet excitons in S_1 are transferred to the triplet excited state T_1. Again, if the rate of phosphorescence and the rate of nonradiative decay from the triplet state is slow, and the energy gap ΔE_{ST} is small enough, rISC occurs after vibrational thermalization followed by emissive decay of S_1 (delayed fluorescence). If the radiative and non-radiative processes are very small, the ISC and rISC processes can occur multiple times, cycling between singlet and triplet state before the delayed fluorescence occurs.[79] While cycling, an equilibrium is formed because the molecule remains in the excited state, before eventually emitting. This equilibrium model assumes:

$$k_F \ll k_{rISC} \tag{2}$$

With the BOLTZMANN statistics, the steady-state population of the excited states can be described (3). The equilibrium constant K expresses the relative population of singlets and triplets. The pre-factor A includes the spin-orbit coupling constant H_{SO} and a factor of 1/3 resulting from spin statistics (one singlet state and three triplet sub-states).

$$K = \frac{[S_1]}{[T_1]} = \frac{k_{rISC}}{k_{ISC}} = A\exp(\frac{-\Delta E_{ST}}{k_B T}) \tag{3}$$

According to equation (3), a high rate of rISC is achieved by either reducing ΔE_{ST} and/or increasing A.

ΔE_{ST} is given in equation (4), which shows a dependency of the exchange energy integral *J*.

$$\Delta E_{ST} = E_{S1} - E_{T1} = 2J \tag{4}$$

Thus, to minimize ΔE_{ST}, the exchange energy integral *J* must also be minimized. *J* is defined in equation (5). Φ and Ψ represent the HOMO and LUMO wavefunctions, respectively, and e is the electron charge. The equation shows that a decrease in the HOMO-LUMO overlap results in a minimization of *J*.[78] As previously discussed, this can be realized by molecular design with D-A structure-like TADFs.

$$J = \iint \phi_{HOMO}(r_1)\, \psi_{LUMO}(r_2) \left(\frac{e^2}{r_1 - r_2}\right) \phi_{HOMO}(r_1)\, \psi_{LUMO}\, dr_1\, dr_2 \qquad (5)$$

Coming back to equation (3). SOC, included in A, is a relativistic effect acting on the angular momentum and the spin. Its defining characteristic is the mixing of orbital and spin degrees of freedom, which allows electronic states of different multiplicities to couple. Further, it's the determining factor for turning prohibited transitions, like ISC and phosphorescence, into permitted processes.[80] The spin-orbit coupling constant, H_{SO}, is proportional to the fourth power of the atomic number, meaning that purely organic TADF molecules consisting only of light elements like Hydrogen, Carbon, Nitrogen and Oxygen are expected to have an inefficient spin-orbit coupling (e.g., $H_{SO(C)} = 32\ cm^{-1}$). In contrast, Platinum or Iridium has a high spin-orbit coupling with $H_{SO(Pt)} = 4481\ cm^{-1}$ and $H_{SO(Ir)} = 3909\ cm^{-1}$, which plays an important role for phosphorescent OLEDs, as explained in Chapter 1.2.3.[81] Since H_{SO} is so small for organic molecules, it was considered negligible. Therefore, the design and synthesis of organic TADF emitters were mainly focused on the minimization of ΔE_{ST} (typically below 0.1 eV).[1] Nevertheless, the assumption that only takes ΔE_{ST} into account is, in some cases, a convenient approach, e.g., analyzing photophysical data. But the key assumption $k_F \ll k_{rISC}$ is not met here.

Many factors contribute to an efficient rate of reverse intersystem crossing. Dias *et al.* reported that ΔE_{ST} is not the only determining factor for k_{rISC}, but H_{SO} has a substantial role, too.[38] If only the transitions of ^1CT and ^3CT are considered, as it was initially, H_{SO} becomes nearly zero.[82] That is because the SOC operator carries the spin magnetic quantum number of the electron and its spatial angular momentum quantum number.[83] A transition of T_1 to S_1 leads to a spin change.

Consequentially to conserve the total angular momentum (product of a spin magnetic quantum number and spatial angular momentum quantum number), the orbital angular momentum must change as compensation.[84-86] Proposed was the utilization of ^{3}LE to strengthen H_{SO}.[83, 86-87] Spin-orbit coupling between ^{1}CT and ^{3}LE is allowed. It was reported that the coupling between ^{1}CT and ^{3}LE enhances H_{SO} and the minimization between these two states plays the determining factor. All three excited states, ^{3}LE, ^{1}CT, and ^{3}CT, have an important role in efficient k_{rISC}. It was found that vibronic coupling between the different states can greatly enhance k_{rISC}. Monkman and coworkers showed that spin-orbit coupling in a donor-acceptor charge transfer molecule between singlet and triplet states is mediated by one of the local triplet states ^{3}LE of the donor or acceptor. They described it as a second-order vibronically coupled mechanism from the ^{3}CT to ^{3}LE to allow the spin-orbit coupling to the ^{1}CT state. Their model predicted that the energy gap between the ^{3}LE and CT states is a critical activation barrier next to ΔE_{ST}, which controls ISC and rISC, and the TADF efficiency.[87] As shown now, more than one energy gap needs to be considered, namely ΔE_{ST} and ^3LE_D to ^{1}CT and ^{1}CT to ^{3}CT. The gaps between the CT states and the LE states are very sensitive to the surrounding environment, so the emitter performance can be easily adjusted by altering the molecules' rigidity or the polarity of the host material.[87]

1.3. Molecular Design – Three Approaches

As previously discussed, a donor-acceptor system to spatially separate the HOMO and LUMO, while still allowing a small overlap often due to a bridging system, is necessary for a small ΔE_{ST}, stable T_1, a sufficient luminescent S_1 with the desired emission color and TADF efficiency. The challenge is to combine suitable donors and acceptors with a fitting connection bridge to balance all these properties. In Figure 2, typical donors and acceptors are listed. Often used donors are carbazoles, acridine derivatives like 9,9-diphenyl-9,10-dihydroacridine (**DPAC**), 9,9-dimethyl-9,10-dihydroacridine (**DMAC**) or 10*H*-phenoxazine (**PXZ**), and diphenyl amines. Benzonitriles, triazines, benzophenones, sulfones, tetrazole, and oxadiazole are commonly used as acceptors.

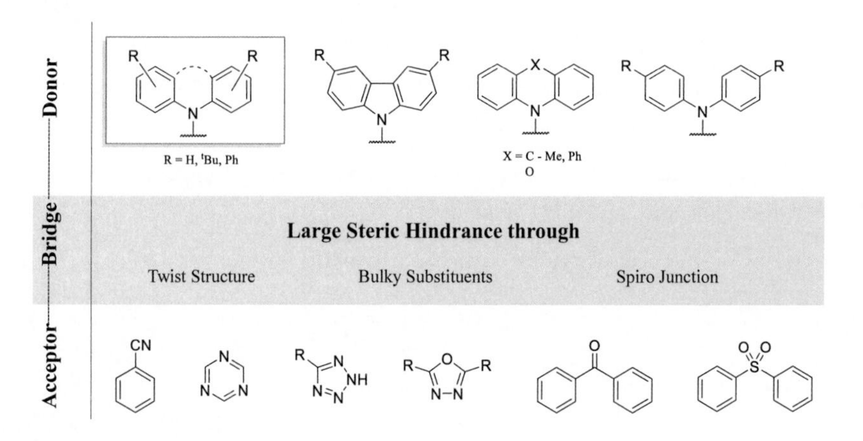

Figure 2 Commonly used donors and acceptors, like carbazole, acridine derivatives, diphenyl amines, benzonitriles, triazines, benzophenones, and more.

The overall approach within the TADF literature is the intramolecular charge transfer effect to design a TADF emitter efficiently. This approach can be divided into the twist-induced charge transfer, the through-space charge transfer, and the multi-resonance effect. The focus is on the first approach; however, the other two principles will also be briefly introduced.

1.3.1. Twist-Induced Charge Transfer

The twist-induced charge transfer approach is probably the most applied in the literature. A donor is connected with an acceptor, D-A or D-A-D, through a bridge that enables a large steric hindrance. This is necessary to sufficiently spatially separate the HOMO, located on the donor, and the LUMO, located on the acceptor. Fine-tuning here enables a slight overlap between HOMO and LUMO, which is important, as discussed in the previous chapters. A large steric can be induced with a spiro junction or bulky substituents leading to a twist.[88] The important parameter here is the dihedral angle between the donor and acceptor. This angle can be adjusted as needed. Among others[89-91], Adachi and coworkers[1] showed the impact of the dihedral angle with their well-known carbazolyl dicyanobenzene series (CDCBs). They synthesized a series of molecules built on a dicyanobenzene acceptor with a varied count and arrangement of carbazole donors (Figure 3).[1] These molecules are distorted so the HOMO, lying on the carbazoles, and the LUMO, located on the benzonitrile(s), are separated. The emitters achieved high TADF efficiency and a wide range of emission colors.

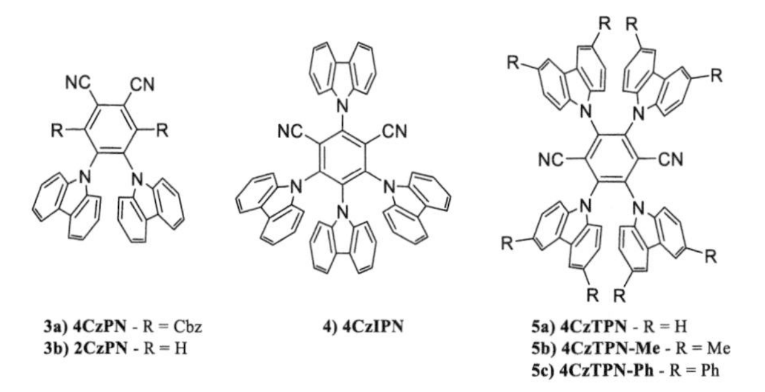

Figure 3 Carbazolyl dicyanobenzene series (CDCB) developed by Adachi and coworkers.

The emission color depends on the donor-acceptor design. Methyl and phenyl groups at the 3- and 6- positions on the carbazoles led to a bathochromic shift of the emission. More donor strength resulted in a shift to longer wavelengths, as opposed to fewer carbazoles like in **2CzPN**, which resulted in a weaker donor ability and hence a shift of the emission to shorter wavelengths. The 4CzIPN was later picked up by Sun *et al.* and found to be the most efficient one in this series. It achieved an EQE of 30% with a green emission color.[92]

A similar system with up to five carbazole donors and one nitrile acceptor was reported.[2] Their series contained emitters with four and five carbazole donors with and without *tert*-butyl groups attached to benzonitrile (Figure 4). All compounds exhibited a blue emission in toluene. The introduction of *tert*-butyl groups resulted in a slight redshift of the emission. Furthermore, a slightly reduced ΔE_{ST} was obtained in the compounds with *tert*-butyl groups with 0.30, 0.24, 0.22, and 0.17 eV for **4CzBN, 4TCzBN, 5CzBN** and **5TCzBN**, respectively.

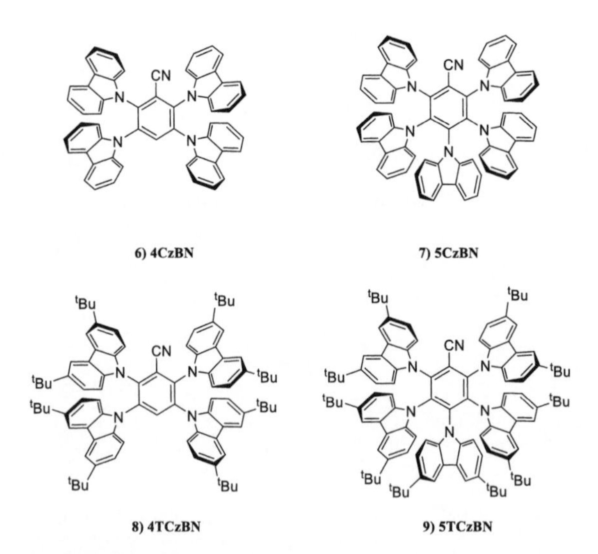

Figure 4 CzBN series, decorated benzonitriles with carbazole derivatives, by Zhang *et al.*

1.3.2. Through-Space Induced Charge Transfer

In this approach, the electronic communication between the donor and acceptor is mediated through space.

Donor and acceptor are disposed in a pseudo facial orientation and linked via a direct σ–bond or through a twisted π–conjugation. The HOMO-LUMO or donor-acceptor separation is created through a space homojunction incorporated in a rigid framework. Therefore, the charge transfer operates through space via aromatic π-bonds or an sp^3-hybridized carbon center.[93] In Figure 5, a selection of through-space TADF emitters is illustrated. Adachi and Liao reported an sp^3-hybrid carbon-centered donor-σ-acceptor TADF emitter, **QAFCN**. Their OLED device achieved an EQE of 18% with a sky-blue emission.[94]

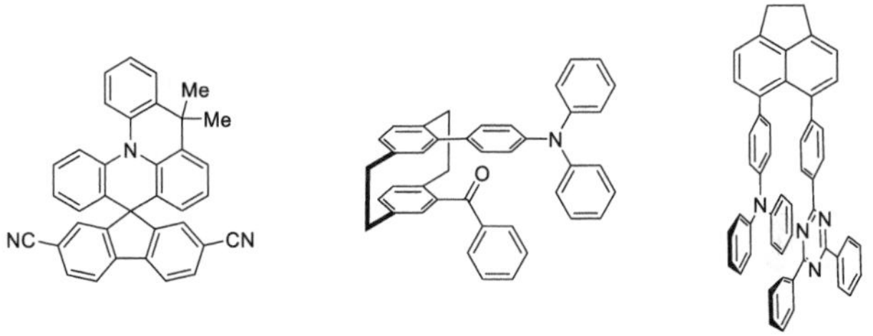

10) QAFCN **11) *cis*-Bz-PCP-TPA** **12) TPA-ace-TRZ**

Figure 5 Through-space charge transfer TADF emitters in the literature.

Spuling *et al.* introduced the first TADF emitter based on the [2.2]paracyclophane scaffold with diphenyl-amine as the donor and benzophenone as the acceptor, ***cis*-Bz-PCP-TPA** (to name only one of the emitters in their report).[95] They achieved a blue luminescence and a short delayed lifetime; however, their emitters lacked photoluminescence quantum yield.

It is often difficult to show that the through-space charge transfer is the dominant contributor to the communication between the donor and acceptor. However, in

2021, Monkman and Zysman-Colman confirmed the specific mechanism of TADF in **TPA-ace-TRZ**, shown in Figure 5.[96]

1.3.3. Multi-Resonance Effect

The multi-resonance effect is known to be the most recent approach in the TADF literature. Hatakeyama reported this new design strategy in 2016, resolving the emission spectra broadening problem.[97] Here, the HOMO-LUMO separation is induced by the opposite resonance effect of nitrogen and boron atoms in a rigid aromatic, polycyclic scaffold (Figure 6).[98] **DABNA-1**, shown in Figure 6, is a well-known multi-resonance TADF emitter with an OLED performance of 13.5% EQE and an emission wavelength of 459 nm.[97]
In 2021, Bin and You reported the orange-red multi-resonance emitter, **CNCz-BNCz** (**14**), with an outstanding EQE of 33.7% in an OLED device (Figure 6).[99] It was long thought that for these emitters the design strategy is limited to the peripheral constellation of the polycyclic N/B/N framework. Furthermore, adjustments on this skeleton could jeopardize the color purity and performance of the emitters. These claims deeply restricted the diversity of multi-resonance emitters.

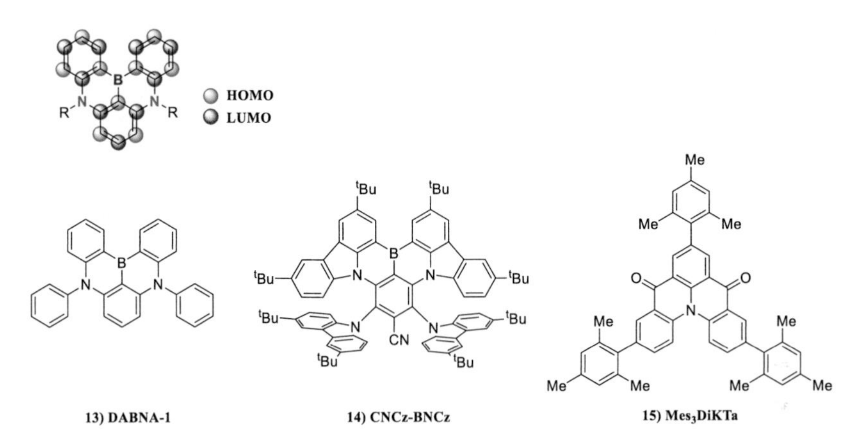

Figure 6 HOMO-LUMO distribution (top left corner). TADF emitter based on the multi-resonance effect. **DABNA-1** (left), **CNCz-BNCz** (center), **Mes3DiKTa** (right). The HOMO-LUMO distribution figure is recreated following the literature.[98]

Despite these challenges, Zysman-Colman published in 2020 a novel multi-resonance TADF emitter, **Mes3DiKTa**, illustrated in Figure 6. The design runs without boron atoms and instead uses benzophenones. It originated from the emitter **DiKTa**, which performed in an OLED with an EQE of 14.7% and an emission wavelength of 468 nm. The introduction of three mesityl groups in **Mes3DiKTa** greatly enhanced the device's performance. The OLED device yielded an EQE of 21.1% with a blue emission at 480 nm.[100-101]

In Summary, great achievements have been gained in each OLED generation. Despite their inefficiency, first-generation OLEDs for blue light are still used in commercially available devices. The efficient, though environmentally challenging, second-generation OLEDs are mostly used for green and red-emitting materials.[102] To change to more acceptable and less hazardous methods, current research focuses on the third generation with TADF materials.
Over the years, many examples of highly efficient OLEDs using metal-free third-generation TADF materials obtained EQEs of over 30% in the blue[103] and

green[104] and close to 13% in the red[105] spectra. With these great advancements, TADF OLEDs supersede first-generation fluorescent OLEDs and become comparable to efficient second-generation phosphorescent OLEDs.

2 Objective

The discovery of TADF was fortunate because it allowed the replacement of inefficient first-generation fluorophores and second-generation phosphors containing concerning heavy metals in their structures. With a 100% theoretical IQE, they can be as efficient as phosphors and simultaneously be designed purely organic. This unique feature makes TADF materials the most promising class for OLED technology.
Focusing solely on the efficiency, OLEDs based on phosphorescent materials are commercially used for generating green and red light. Blue light applications, however, still rely on inefficient fluorescent materials. As a contribution to better solutions for blue light generation, this work focused on designing purely organic blue TADF emitters.

The first part targeted the modification of the tristriazolotriazine acceptor core, which was, among others, investigated in our group previously.[106-108] Herein, the enlargement of the acceptor core to form the novel acceptors monotriazolotriazine and ditriazolotriazine is intended. Both acceptors are decorated with different TADF donors. Efficient and commonly used acridine derivatives are chosen to find the most efficient donor-acceptor pair in this category. In a further approach, the dihedral angle between the donor and acceptor is enhanced by inserting different functional groups into the spacer unit. The synthesized TADF emitters are characterized by standard organic compound methods and are further investigated with photophysical measurements to explore their TADF character. The promising compounds are embedded in films with different hosts to get their best TADF properties. The molecules are calculated using density functional theory (DFT) methods to underline the experiments theoretically.

The second project is designed based on the outstanding results of Zhang *et al.*[2] The acceptor of a five 3,6-di-*tert*-butyl-9*H*-carbazole (**tCz**) donor decorated with a benzonitrile is transformed into a tetrazole or oxadiazole acceptor. The derivatization of the tetrazole or oxadiazole group is targeted to explore the tolerance for different functional groups and steric challenges. A modular approach is designed to allow accessible adjustments of photophysical properties. The distinctive charge transfer character is actively impacted by changing the chemical structure using electron-donating or withdrawing groups with different strengths and steric. Additionally, the importance of the dihedral angle is studied by incorporating a phenylene spacer between the donor unit and the acceptor. After that, the donor system is changed to four or two **tCz** donors and a D-A-A-D-scaffold is approached. A modulative synthetic route is established to connect different donor systems via two oxadiazole acceptors with altering bridging units. The impact of the connecting part between the acceptors and different donor combinations is investigated. Besides, the donor strength, count, and arrangement are considered. Again, all molecules are characterized structure-wise and through basic photophysical measurements to assess their TADF properties. Those with a sufficient TADF character are tested in films with different hosts to maximize their TADF outcome.

3 Results and Discussion

3.1. Mono- and Di[1,2,4]-triazolo[1,3,5]-triazine (MTT & DTT)

3.1.1. Tris[1,2,4]-triazolo[1,3,5]-triazine (TTT)

In 1961, Huisgen synthesized the discotic-shaped tris[1,2,4]-triazolo[1,3,5]-triazine (TTT) using cyanuric chloride and three equivalents of tetrazole. Their proposed mechanism is shown in Scheme 6.[109] The tetrazole nucleophilic attacks the carbon of cyanuric chloride. What follows is the cleavage of Nitrogen and ring closure. This repeats three times to substitute all three chlorides. The TTT can occur in two isomers, namely B-TTT and L-TTT. Whereas the B-TTT can be synthesized with the previous described route, the L-TTT can be obtained with B-TTT and high temperatures of up to 350 °C.[110]

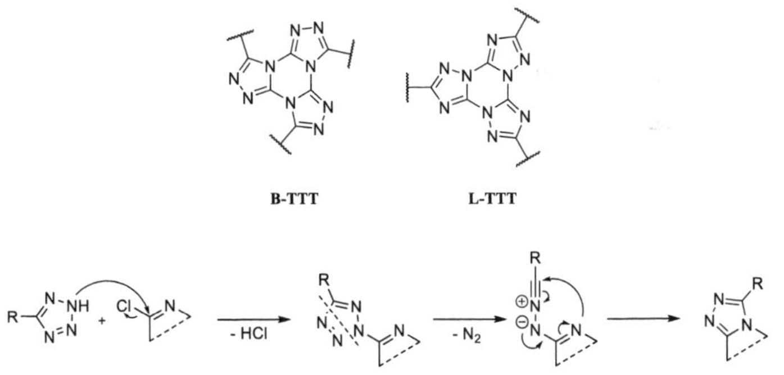

B-TTT L-TTT

- HCl - N_2

Scheme 6 Tris[1,2,4]-triazolo[1,3,5]-triazine displayed in its two isomers, the B-TTT and L-TTT. Mechanism of the synthesis of the B-TTT core (below).

Long only applied for its liquid crystal properties [111-112], the TTT recently appeared in the TADF OLED research. Several groups picked up the core including Pathak *et al.*[107], Wang *et al.*[108], and the Bräse group[106]. For instance, the EQE of one of the investigated TTT emitters using **DMAC** as donors, namely **TTT-DMAC,** ranged from 1.9% reported by Pathak *et al.*, to 5.8% by our group to 9.73% by Wang *et al.*

The first project of this work created the mono[1,2,4]-triazolo[1,3,5]-triazine (MTT) and di[1,2,4]-triazolo[1,3,5]-triazine (DTT) (Figure 7). This class represents a new type of triazolotriazine acceptor. The idea was to design, synthesize and characterize these new TADF emitters that originated from the TTT core. Instead of three "triazole-fused" tetrazoles, two are used for DTT and one for MTT. The synthetic procedures followed the protocol established within our group.[106] During the work of this thesis, Su *et al.* published three TADF molecules based on the L-MTT isomer.[113] Nevertheless, their synthetic approach differentiates from ours. They synthesized the triazolotriazine ring using 2-chloro-4,6-diphenyl-[1,3,5]-triazine to build up the B-isomer. To obtain the L-isomer, they used the DIMROTH rearrangement at high temperatures. Lastly, the donor, in their case carbazole derivatives, were attached using BUCHWALD-HARTWIG reaction conditions.

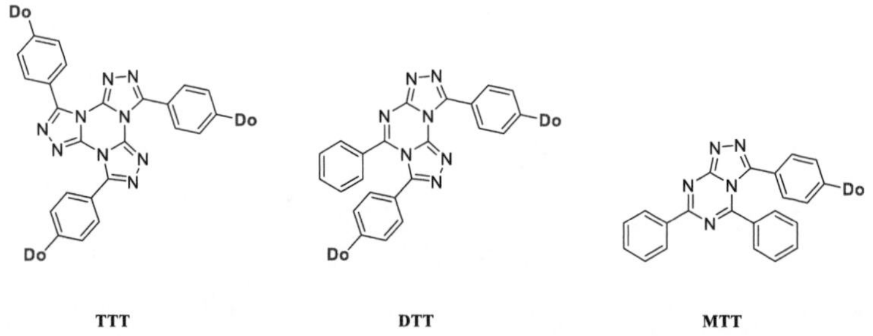

TTT DTT MTT

Figure 7 Core structures of tris[1,2,4]-triazolo[1,3,5]-triazine (TTT), di[1,2,4]-triazolo[1,3,5]-triazine (DTT) and mono[1,2,4]-triazolo[1,3,5]-triazine (MTT).

3.1.2. Syntheses

3.1.2.1. Donors

A variety of acridine derivatives such as 9,9-diphenyl-9,10-dihydroacridine, 9,9-dimethyl-9,10-dihydroacridine and 10*H*-phenoxazine were tested as donors to decorate the MTT and DTT core (Figure 8). **DMAC** and **PXZ** are commercially affordable as opposed to **DPAC**, which was synthesized following literature procedures.[114-115]

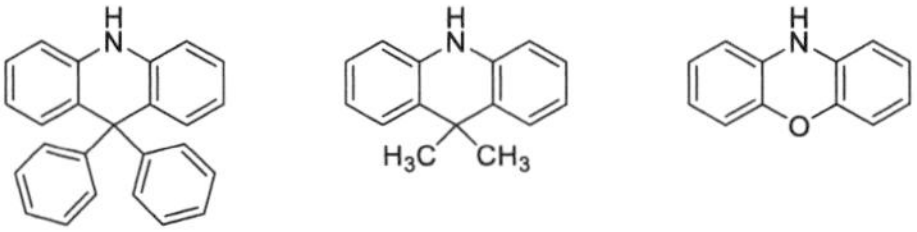

16) DPAC **17) DMAC** **18) PXZ**

Figure 8 Selected acridine derivatives **DPAC**, **DMAC**, and **PXZ** to use as donors for the MTT and DTT emitters.

Synthesis of 9,9-Diphenyl-9,10-dihydroacridine (Scheme 7)

Starting from methyl 2- bromobenzoate (**19**), the first step contained a BUCHWALD-HARTWIG amination with aniline (**20**) using 1,1′-ferrocenediyl-bis(diphenylphosphine), bis(dibenzylideneacetone)-palladium and cesium carbonate as ligand, catalyst, and base, respectively. Methyl 2-(phenylamino)benzoate (**21**) was isolated in a 90% yield. Subsequently, **21** was reacted with phenyl-magnesium chloride (**22**) in a GRIGNARD reaction to obtain **23** in a 64% yield. Lastly, **23** was cyclized under acidic conditions with hydrochloric acid. After purification **DPAC** (**16**) was isolated in 82% yield. The overall yield over these three steps amounted to 47%.

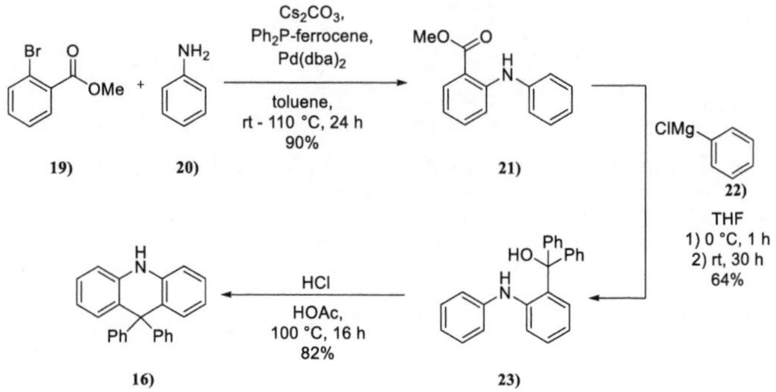

Scheme 7 Synthesis of 9,9-diphenyl-9,10-dihydroacridine through three steps: 1) BUCHWALD-HARTWIG, 2) GRIGNARD, 3) FRIEDEL-CRAFTS CYCLIZATION.

3.1.2.2. Nitrile Precursors

Different benzonitriles, which later acted as the spacer in the TADF molecule, were connected to the selected donors. The first approach installed an unsubstituted phenyl group between the donor and acceptor. Different spacers with altering electronic or steric effects were installed later to investigate the dihedral angle's importance. A nucleophilic aromatic substitution reaction (S_NAr) was a suitable route to synthesize six nitrile precursors, successfully giving fair to good yields. In Scheme 8, all synthesized nitrile-spacer-donor pairs are listed. The base is added to deprotonate the amine function of the donor to enhance the nucleophilicity. In the selection of the phenylene spacer, **PXZ-BN (25a)** achieved the highest yield at 83%, followed by **DPAC-BN (25c)** at 76% and **DMAC-BN (25b)** at 47%.

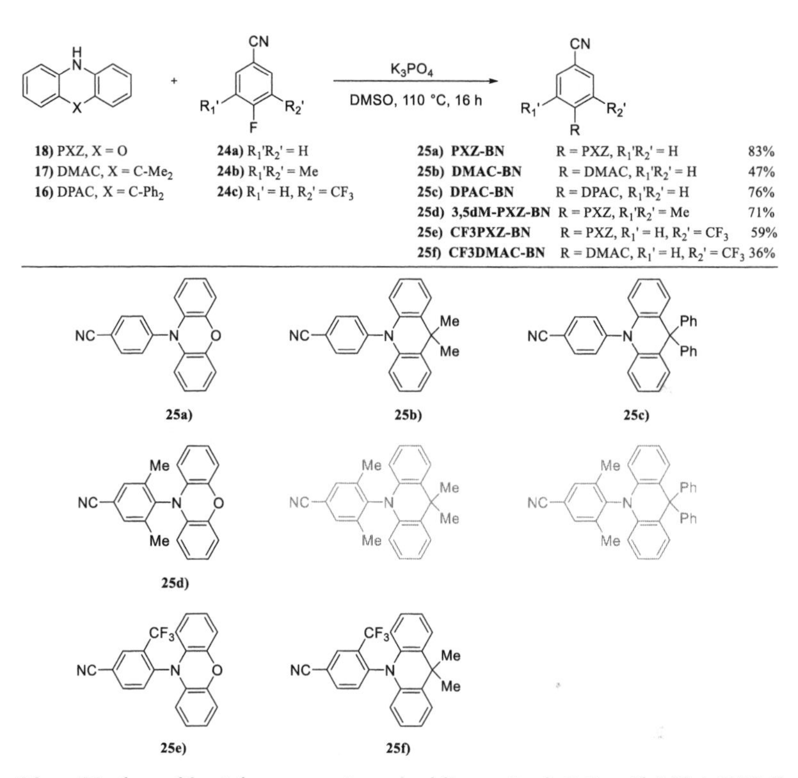

Scheme 8 Syntheses of the nitrile precursors via a nucleophilic aromatic substitution with K_3PO_4 in DMSO. The two compounds in grey are the ones that didn't show conversion.

To enhance the dihedral angle, 4-fluoro-3,5-dimethylbenzonitrile (**24b**) was used to install methyl groups in the 3- and 5-position of the spacer. The reaction with the donor **PXZ** led to the desired compound **3,5dM-PXZ-BN** (**25d**) in 71% yield. **DPAC** and **DMAC** could not be successfully attached to **24b** neither with the described nor another route for both using different conditions such as higher temperatures and longer reaction times. For the reaction with **DMAC** BUCHWALD-HARTWIG conditions were tested with Cs_2CO_3, Ph_2P-ferrocene, and $Pd(dba)_2$ or NaO^{t}Bu, P(^{t}Bu)$_3$, and $Pd(OAc)_2$. Both approaches did not lead to the desired product. Through an alternative route to enhance the dihedral angle, **DMAC** was tried

to be attached under S_NAr conditions to the 2- and 6- positions using 2-bromo-5-fluoro-1,3-dimethylbenzene. The bromide can later be transformed into nitrile. Again, no conversion was observed. To explain why **DMAC** and **DPAC** can be attached to 4-fluorobenzonitrile (**24a**) but not to 4-fluoro-3,5-dimethylbenzonitrile and why **PXZ** succeeds in both cases, two aspects must be considered. First, **24a** shows a different reactivity in an S_NAr than **24b**. In an S_NAr reaction compound **24a** directs its new substituent in *ortho*/*para*-position. In *para*-position the fluoride is bound to give an appropriate starting point for this reaction. Compound **24b** directs in *ortho*/*para*-position; however, the methyl groups direct in *meta*-position. Considering the directing factors, compound **24b** is less reactive in this reaction constellation. Nevertheless, the reaction using **PXZ** succeeded with a good yield of 71%.
The second aspect to be considered is that **PXZ** is a much stronger donor than **DPAC** and **DMAC**. With its stronger nucleophilic character, it can overcome the unfavored starting situation with **24b** being less reactive than **24a**. This observation is found in compounds **25a-c** as well. The reaction with the **PXZ**-donor to obtain **25a** achieved the highest yield. **DPAC-BN** follows because it is less nucleophilic than **PXZ**. The last one in this sequence with the lowest yield is the **DMAC-BN** because of the relatively low nucleophilicity. Next to the methyl groups in the spacer, another functional group was investigated. **CF3PXZ-BN** and **CF3DMAC-BN** with a trifluoromethyl group in the spacer were obtained in 59% and 36% yield, which is lower than the nitriles with the unsubstituted phenyl group. The steric hindrance of the trifluoromethyl group can be issued here. Comparing compounds **25e** and **25f**, the effect of the strong nucleophilicity of **PXZ** is observed again, leading to higher yields than the **DMAC** nitrile.

3.1.2.3. Tetrazole Precursors

The next step to yield the desired MTT- and DTT-TADF emitters is transforming the nitriles into tetrazoles. A straightforward synthetic approach was utilized using sodium azide in a 1,3-dipolar cycloaddition (Scheme 9). The reaction scale was limited because an excess of 3.00 equivalents of sodium azide was used, and the tetrazole compounds were precipitated with hydrochloric acid during the workup.

The isolated products were obtained in excellent to quantitative yields with 99% for **PXZ-T**, 95% for **DMAC-T** and **DPAC-T**, and >99% for **CF3DMAC-T** (**26a-c** and **f**). Both **PXZ** tetrazoles **3,5dM-PXZ-T** and **CF3PXZ-T** containing a substituted phenylene spacer resulted in 82% and 74% lower yields.

It is noted that all tetrazole structures in this work were drawn in the 2*H* form. A tetrazole can consist of three tautomers, the 1*H*-, 2*H*-, and 5*H*-tetrazole, of which the latter is not aromatic. In solution, the 1*H*- and 2*H* form lie in an equilibrium that favors one tautomer depending on the polarity of the solvent. In polar solvents, the 1*H* form is favored because it is the more polar tautomer, as opposed to the 2*H* form.[116-117] However, for simplicity reasons, it was agreed to draw the 2*H*-form for all tetrazoles.

NaN$_3$, NH$_4$Cl
DMF, 130 °C, 16 h

25a) PXZ-BN	R = PXZ, $R_1'R_2'$ = H	**26a) PXZ-T**	R = PXZ, $R_1'R_2'$ = H	99%
25b) DMAC-BN	R = DMAC, $R_1'R_2'$ = H	**26b) DMAC-T**	R = DMAC, $R_1'R_2'$ = H	95%
25c) DPAC-BN	R = DPAC, $R_1'R_2'$ = H	**26c) DPAC-T**	R = DPAC, $R_1'R_2'$ = H	95%
25d) 3,5dM-PXZ-BN	R = PXZ, $R_1'R_2'$ = Me	**26d) 3,5dM-PXZ-T**	R = PXZ, $R_1'R_2'$ = Me	82%
25e) CF3PXZ-BN	R = PXZ, R_1' = H, R_2' = CF_3	**26e) CF3PXZ-T**	R = PXZ, R_1' = H, R_2' = CF_3	74%
25f) CF3DMAC-BN	R = DMAC, R_1' = H, R_2' = CF_3	**26f) CF3DMAC-T**	R = DMAC, R_1' = H, R_2' = CF_3	>99%

26a) **26b)** **26c)**

26d)

26e) **26f)**

Scheme 9 Syntheses of the tetrazole precursors (**26a-f**) using sodium azide in a 1,3-dipolar cycloaddition.

3.1.2.4. Di[1,2,4]-triazolo[1,3,5]-triazine

The tetrazole precursors were reacted with 4-dichloro-6-phenyl-[1,3,5]-triazine (**27**) and 2,6-lutidine, which acts as the base to deprotonate the NH tetrazole (Scheme 10). Six DTT emitters have been obtained in fair to good yields. The yields are generally lower than those from the corresponding TTT compounds.[106, 118] The TTT core is synthesized using cyanuric chloride, a triazine bearing three chlorides. This molecule has a low electron density in its core, which makes it very reactive in nucleophilic substitution reactions. In contrast,

27 has one chloride less, which makes it less electrophilic and, therefore, less reactive. The yields ranged from 33%, 35%, 59% and 68% to 94% for **DPAC-DTT** (**28c**), **3,5dM-PXZ-DTT** (**28d**), and **CF3DMAC-DTT** (**28f**), **DMAC-DTT** (**28b**), **PXZ-DTT** (**28a**), and **CF3PXZ-DTT** (**28e**), respectively.

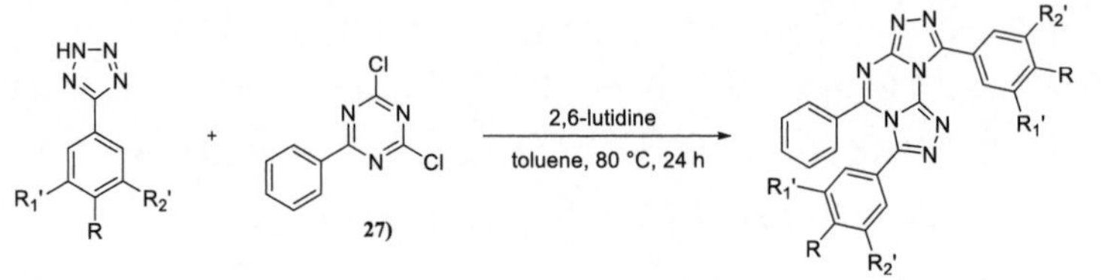

26a) **PXZ-T**	R = PXZ, $R_1'R_2'$ = H	
26b) **DMAC-T**	R = DMAC, $R_1'R_2'$ = H	
26c) **DPAC-T**	R = DPAC, $R_1'R_2'$ = H	
26d) **3,5dM-PXZ-T**	R = PXZ, $R_1'R_2'$ = Me	
26e) **CF3PXZ-T**	R = PXZ, R_1' = H, R_2' = CF_3	
26f) **CF3DMAC-T**	R = DMAC, R_1' = H, R_2' = CF_3	
28a) **PXZ-DTT**	R = PXZ, $R_1'R_2'$ = H	63%
28b) **DMAC-DTT**	R = DMAC, $R_1'R_2'$ = H	59%
28c) **DPAC-DTT**	R = DPAC, $R_1'R_2'$ = H	33%
28d) **3,5dM-PXZ-DTT**	R = PXZ, $R_1'R_2'$ = Me	35%
28e) **CF3PXZ-DTT**	R = PXZ, R_1' = H, R_2' = CF_3	94%
28f) **CF3DMAC-DTT**	R = DMAC, R_1' = H, R_2' = CF_3	35%

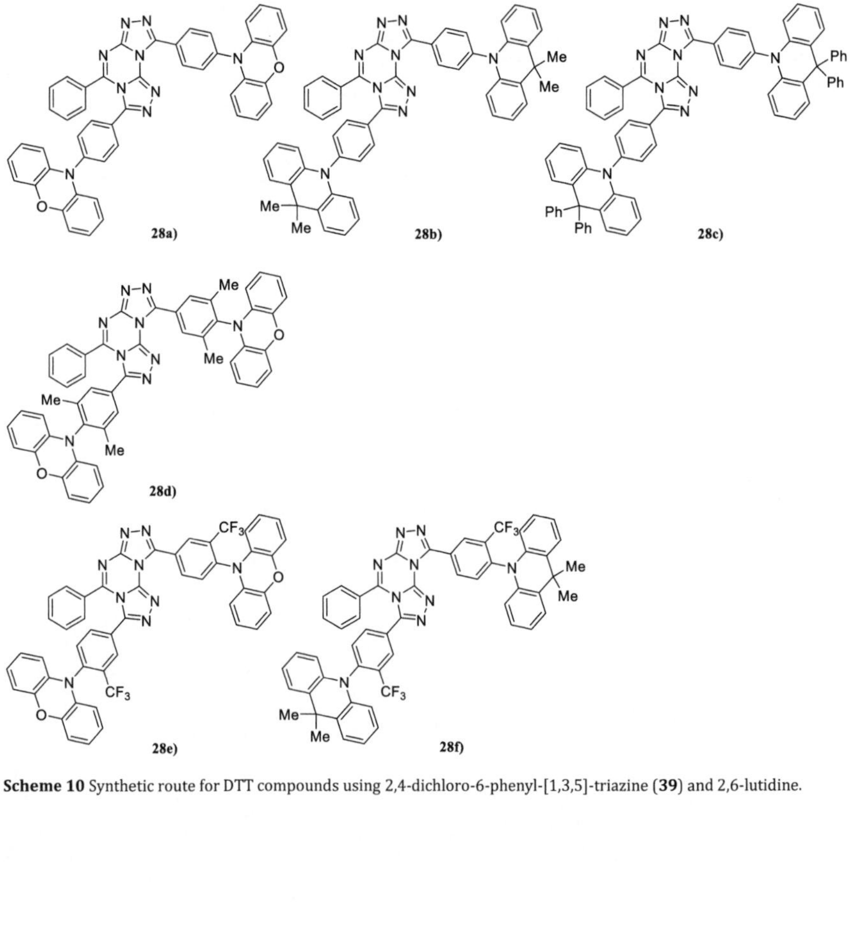

Scheme 10 Synthetic route for DTT compounds using 2,4-dichloro-6-phenyl-[1,3,5]-triazine (**39**) and 2,6-lutidine.

3.1.2.5. Mono[1,2,4]-triazolo[1,3,5]-triazine

Using the same protocol as for DTT syntheses (condition A), six MTT emitters were successfully synthesized. MTT emitter **30d** was obtained using a different route (condition B). All synthesized MTTs are shown in Scheme 11.
Compared to TTT and DTT compounds, the yields of the isolated products were even lower. This observation is not surprising since 2-chloro-4,6-diphenyl-[1,3,5]-triazine (**29a**) shows a poorer electrophilic reactivity. In addition, these compounds are less soluble than TTT and DTT, contributing to the low yields. **DMAC-MTT** (**30b**) was obtained with a 26% yield because the methyl groups enhance the solubility. In contrast, **PXZ-MTT** (**30a**) and **DPAC-MTT** (**30c**) only yielded 11% and 7%, respectively, probably because they are less soluble. The same trend can be observed in compounds **CF3PXZ-MTT** (**30e**), **CF3DMAC-MTT** (**30f**) and **PXZ-MTT-Me** (**30g**). The CF_3 and the methyl groups increase the compound's solubility and help achieve higher yields.
Because of the low yields, another route using condition B) was tried. 2-Bromo-4,6-diphenyl-1,3,5-triazine was reacted with **29b** under BUCHWALD-HARTWIG conditions. Indeed, this route led to an enhanced conversion of **3,5dM-PXZ-MTT** (**30d**) with a yield of 73%. Therefore, this route was tried with other precursors like **26a**. However, the obtained yields were even lower or gave no conversion. For that reason, this approach was not further pursuit.

A surprising finding was observed through the analysis of the crystal structures of **30d** and **30g**, which have the **PXZ** donor incorporated. The crystal structures were characterized by single X-ray diffraction revealing that the MTT emitters are arranged in another isomer, the L-MTT, and not the expected B-isomer (Figure 9).

A) 2,6-lutidine, toluene, 80 °C, 24 h
B) Cs_2CO_3, $Pd(dba)_2$, Ph_2P-ferrocene, toluene, 110 °C, 16 h

26a) PXZ-T	R = PXZ, $R_1'R_2'$ = H	**29a)** X = Cl, R" = H	**30a) PXZ-MTT**	R = PXZ, $R_1'R_2'$ = H, R" = H	11% A)
26b) DMAC-T	R = DMAC, $R_1'R_2'$ = H	**29b)**[B] X = Br, R" = H	**30b) DMAC-MTT**	R = DMAC, $R_1'R_2'$ = H, R" = H	26% A)
26c) DPAC-T	R = DPAC, $R_1'R_2'$ = H	**29c)*** R" = Me	**30c) DPAC-MTT**	R = DPAC, $R_1'R_2'$ = H, R" = H	7% A)
26d) 3,5dM-PXZ-T	R = PXZ, $R_1'R_2'$ = Me		**30d) 3,5dM-PXZ-MTT**	R = PXZ, $R_1'R_2'$ = Me, R" = H	73% B)
26e) CF3PXZ-T	R = PXZ, R_1' = H, R_2' = CF_3		**30e) CF3PXZ-MTT**	R = PXZ, R_1' = H, R_2' = CF_3, R" = H	27% A)
26f) CF3DMAC-T	R = DMAC, R_1' = H, R_2' = CF_3		**30f) CF3DMAC-MTT**	R = DMAC, R_1' = H, R_2' = CF_3, R" = H	76% A)
			30g) PXZ-MTT-Me	R = PXZ, $R_1'R_2'$ = H, R" = Me	29% A)

30a) **30b)** **30c)**

30d)[B]

30e) **30f)**

30g)*

Scheme 11 Syntheses of seven MTT compounds using two different conditions. Condition A) **29a** or **29c**, 2,6-lutidine, toluene, 80 °C, 24 h. Condition B) **29b**, Cs_2CO_3, $Pd(dba)_2$, Ph_2P-ferrocene, toluene, 110 °C, 16 h. The MTT molecules are arranged in the L-form.

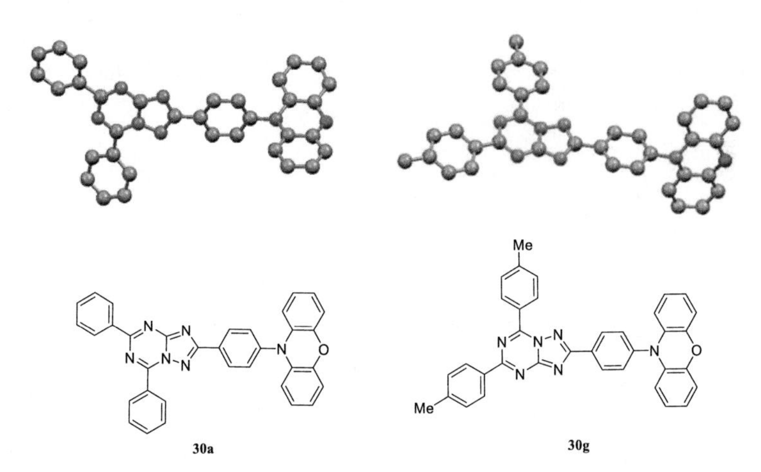

Figure 9 Crystal structures of **PXZ-MTT (30a)** and **PXZ-MTT-Me (30g)** revealed the L-MTT isomer.

It appears that the L-isomer is a result of the DIMROTH rearrangement.[119] It is long known that polyazaindolizine systems can undergo the DIMROTH rearrangement under basic[119-120] or acidic catalysis[121].

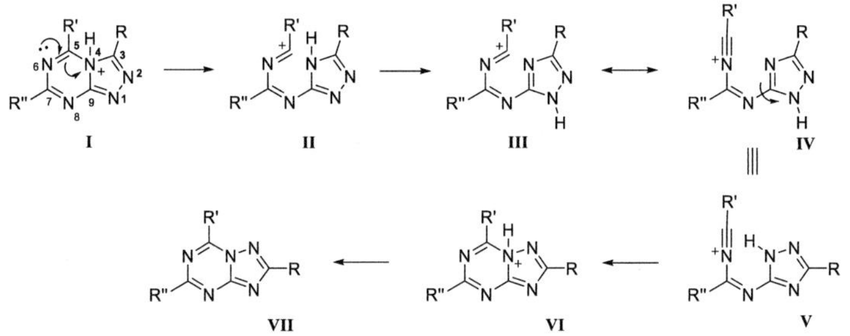

Scheme 12 Acid-catalyzed mechanism of the DIMROTH rearrangement of B-monotriazolotriazine to L-monotriazolotriazine.[121]

The syntheses of the MTT molecules were performed under basic conditions with 2.00 equivalents of 2,6-lutidine at 80 °C. The excess base was considered to enhance the reaction. Under these conditions, the first intuition assumes that the arrangement occurs base-catalyzed. According to Guerret *et al.*, the base-induced rearrangement's first step is the base's nucleophilic attack at the 5-position.[119] 2,6-Lutidine was intentionally chosen because it is a non-nucleophilic base due to the steric hindrance of the methyl groups. Therefore, this assumption is not plausible.

A more likely explanation for the formation of the L-MTT-isomers is the acidic catalyzed DIMROTH rearrangement (Scheme 12).[121] 2,6-Lutidine deprotonates in the first step of the reaction the tetrazole starting material. As a weak acid due to the loss of a proton, it can protonate the nitrogen in the 4-position (**I**). A ring opening of intermediate **II** with subsequent tautomerization (H-shift) leading to **III** follows. Afterwards, the triazole group rotates, and the ring closes (**VI**). After deprotonation, the L-isomer (**VII**) is formed.

To avoid excessive base, the procedure was reviewed to use a 1.00 equivalent base. With an equal amount of tetrazole and base, perhaps no rearrangement occurs. The reaction with the donor **PXZ** was repeated, and yellow crystals were obtained. The analysis using single crystal X-ray diffraction revealed that the molecule is still arranged in the L-form, despite the change of reaction conditions. The variation of the base concentration, therefore, does not lead to the B-isomer. Since no crystal structures of the other MTTs could be obtained and no evidence could be brought that contradicts the L-isomer finding, it was assumed that all MTT molecules are present in the L-form.

3.1.3. DFT Calculations

The DTT and MTT compounds with the unsubstituted and methyl-substituted phenylene spacer have been calculated through the collaboration with Ettore Crovini from the Eli Zysman-Colman group. They determined the frontier orbitals distribution, HOMO-LUMO levels, and the energies of the excited states using DFT calculations adopting the Gaussian 09 revision D.018 suite. The ground states of the molecules were calculated in the gas phase to determine their geometries. Then, the optimized structures were used to calculate the excited states by applying time-dependent density functional theory (TD-DFT) using the Tamm-Dancoff Approximation (TDA).

DFT Calculations for DTTs

The energy levels and frontier orbitals for the DTT emitters are illustrated in Figure 10. As expected, the HOMO is located on the donor groups, although with a small overlap with the phenylene spacers. The emitters have two donor groups, and multiple donor-located molecular orbitals are formed. Therefore, the HOMO is partially delocalized over all donor groups (HOMO, HOMO-1, HOMO-2, etc.). The LUMO is mostly located on the triazolotriazine acceptor core and outskirts slightly to the phenylene spacers, hence creating a sufficient overlap of the frontier orbitals. A dihedral angle change by introducing methyl substituents to the spacer does not significantly change the calculated energy levels, hence the TADF properties. **PXZ-DTT** and **3,5dM-PXZ-DTT** showed a comparable optical band gap (energy gap between HOMO and LUMO) with 2.42 eV and 2.43 eV, respectively. The S_1 levels are calculated to be 2.02 eV and 2.18 eV estimating a yellow and orange emission color, respectively. The ΔE_{ST} value is 0.03 eV for **PXZ-DTT**, making it a promising TADF candidate. **3,5dM-PXZ-DTT** was calculated with an ΔE_{ST} of 0 eV. The strong donor phenoxazine stabilizes the LUMO with -2.43 eV for both **PXZ-DTTs**, as opposed to **DPAC-DTT** with a LUMO level of -2.13 eV. **DPAC-**

DTT showed the largest optical band gap with 3.13 eV. With an S_1 energy level of 2.71 eV, the emitting color of **DPAC-DTT** is estimated to be blue. The energy gap between S_1 and T_1 is smaller with 0.01 eV, as opposed to 0.03 eV for **PXZ-DTT**. **DMAC-DTT** was calculated with a blue emission with an S_1 level of 2.50 eV, although not as blue as **DPAC-DTT**. Its ΔE_{ST} is calculated to 0 eV. Considering their small energy gaps, all synthesized DTT emitters seem to have promising TADF properties. Having the weakest donor, **DPAC-DTT**, followed by **DMAC-DTT**, is the bluest emitter in this series.

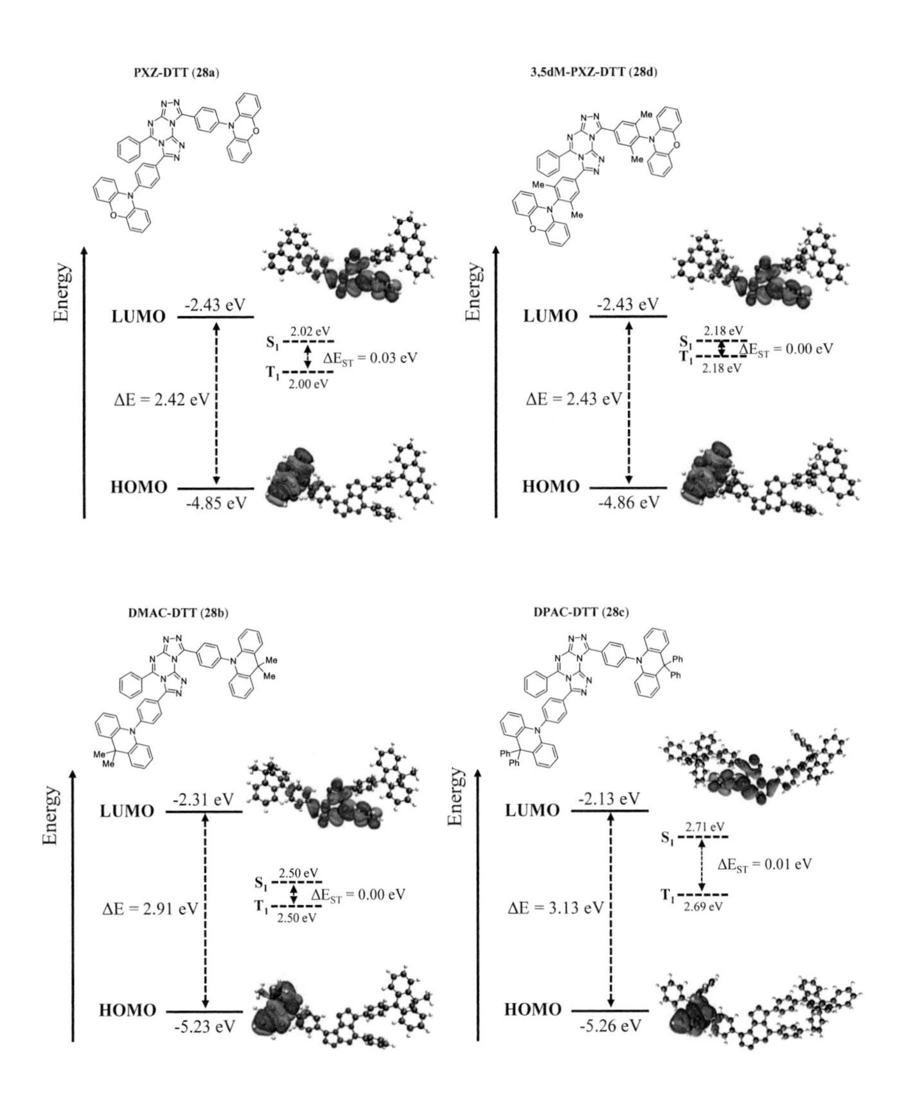

Figure 10 Visualization of the HOMO-LUMO orbitals and DFT calculations of **PXZ-DTT** and **3,5dM-PXZ-DTT** (top) and **DMAC-DTT** and **DPAC-DTT** (bottom). The DFT calculations were run with the PBE1PBE functional and the 6-31G(d,p) basis set. For simplicity reasons, the energy gaps with 0 eV between S_1 and T_1 were drawn with a gap. Results provided by Ettore Crovini.

DFT Calculations for MTTs

The MTT compounds were calculated with DFT methods as well. The simulations of the L-MTTs are visualized in Figure 11. In all four molecules, the HOMO is located on the donors, whereas the LUMO is located on the acceptor and outskirts to the phenylene spacer.

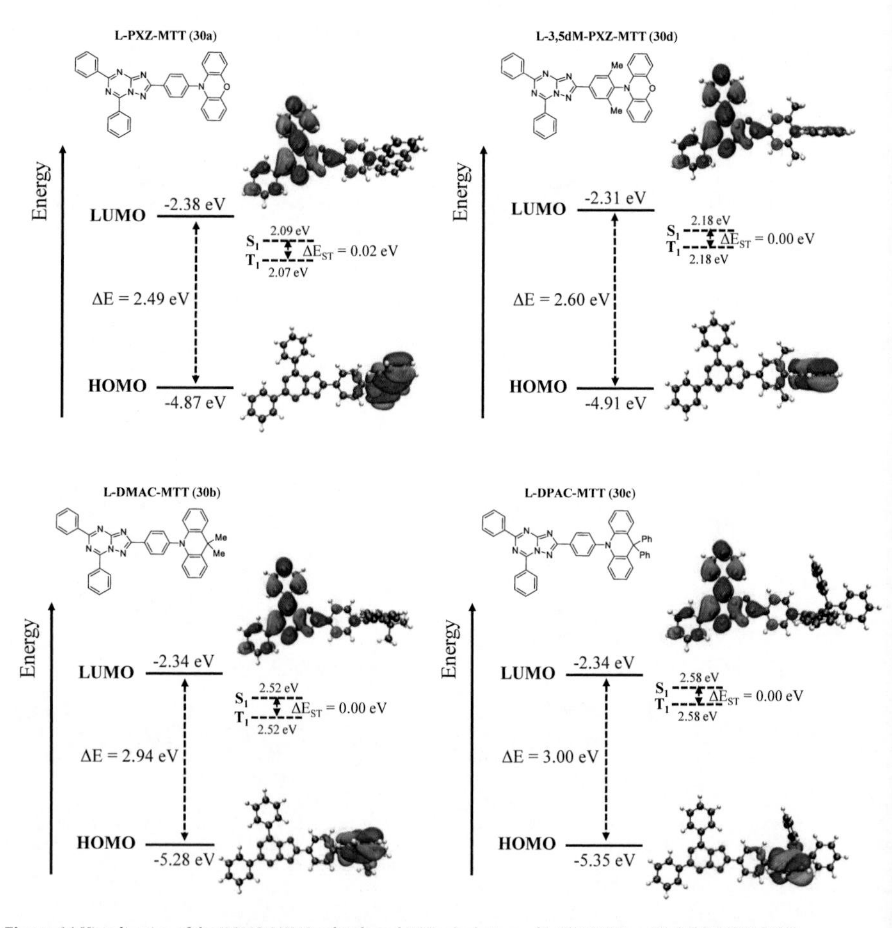

Figure 11 Visualization of the HOMO-LUMO orbitals and DFT calculations of **L-PXZ-MTT** and **L-3,5dM-PXZ-MTT** (top) and **L-DMAC-MTT** and **L-DPAC-MTT** (bottom). The DFT calculations were run with the PBE1PBE functional

and the 6-31G(d,p) basis set. For simplicity reasons, the energy gaps with 0 eV between S_1 and T_1 were drawn with a gap. Results provided by Ettore Crovini.

The impact of the dihedral angle is more evident than in the DTT molecules. **L-PXZ-MTT** showed a HOMO energy level of -4.87 eV and a LUMO level of -2.38 eV. Emitter **L-3,5-dM-PXZ-MTT**, with the extended methylated phenylene spacer, was calculated to have a larger optical band gap of 2.60 eV with HOMO and LUMO levels of -4.91 eV and -2.31 eV, respectively. The optical band gap of the **L-DMAC-MTT** and **L-DPAC-MTT** is larger than the **PXZ** compounds with 2.94 eV and 3.00 eV, respectively. As expected, the LUMO levels are determined to show roughly the same properties for all four molecules. The HOMOs with -5.28 eV and -5.35 eV for **L-DMAC-MTT** and **L-DPAC-MTT**, respectively, differentiate from the **PXZ** compounds. The singlet-triplet energy gap is calculated to be 0 eV for **L-DMAC-MTT**, **L-DPAC-MTT**, and **L-3,5-dM-PXZ-MTT** and 0.02 eV for **L-PXZ-MTT**. The **DMAC** and **DPAC** emitters were calculated to show a bluer emission than the **PXZ** compounds, as was the case for the DTT series.

In addition, the B-MTTs were calculated with DFT methods. The calculated molecular orbitals and excited state energy levels of the B-MTTs are displayed in Figure 12.

All B-MTT emitters showed that the LUMO is located on the triazolotriazine acceptor. Whereas for all emitters, the HOMO is mainly located on the donor groups, although slightly interfering in the phenylene spacers.

B-DPAC-MTT was determined to show the bluest emission in this series with an S_1 value of 2.77 eV and an optical band gap of 3.22 eV, with a HOMO value of -5.52 eV and a LUMO value of -2.20 eV. The calculated ΔE_{ST} of 0.19 eV was the highest in this series. Slightly red-shifted was **B-DMAC-MTT** with an S_1 level of 2.69 eV and an ΔE of 3.13 eV. The ΔE_{ST} was calculated to 0.10 eV, which is suitable for TADF. The **PXZ** compounds, **B-PXZ-MTT** and **B-3,5dM-PXZ-MTT**, were calculated to emit with a red shift compared to the **DMAC** and **DPAC** emitters. A minute change is noted in comparing the **B-PXZ-MTT** with the methyl-substituted compound **B-3,5dM-PXZ-MTT**. While the HOMO is barely affected, the LUMO is slightly destabilized from -2.37 eV to -2.29 eV for **B-3,5dM-PXZ-MTT**. With an S_1 level of 2.23 eV with an ΔE of 2.66 eV and an S_1 level of 2.35 eV with an ΔE 2.78 eV, the emitted color is estimated to be yellow and green for **B-PXZ-MTT** and **B-3,5dM-PXZ-MTT**, respectively. **B-PXZ-MTT** shows an excellent ΔE_{ST} of 0.01 eV, and the ΔE_{ST} of **B-3,5dM-PXZ-MTT** was calculated to be 0 eV.

Comparing **B-PXZ-MTT** with **PXZ-DTT**, a small stabilization of the LUMO of 0.06 eV is notable for the DTT emitter due to the second **PXZ** donor. This indicates that the monotriazolotriazine and the ditriazolotriazine are similar strong acceptors. However, comparing the DTTs and MTTs with their associated TTT parent molecules (**DMAC-TTT**[106], **PXZ-TTT**[107]), it seems that the DTT and MTT core are stronger acceptors than the TTT. An explanation might be the complete surrounding of triazolo groups enclosing the triazine with a strong resonance effect.

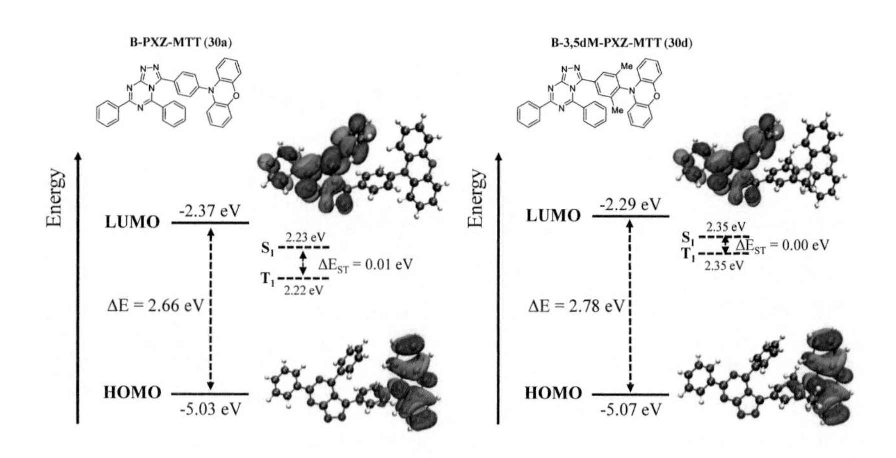

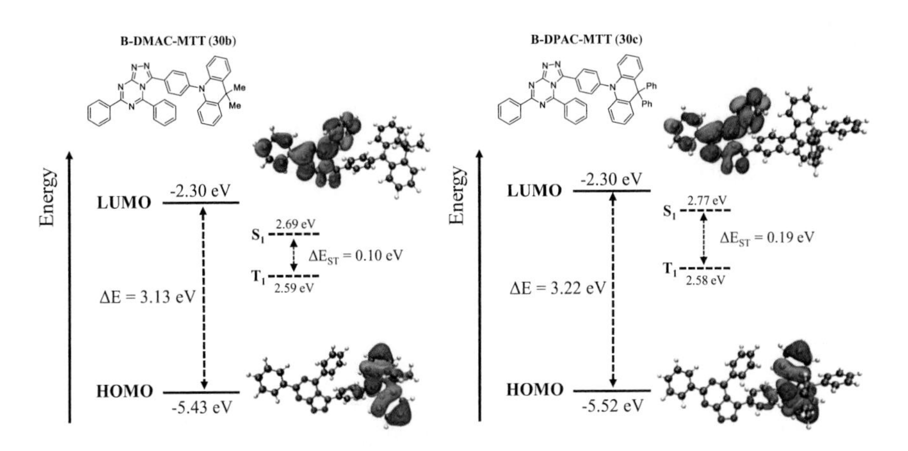

Figure 12 Visualization of the HOMO-LUMO orbitals and DFT calculations of **B-PXZ-MTT**, **B-3,5dM-PXZ-MTT**, **B-DMAC-MTT**, and **B-DPAC-MTT**. The DFT calculations were run with the PBE1PBE functional and the 6-31G(d,p) basis set. For simplicity reasons, the energy gaps with 0 eV between S_1 and T_1 were drawn with a gap. Results provided by Ettore Crovini.

In the following, the L- and B-isomers will be compared. The optical band gap is higher in the B-form for all four molecules. The LUMO levels are similar for the B- and L-form. The HOMO levels, however, lie energetically deeper in the B-isomers, indicating that the donors are more affected by this change in the chemical environment. The ΔE_{ST} values are for the **PXZ** compounds roughly the same for both isomers. In contrast, ΔE_{ST} increases significantly for **DMAC-MTT** and **DPAC-MTT** when going from the L-form to the B-form.

3.1.4. Photophysical Characterization

The photophysical measurements of the original MTT and DTT series (meaning those with an unsubstituted phenyl spacer) were carried out by Ettore Crovini from the Eli Zysman-Colman group through a collaboration. The expanded systems were measured later by Idoia Garin from the research group of Prof. Lemmer at KIT. The emitters from the original MTT and DTT series have been fully characterized. However, the remaining emitters were not fully photophysically characterized since these measurements are very time-consuming.

Cyclic Voltammetry

Cyclic voltammetry (CV) measurements were conducted to evaluate the HOMO and LUMO levels of the MTT and DTT emitters. The measurements were compared with the theoretical values and summarized in Table 1 and shown Figure 13.

Table 1 Comparison of the theoretical HOMO-LUMO levels with the experimental values obtained via CV. The materials were measured in a dichloromethane solution. The measured HOMO and LUMO values were obtained from the redox potentials from the DPV, $E_{HOMO/LUMO}$= -($E_{ox/red}$ + 4.8), where $E_{ox/red}$ was obtained from the DPV corrected vs. Ferrocene.

Entry	Emitter material	#	$HOMO_{calcd}$ [eV]	$LUMO_{calcd}$ [eV]	$HOMO_{meaus}$ [eV]	$LUMO_{meaus}$ [eV]
1	**PXZ-DTT**	**28a**	-4.85	-2.43	-5.55	-3.41
2	**DMAC-DTT**	**28b**	-5.23	-2.31	-5.79	-3.40
3	**DPAC-DTT**	**28c**	-5.26	-2.13	-5.93	-3.45
4	**PXZ-MTT**	**30a**	-5.03	-2.37	-5.13	-2.99
5	**DMAC-MTT**	**30b**	-5.43	-2.30	-5.33	-2.99
6	**DPAC-MTT**	**30c**	-5.52	-2.30	-5.44	-2.98

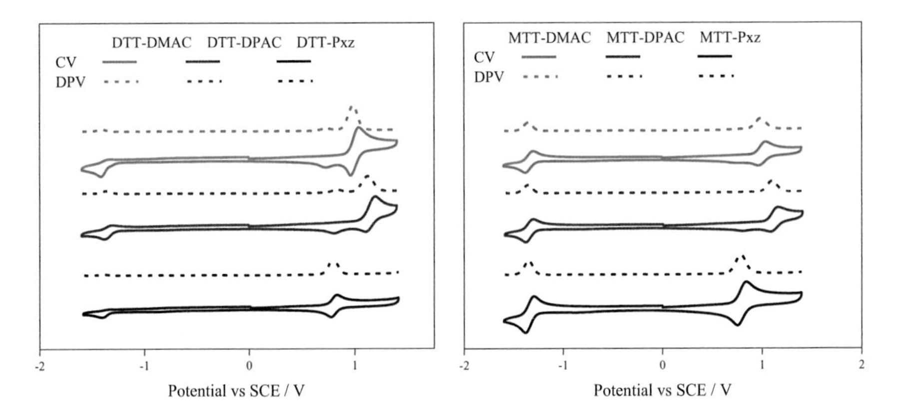

Figure 13 Cyclic voltammetry (CV) and differential pulse voltammetry (DPV) plots for **DMAC-**, **DPAC-**, **PXZ-DTT**, and **-MTT** samples. All samples were measured in a dichloromethane solution. The redox potentials are measured relative to a saturated calomel electrode (SCE) with a ferrocene/ferrocenium (Fc/Fc+) redox couple as the internal standard. Results provided by Ettore Crovini.

Deviations between the theory and experiments are common because the calculations were carried out for molecules in the gas phase, and the CVs are measured in a dichloromethane solution. Interestingly, the DTT compounds show higher deviations than the MTTs. For instance, the HOMO of **PXZ-MTT** was calculated at -5.03 eV and measured at -5.13 eV. In contrast, the HOMO of the corresponding DTT compound **PXZ-DTT** was calculated to be -4.85 eV and has a HOMO energy level of -5.55 eV. This accounts for a difference of 0.7 eV for calculation and CV. Almost the same deviation is seen for **DPAC-DTT**, with a difference of 0.67 eV between the calculated HOMO (-5.26 eV) and the measured HOMO (-5.93 eV). These significant differences in the theory and CVs between the calculated MTTs and DTTs is caused by a more complex structure of the DTTs.

Photophysical Measurements of DTTs

Emission and excitation spectra and time-resolved photoluminescence decay measurements of all original DTT emitters were measured in neat thin films and as dopants in films. The host materials and the dopant concentrations were varied to find the best guest-host combination regarding the TADF properties. With these results in hand, the photoluminescence quantum yields (PLQY or Φ_{PL}) were calculated.

A screening with commonly used host materials for blue emitters like N,N′-dicarbazolyl-3,5-benzene (mCP)[122-123] and poly (methyl methacrylate) (PMMA)[124-125] was carried out. The doping concentration was set to 10% for PMMA and varied from 3% to 20% for mCP.

The screening results for **PXZ-DTT** in neat and in films with the host materials using different concentrations are shown in Table 2.

Table 2 Photoluminescence quantum yields in air and nitrogen of **PXZ-DTT** in neat thin films and doped in different concentrations in mCP and PMMA with an excitation wavelength of 340 nm.

Entry	Host material	Doping concentration [wt%]	Φ_{PL} air [%]	Φ_{PL} nitrogen [%]
1	**mCP**	3	36.9	49.0
2	**mCP**	5	36.3	45.2
3	**mCP**	10	33.4	40.2
4	**mCP**	15	24.0	27.9
5	**mCP**	20	17.6	17.9
6	**PMMA**	10	14.9	17.1
7	-	neat	1.9	1.9

The films of all DTT compounds showed an enhanced PLQY when measured under a nitrogen atmosphere instead of air. **PXZ-DTT** doped in mCP (3wt%) delivered the best PLQY with 36.9% and 49.0% in air and nitrogen, respectively. The PLQY decreased by increasing the doping concentration in mCP. The PLQY was much lower when doping in PMMA instead of mCP, giving a PLQY of 14.9% in air and 17.1% under nitrogen. Without a host, the emitter performed poorly with a PLQY of 1.9% in air and nitrogen.

The screening results for **DPAC-DTT** are shown in Table 3. The PLQY doped in mCP gave the best result for 3wt% with 32.2% in air and 33.5% under nitrogen. The PLQY decreases by increasing the concentration from 3wt% to 15wt% and then increases again with 20wt% to 31.7% in air and 32.1% under nitrogen, almost as much as with 3wt%. The performance in mCP gave the best PLQY, same as for **PXZ-DTT**, and lowered when using PMMA with 21.9% and 20.0% in air and under nitrogen, respectively. The PLQY without a host amounted to 13.7% in air and 15.6% under nitrogen.

Table 3 Photoluminescence quantum yields in air and nitrogen of **DPAC-DTT** in neat thin films and doped in different concentrations in mCP and PMMA with an excitation wavelength of 340 nm.

Entry	Host material	Doping concentration [wt%]	Φ_{PL} air [%]	Φ_{PL} nitrogen [%]
1	**mCP**	3	32.2	33.5
2	**mCP**	5	31.5	31.5
3	**mCP**	10	29.7	29.9
4	**mCP**	15	26.8	27.8
5	**mCP**	20	31.7	32.1
6	**PMMA**	10	21.9	20.0
7	-	neat	13.7	15.6

Table 4 shows the measurement results of **DMAC-DTT**. A doping concentration of 5wt% in mCP gave the best performance in air with a PLQY of 26.5%. The best result under nitrogen was observed with a doping concentration of 3wt% mCP with 31.0%.

Table 4 Photoluminescence quantum yields in air and nitrogen of **DMAC-DTT** in neat thin films and doped in different concentrations in mCP and PMMA with an excitation wavelength of 340 nm.

Entry	Host material	Doping concentration [wt%]	Φ_{PL} air [%]	Φ_{PL} nitrogen [%]
1	**mCP**	3	26.0	31.0
2	**mCP**	5	26.5	30.4
3	**mCP**	10	20.5	24.5
4	**mCP**	15	16.8	20.6
5	**mCP**	20	23.5	29.4
6	**PMMA**	10	14.1	17.6
7	-	neat	10.1	12.7

Again, the performance reduces with raising the doping concentration, except for 20wt% mCP, which gave 23.5% in air and 29.4% under nitrogen. Hence, 20wt% was almost as sufficient as 3wt%. Following the previous trend, doping in PMMA gave less performance with 14.1% and 17.6% in air and nitrogen, respectively. Measuring the emitter material **DMAC-DTT** solely, a PLQY of 10.1% in air and 12.7% under nitrogen was achieved.

Comparing all three DTT emitters, doping in mCP gave the best results in the following order: **PXZ-DTT**, **DPAC-DTT**, and **DMAC-DTT**. Using **PXZ-DTT** in a neat film showed a very low performance compared to **DPAC-DTT** and **DMAC-DTT**. Doping in PMMA gave less performance for all three emitters as doping in mCP.

Next, emission and excitation spectra and time-resolved spectroscopy were measured using all emitter-host partners with their best-performed doping concentration. Besides, the time-resolved photoluminescence decay curves for the emitter host pairs showed the best performance are displayed.

The complete photophysical characterization of **PXZ-DTT** is summarized in Table 5. The emission maxima shifted bathochromically from 530 nm when doped with mCP to 597 nm using PMMA (Figure 14). The prompt emission lifetime increased from 27.1 ns in mCP (3wt%) to 31.7 ns in PMMA (10wt%). The delayed emission lifetime is small, with 2.89 µs in mCP (3wt%), as opposed to 2.26 ms in PMMA (10wt%). For the neat compound, no data could be measured because the sample was too little emissive.

Table 5 Photophysical data of **PXZ-DTT** measured in neat thin films and doped in different hosts. τ_{PF} is the lifetime of the prompt fluorescence, and τ_{DF} is the lifetime of the delayed fluorescence.

Entry	Host	λ_{PL} [nm]	τ_{PF} [ns]	τ_{DF} [µs]
1	**mCP** (3wt%)	530	27.1	2.89
2	**PMMA** (10wt%)	597	31.7	2260
3	- (neat)	-	-	-

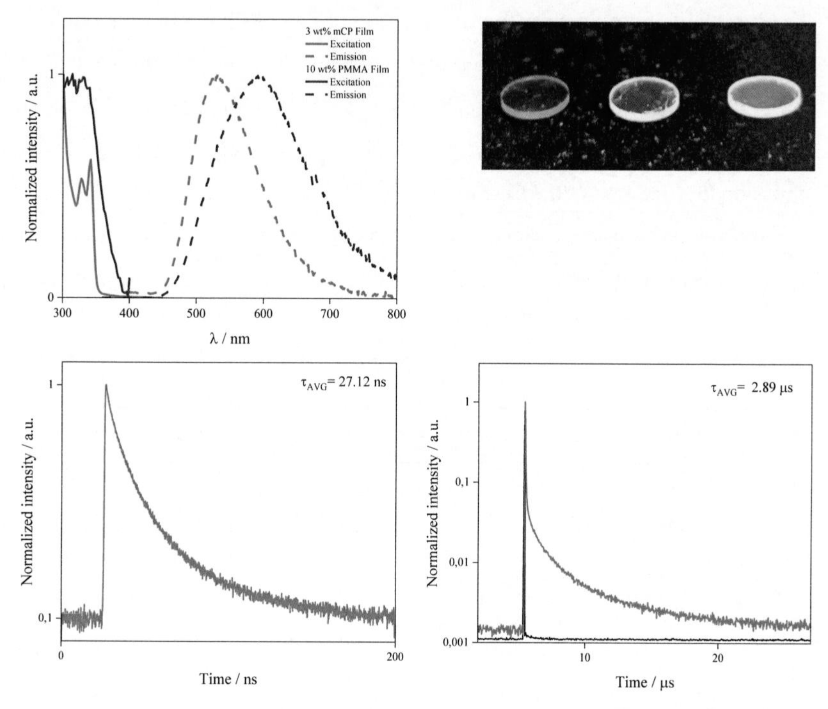

Figure 14 Emission and excitation spectra of **PXZ-DTT** in mCP film (3wt%, red) and PMMA film (10wt%, blue) (top left). A photo of the doped emitter in the following order from left to right: neat, PMMA, mCP (top right). Time-resolved spectroscopy of **PXZ-DTT** (3wt% mCP, red) with prompt decay (bottom left) and delayed decay (bottom right). The instrument response factor (IRF) is shown in the black curve. Results provided by Ettore Crovini.

The photophysical properties of **DPAC-DTT** are summarized in Table 6. The doping in mCP gave the bluest emission with 483 nm and shifted bathochromically when using PMMA with 502 nm and the neat sample with 575 nm (Figure 15). Figure 15 shows the time-resolved photoluminescence decay curves for the host material mCP (3wt%). The prompt fluorescence lifetime is small for mCP with 9.80 ns and increases when using PMMA with 11.9 ns and neat film with 19.7 ns. The delayed fluorescence lifetime is shortest for the host mCP with 3.39 μs and increases when using a neat film with 3.91 μs and PMMA with 4.59 μs. In summary, doping **DPAC-DTT** in mCP enhances its properties and gives the desired blue emission.

Table 6 Photophysical data of **DPAC-DTT** measured in neat thin films and doped in different hosts. τ_{PF} is the lifetime of the prompt fluorescence, and τ_{DF} is the lifetime of the delayed fluorescence.

Entry	Host	λ_{PL} [nm]	τ_{PF} [ns]	τ_{DF} [μs]
1	**mCP** (3wt%)	483	9.80	3.39
2	**PMMA** (10wt%)	502	11.9	4.59
3	- (neat)	575	19.7	3.91

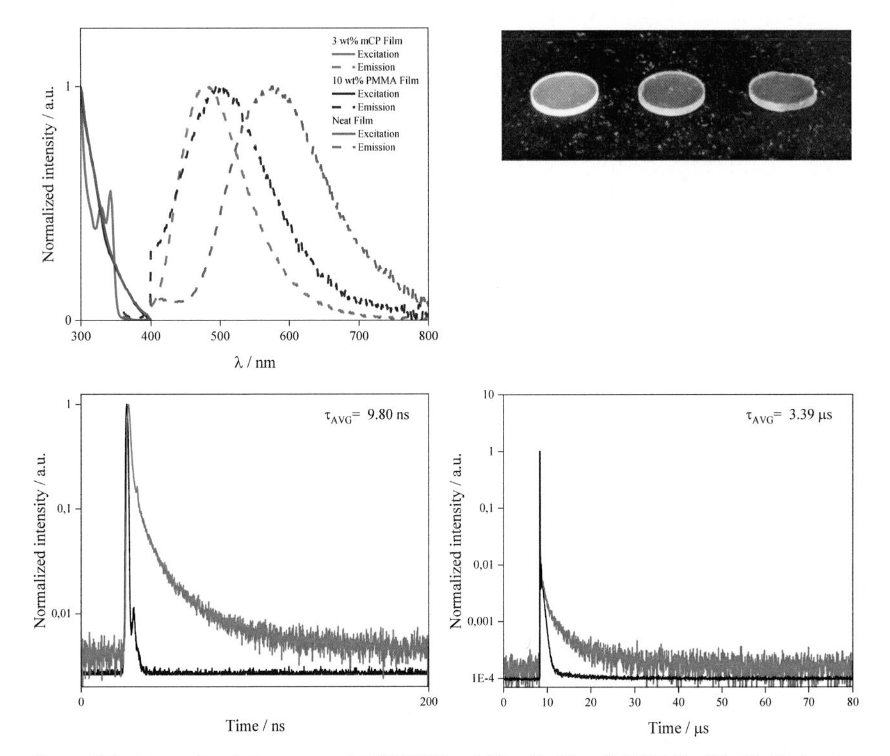

Figure 15 Emission and excitation spectra of **DPAC-DTT** in mCP film (3wt%, red), PMMA film (10wt%, blue), and neat film (green) (top left). A photo of the doped material in the following order from left to right: neat, PMMA, mCP (top right). Time-resolved spectroscopy of **DPAC-DTT** (3wt% mCP) with the prompt decay (bottom left) and the delayed decay (bottom right). IRF (black curve). Results provided by Ettore Crovini.

At last, the photophysical measurements of **DMAC-DTT** are shown in Table 7 and Figure 16. Following the trend of the other DTTs, the **DMAC-DTT** shows the bluest emission of 488 nm in mCP (3wt%). The emission color shifts to the red spectra using PMMA with 533 nm and neat film with 563 nm. The prompt fluorescence lifetime is short with 13.6 ns for the mCP host, and increases when using PMMA with 18.1 ns and neat with 58.3 ns. The delayed fluorescence lifetime is highest in mCP with 3.22 μs and decreases using the neat emitter with 2.10 μs and 2.46 μs in PMMA.

Table 7 Photophysical data of **DMAC-DTT** measured in neat thin films and doped in different hosts. τ_{PF} is the lifetime of the prompt fluorescence, and τ_{DF} is the lifetime of the delayed fluorescence.

Entry	Host	λ_{PL} [nm]	τ_{PF} [ns]	τ_{DF} [μs]
1	**mCP** (3wt%)	488	13.6	3.22
2	**PMMA** (10wt%)	533	18.1	2.46
3	- (neat)	563	58.3	2.10

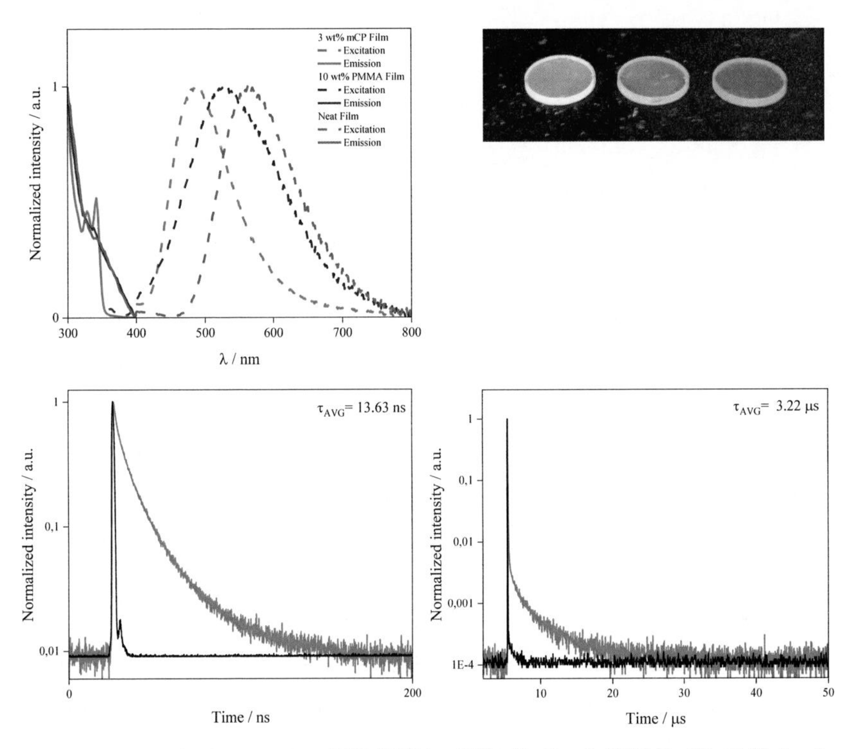

Figure 16 Emission and excitation spectra of **DMAC-DTT** in mCP film (3wt%, red), PMMA film (10wt%, blue), and neat film (green) (top left). A photo of the doped material in the following order from left to right: neat, PMMA, mCP (top right). Time-resolved spectroscopy of **DMAC-DTT** (3wt% mCP) with prompt decay (bottom left) and delayed decay (bottom right). IRF (black curve). Results provided by Ettore Crovini.

To summarize, all emitters performed best when using mCP as a host. The **DPAC-DTT** emitter showed the bluest fluorescence with 483 nm in the originating DTT series containing **PXZ-**, **DPAC-**, and **DMAC-DTT**. Closely follows **DMAC-DTT** with 488 nm. **PXZ-DTT** is significantly red-shifted owing to its strong **PXZ** donors. All emitters showed a short delayed lifetime in the μs range, which enhances the TADF step.

The original DTT system was expanded with methyl and trifluoromethyl groups in the phenylene spacers between the donor and acceptor. Methyl and trifluoromethyl groups, according to the literature, can enhance TADF and, in the case of CF_3, blueshift the emission.[126] Li *et al.* designed a series of blue TADF emitters and investigated the influence of different functional groups in the spacer. They incorporated methyl and trifluoromethyl at different positions in the phenylene spacer. They found that the CF_3 group, which deepens the LUMO with its electron-withdrawing character, reduces ΔE_{ST} and shortens τ_{DF}. Furthermore, they learned that a functional group like CH_3 and CF_3 close to the donor side, rather than to the acceptor, leads to the desired highly twisted structure, which is necessary to sufficiently separate HOMO and LUMO. In summary, they learned that incorporating substituents into the spacer, such as the CF_3 group close to the donor, can make the difference of an initially non-TADF becoming a TADF molecule.

To adopt their findings, the spacer of the DTT core with **PXZ** and **DMAC** as donors was modified using methyl and trifluoromethyl groups in the position closest to the donor. The photophysical measurements for compounds **CF3-PXZ-DTT** and **CF3-DMAC-DTT** were carried out through another collaboration with Idoia Garin from the research group of Prof Lemmer. The compounds were measured in a toluene solution. The results of the photophysical measurements are shown in Figure 17. Table 8 shows a summary and comparison with the original **PXZ-** and **DMAC-DTT**.

Table 8 Photophysical results of **CF3-PXZ-DTT, CF3-DMAC-DTT, PXZ-DTT,** and **DMAC-DTT** measured in toluene solution. The emission maxima were averaged from the prompt and delayed emission, except for **28e**.

Entry	Emitter	#	λ_{PL} [nm]	τ_{PF} [ns]	τ_{DF} [µs]
1	**CF3-PXZ-DTT**	**28e**	439_{PF} / 613_{DF}	1.50	0.02
2	**CF3-DMAC-DTT**	**28f**	527	17.5	0.8
3	**PXZ-DTT**	**28a**	602_{PF}	35.6	-

4	**DMAC-DTT**	**28b**	538	28.2	34.7

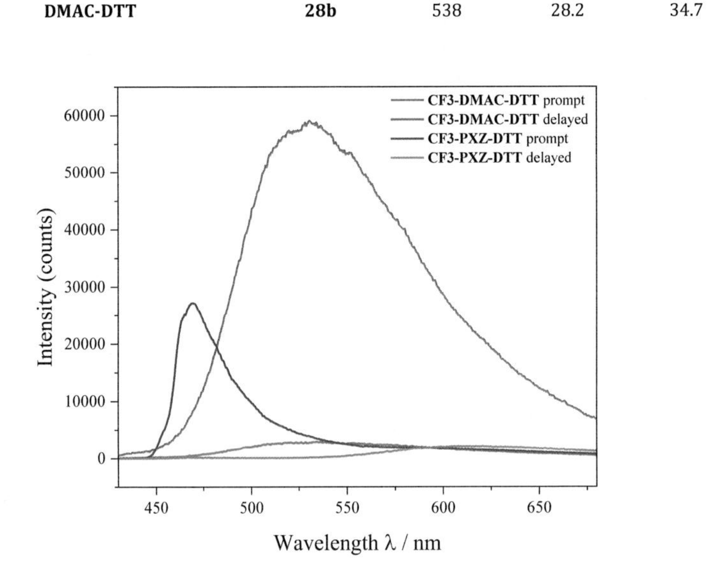

Figure 17 Emission spectra of the DTT emitters **28e-f** with the trifluoromethylated spacer. Results provided by Idoia Garin.

CF3-PXZ-MTT showed a rather low prompt fluorescent intensity compared to **CF3-DMAC-MTT**. Both emitters only exhibit a weak delayed emission. Usually, the emission maxima of the prompt and the delayed fluorescence are similar and only with small deviations. However, **CF3-PXZ-DTT** showed a significant shift between prompt and delayed components with a blue emission of 439 nm prompt fluorescence and a red emission of 613 nm delayed fluorescence.

The extended DTTs, **CF3-PXZ-DTT** and **CF3-DMAC-DTT**, were measured in a toluene solution. For a reasonable comparison, the original DTT emitters with **PXZ** and **DMAC** were also measured in toluene. The **DMAC-DTT** shows an emission maximum of 538 nm. By installing a CF_3 group in the *ortho*-position to the donor, a blueshift in the emission occurs at 527 nm. Furthermore, the delayed lifetime decreases from 34.7 μs to 800 ns for **DMAC-DTT** and **CF3-DMAC-DTT**, respectively. **PXZ-DTT**, measured in toluene, showed an emission maximum at 602 nm with a prompt lifetime of 35.6 ns. A delayed component was not observed. Following the trend Li *et al.* proposed, the emission of **CF3-PXZ-DTT** is blue-shifted with 439 nm and the prompt lifetime is shorter with 1.50 ns. A short delayed lifetime of 0.02 μs was observed.

Photophysical Measurements of MTT

The MTT emitters were photophysically characterized with the same methods as the DTT emitters.

First, the host screening for the original MTT series, containing **DMAC-**, **DPAC-**, and **PXZ-MTT**, was carried out. **DMAC-MTT** was tested in mCP, PMMA, 9-(4-*tert*-butylphenyl)-3,6-bis(triphenylsilyl)-9*H*-carbazole (CzSi), bis[2-(diphenylphosphino)phenyl] ether oxide (DPEPO) and without a host. The results are shown in Table 9.

Table 9 Host screening for **DMAC-MTT**. Photoluminescence quantum yields in air and nitrogen of **DMAC-MTT** in neat thin films and doped in different concentrations in mCP, CzSi, DPEPO and PMMA with an excitation wavelength of 340 nm.

Entry	Host material	Doping concentration [wt%]	Φ_{PL} air [%]	Φ_{PL} nitrogen [%]
1	**mCP**	3	40.7	56.8
2	**mCP**	5	52.3	59.1
3	**mCP**	10	56.2	62.1
4	**mCP**	15	49.9	53.7
5	**mCP**	20	54.7	57.1
6	**CzSi**	3	31.8	51.5
7	**CzSi**	5	36.0	52.9
8	**CzSi**	10	40.8	53.3
9	**CzSi**	15	40.7	49.0
10	**CzSi**	20	36.6	43.4
11	**DPEPO**	3	31.5	37.9

12	**DPEPO**	5	32.6	35.9
13	**DPEPO**	10	29.7	32.1
14	**DPEPO**	15	33.3	35.6
15	**DPEPO**	20	38.4	42.2
16	**PMMA**	10	27.1	31.6
17	-	neat	20.7	22.9

As expected, the PLQYs were higher when measured under a nitrogen atmosphere. In line with the results of the DTT series, the host mCP showed the best results. The highest PLQY, with 56.2% in air and 62.1% under nitrogen, was achieved when using 10wt% mCP. Using CzSi as a host the best PLQY with 53.3% under nitrogen and 40.8% in air was achieved with a host concentration of 10wt%. DPEPO achieved a poorer PLQY compared to the two above. The best outcome was achieved with 20wt% with 42.2% and 38.4% under nitrogen and in air, respectively. PMMA showed the poorest PLQY in this host screening series, with 20.7% in air and 22.9% under nitrogen.

The host screening of compound **DPAC-MTT** is shown in Table 10. Following the trend, mCP came off best, with 57.0% under nitrogen and 46.0% in air (15wt%). PMMA followed with 35.0% and 28.1% under nitrogen and air, respectively. As expected, the neat sample performed the poorest, with 28.0% under nitrogen and 19.1% in air.

Table 10 Photoluminescence quantum yields in air and nitrogen of **DPAC-MTT** in neat thin films and doped in different concentrations in mCP and PMMA with an excitation wavelength of 340 nm.

Entry	Host material	Doping concentration [wt%]	Φ_{PL} air [%]	Φ_{PL} nitrogen [%]
1	**mCP**	3	24.6	42.5
2	**mCP**	5	32.6	44.5
3	**mCP**	10	39.9	51.6

4	**mCP**	15	46.0	57.0
5	**mCP**	20	44.4	54.2
6	**PMMA**	10	28.1	35.0
7	-	neat	19.1	28.0

The results of **PXZ-MTT**, which is the last in the series of the origin MTTs, are shown in Table 11. A host concentration of 5wt% mCP gave the highest PLQY, 57.3% and 55.3% under nitrogen and air, respectively. The emitter performed poorly when embedded in PMMA, with only ~10%. A low PLQY of ~2% was measured using the emitter solely without a host. It was shown again that mCP led to the best performance.

Table 11 Photoluminescence quantum yields in air and nitrogen of **PXZ-MTT** in neat thin films and doped in different concentrations in mCP and PMMA with an excitation wavelength of 340 nm.

Entry	Host material	Doping concentration [wt%]	Φ_{PL} air [%]	Φ_{PL} nitrogen [%]
1	**mCP**	3	53.8	56.0
2	**mCP**	5	55.3	57.3
3	**mCP**	10	37.9	38.6
4	**mCP**	15	31.2	31.6
5	**mCP**	20	22.5	22.4
6	**PMMA**	10	10.1	10.3
7	-	neat	2.10	2.20

Additionally, the excitation and emission spectra and the time-resolved photoluminescence decay were measured for all original MTT emitters.

DMAC-MTT (10wt%) in mCP had an emission maxima of 530 nm, resulting in a blue-greenish emission. The emission is bathochromically shifted using PMMA as a host with 540 nm and the neat sample with 592 nm, emitting yellow light. The data for **DMAC-MTT** is shown in Table 12 and Figure 18.

Table 12 Photophysical data of **DMAC-MTT** measured in neat thin films and doped in different hosts. τ_{PF} is the lifetime of the prompt fluorescence, and τ_{DF} is the lifetime of the delayed fluorescence.

Entry	Host	λ_{PL} [nm]	τ_{PF} [ns]	τ_{DF} [µs]
1	**mCP** (10wt%)	530	40.6	5.91
2	**PMMA** (10wt%)	540	38.9	4.41
3	- (neat)	592	230	3.93

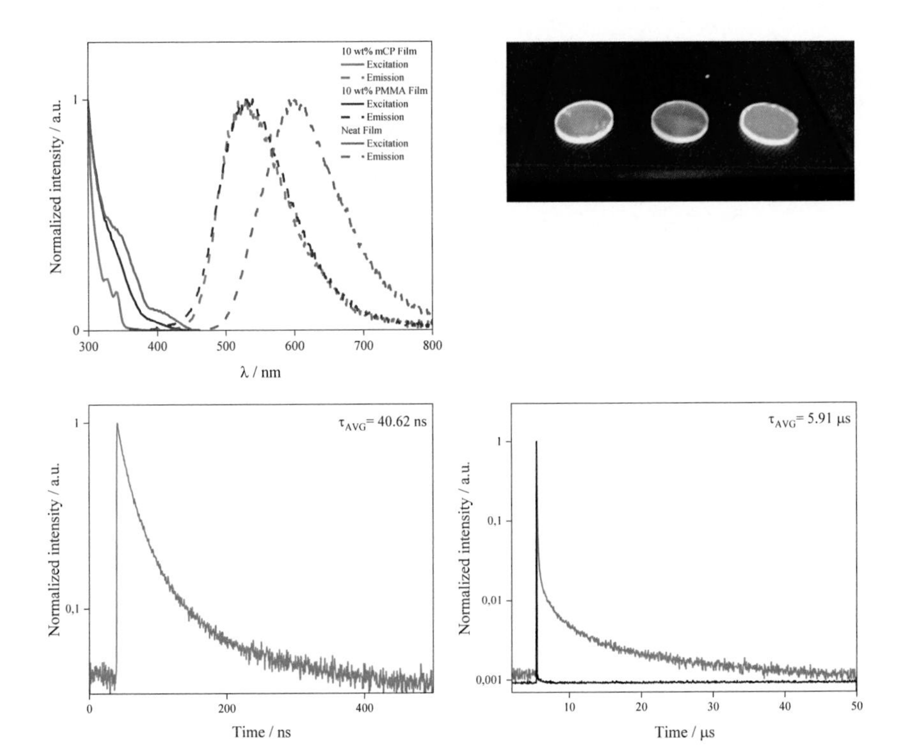

Figure 18 Emission and excitation spectra of **DMAC-MTT** in mCP film (10wt%, red), PMMA film (10wt%, blue), and neat film (green) (top left). A photo of the doped material in the following order from left to right: neat, PMMA, mCP (top right). Time-resolved spectroscopy of **DMAC-MTT** (10wt% mCP) with prompt decay (bottom left) and delayed decay (bottom right). IRF (black curve). Results provided by Ettore Crovini.

The delayed component τ_{DF} was the longest with 5.91 µs in mCP (10wt%) compared to 4.41 µs in PMMA and 3.93 µs in neat film. Nevertheless, the difference is rather small, and all three are in an appropriate range for TADF. The prompt lifetime is in the usual ns range for prompt fluorescence with 40.6 ns, 38.9 ns, and 230 ns for mCP, PMMA, and neat film, respectively.

Measuring the emission spectra of **DPAC-MTT** in mCP (15wt%) gave a maximum of 489 nm, exhibiting a blue emission (Table 13 and Figure 19). As expected, a red shift in the emission is observed with 513 nm (PMMA) and 532 nm (neat film). The prompt component is in the ns range with 29.7 ns, 26.9 ns, and 205 ns in mCP, PMMA and neat film, respectively. The delayed fluorescence exhibited a long lifetime of 881 μs in mCP and 1.08 ms in PMMA. The neat film did not show a delayed component, hence no TADF.

Table 13 Photophysical data of **DPAC-MTT** measured in neat thin films and doped in different hosts. τ_{PF} is the lifetime of the prompt fluorescence, and τ_{DF} is the lifetime of the delayed fluorescence.

Entry	Host	λ_{PL} [nm]	τ_{PF} [ns]	τ_{DF} [μs]
1	**mCP** (15wt%)	489	29.7	881
2	**PMMA** (10wt%)	513	26.9	1078
3	- (neat)	532	205	-

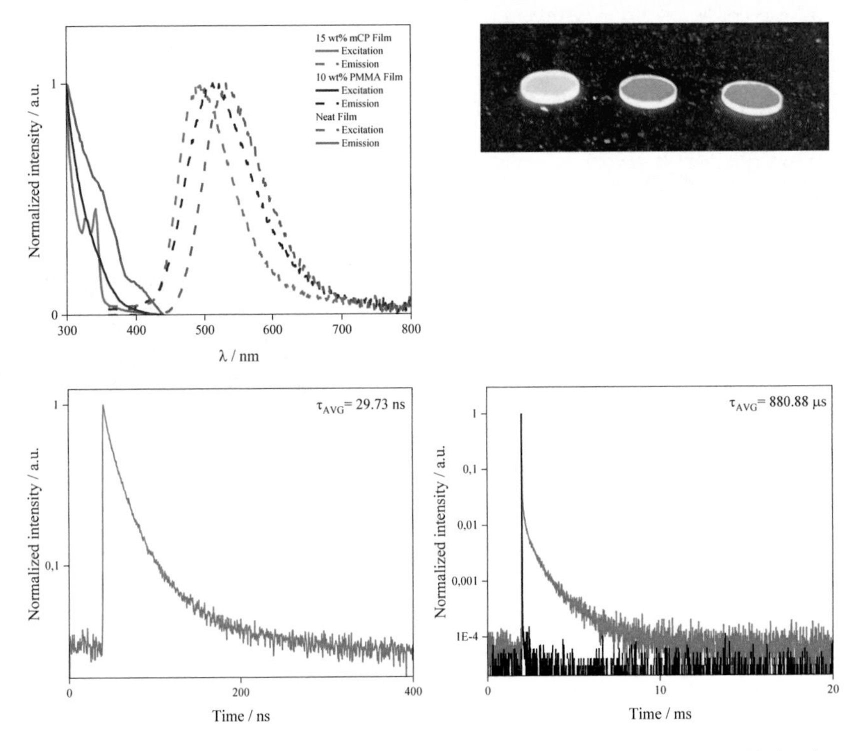

Figure 19 Emission and excitation spectra of **DPAC-MTT** in mCP film (15wt%, red), PMMA film (10wt%, blue), and neat film (green) (top left). A photo of the doped material in the following order from left to right: neat, PMMA, mCP (top right). Time-resolved spectroscopy of **DPAC-MTT** (15wt% mCP) with prompt decay (bottom left) and delayed decay (bottom right). IRF (black curve). Results provided by Ettore Crovini.

The last emitter in the original MTT series is the **PXZ-MTT**. The photophysical data and the photoluminescence spectrum are shown in Table 14 and Figure 20. When doped in mCP (5wt%), the emission resulted in two maxima with 540 and 580 nm. Replacement with PMMA gave an emission maximum of 596 nm, slightly red-shifted to the emission with mCP dopant. The prompt and delayed fluorescence lifetimes for the emitter doped in mCP and PMMA had similar rates with 46.5 and 52.2 ns prompt and 3.60 and 2.82 µs delayed, respectively. The neat film was too little emissive to measure any decay curve data.

Table 14 Photophysical data of **PXZ-MTT** measured in neat thin films and doped in different hosts. τ_{PF} is the lifetime of the prompt fluorescence, and τ_{DF} is the lifetime of the delayed fluorescence.

Entry	Host	λ_{PL} [nm]	τ_{PF} [ns]	τ_{DF} [µs]
1	**mCP** (5wt%)	540-580	46.5	3.60
2	**PMMA** (10wt%)	596	52.2	2.82
3	- (neat)	592	-	-

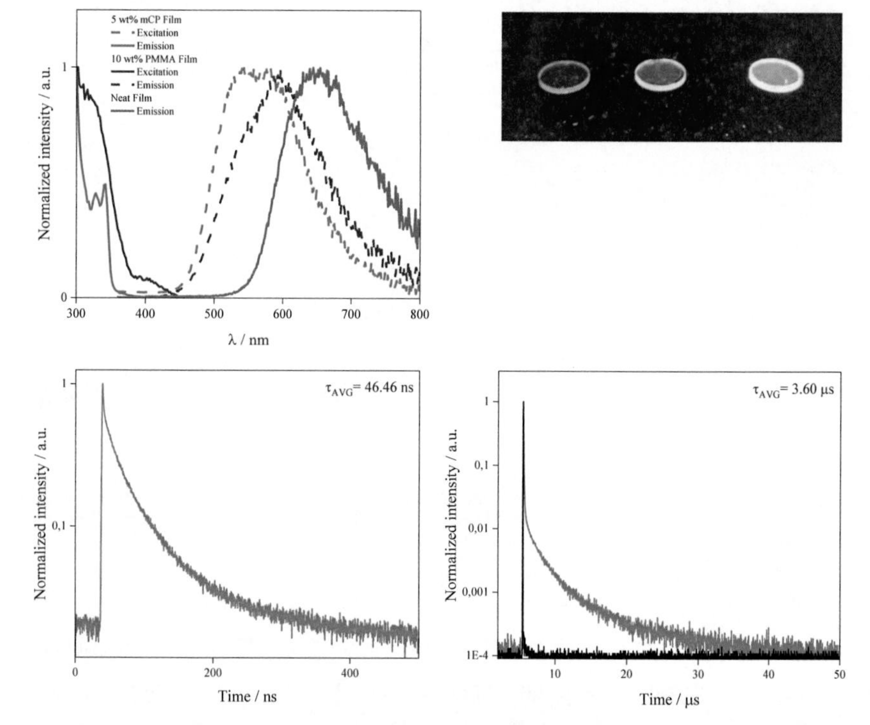

Figure 20 Emission and excitation spectra of **PXZ-MTT** in mCP film (5wt%, red), PMMA film (10wt%, blue), and neat film (green) (top left). A photo of the doped material in the following order from left to right: neat, PMMA, mCP (top right). Time-resolved spectroscopy of **PXZ-MTT** (5wt% mCP) with prompt decay (bottom left) and delayed decay (bottom right). IRF (black curve). Results provided by Ettore Crovini.

As for the DTT series, the MTTs with **PXZ** and **DMAC** were decorated with methyl and trifluoromethyl groups in the phenylene spacer. Furthermore the **PXZ-MTT** was additioned with methyl groups on the phenyl groups bearing no donor, giving **PXZ-MTT-Me**. The photophysical measurements of **30e-g** were carried out by Idoia Garin, and **30a-b** were carried out by Ettore Crovini. The results are shown in Table 15 and Figure 21.

Table 15 Measurements of **30a-b**, **-e-g** in toluene solution. The emission maxima were averaged from the prompt and delayed emission.

Entry	Emitter	#	λ_{PL} [nm]	τ_{PF} [ns]	τ_{DF} [ms]
1	**CF3-PXZ-MTT**	**30e**	586	29.5	0.0005
2	**CF3-DMAC-MTT**	**30f**	527	35.5	4
3	**PXZ-MTT-Me**	**30g**	585	32.0	0.00055
4	**DMAC-MTT**	**30b**	538	-	0.0109
5	**PXZ-MTT**	**30a**	596	26.6	1270

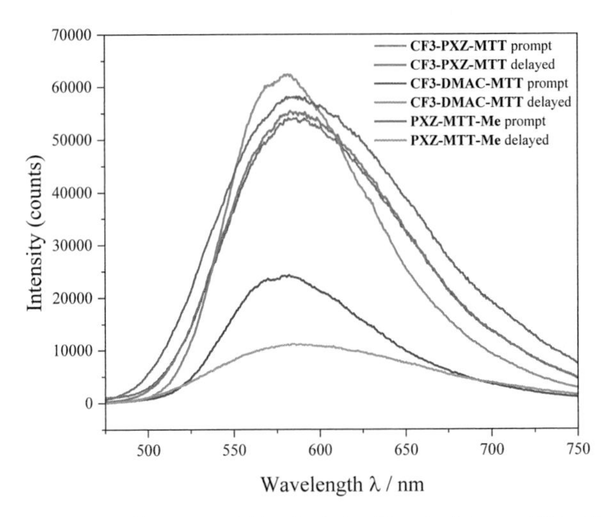

Figure 21 Emission spectra with prompt and delayed emission of the extended spacer MTT emitters **30e-g**. Results provided by Idoia Garin.

The emission wavelengths ranged from 527 to 585 and 586 nm for **CF3-DMAC-MTT**, **PXZ-MTT-Me**, and **CF3-PXZ-MTT**, respectively. The emission maxima and

fluorescent lifetimes for the **PXZ-MTT** emitters **30e** and **30g** were similar with only small deviations. The intensity of the delayed emission is significantly weaker for **PXZ-MTT-Me**, whereas a similar intensive prompt and delayed fluorescence is shown for **CF3-PXZ-MTT**. The delayed fluorescence of **CF3-DMAC-MTT** was significantly stronger than its prompt emission, which is favorable for TADF applications.
Photoluminescent and time-resolved spectra were measured for **PXZ-MTT** and **DMAC-MTT** to compare them with **30e-f**. The pattern repeats, showing that the emission maximum hypsochromically shifts from 538 nm for **DMAC-MTT** to 527 nm for **CF3-DMAC-MTT** and from 596 nm for **PXZ-MTT** to 586 nm for **CF3-PXZ-MTT**. The delayed fluorescence lifetime decreases significantly from 1270 ms for **PXZ-MTT** to 0.0005 ms for **CF3-PXZ-MTT**. Except for an increase in the lifetime of the delayed component with 0.0109 ms for **DMAC-MTT** to 4 ms for **CF3-DMAC-MTT**, the overall trend Li *et al.* proposed can be observed.

Solvatochromism

DMAC-MTT and **DMAC-DTT** were studied for solvatochromism (Figure 22). Solvents like Methylcyclohexane, toluene and tetrahydrofuran were tested for both and dichloromethane for **DMAC-MTT**. The emission maxima for **DMAC-MTT** shifted bathochromically from 448 nm (Me-Cy-Hex), 533 nm (toluene), and 688 nm (THF) to 714 nm (DCM). The same order was observed for **DMAC-DTT** with 507 nm (Me-Cy-Hex), 525 nm (toluene) to 656 nm (THF).

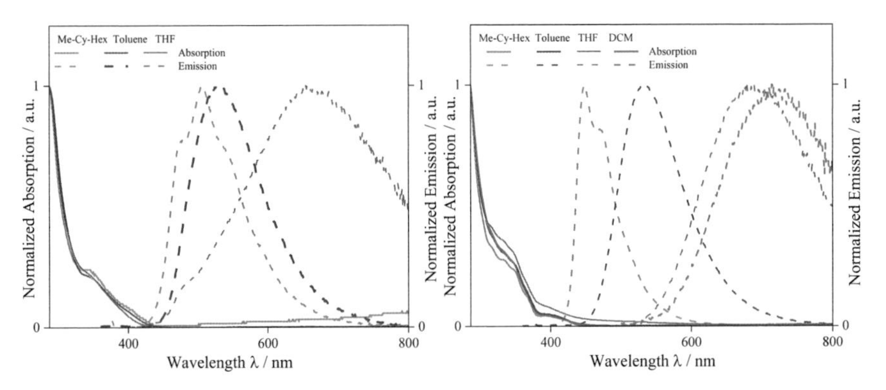

Figure 22 Solvatochromism study of **DMAC-DTT** (left) and **DMAC-MTT** (right) using different solvents, like methylcyclohexane, toluene, tetrahydrofuran, and dichloromethane. Results provided by Ettore Crovini.

Temperature-dependent Time-resolved Decay Measurements

The temperature-dependent time-resolved decay curves were measured for **DMAC-MTT** at 77 K, 200 K, 250 K, and 300 K. The results are shown in Figure 23. The delayed fluorescence decay data revealed an increase of the intensity of the delayed emission by increasing the temperature. This intensified TADF decay is characteristic for molecules showing TADF. Furthermore **DMAC-DTT** was measured at 77 K, 250 K, and 300 K (Figure 24). The spectra are split up in two because the curve at 250 K exactly overlaps with the one at 300 K. The temperature dependency, characteristic for TADF, can also be overserved here.

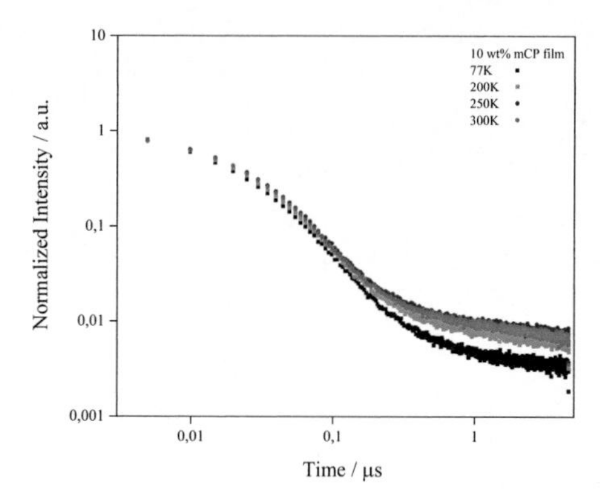

Figure 23 Delayed fluorescent decay data for **DMAC-MTT**, in 10wt% doped mCP films, measured at different temperatures. Results provided by Ettore Crovini.

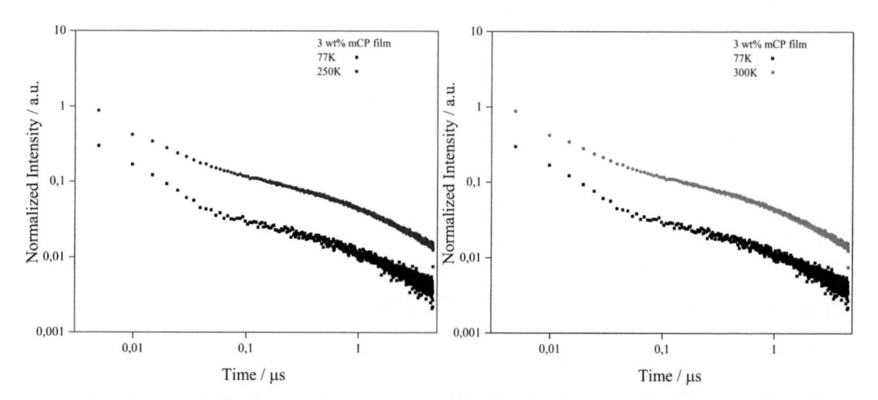

Figure 24 Delayed fluorescent decay data for compound **DMAC-DTT** (in 3wt% doped mCP films), measured at different temperatures. Results provided by Ettore Crovini.

To summarize, a wide range of TADF emitters based on the MTT and DTT acceptor core with different donors and alternations in the phenylene spacer were synthesized, theoretically calculated and photophysically characterized. **DMAC-DTT** and **DPAC-DTT** emit in the desired blue region and exhibit similar TADF properties in doped films. In the MTT class, **DMAC-MTT** performed best in all aspects, with its blue emission and good TADF properties. **DMAC-DTT** is the best TADF emitter in the MTT and DTT series considering blue emission coupled with good TADF properties. It showed a bluer emission and superseded **DMAC-MTT** with its shorter prompt and delayed decay rates of 13.6 ns and 3.22 µs, as opposed to 40.6 ns and 5.91 µs for **DMAC-MTT**. A possible explanation could be that the presence of two **DMAC** donors enhances the TADF properties. Furthermore, it was shown that the emission could be shifted hypsochromically when installing a trifluoromethyl group in the *ortho*-position of the donor, like in **CF3-PXZ-MTT/-DTT** and **CF3-DMAC-MTT/-DTT**.

Thermal Stability

A thermogravimetric analysis (TGA) was conducted exemplarily with **PXZ-DTT** for the triazolotriazine compounds as a representative to explore the thermal stability of this molecule class. The TGA measurement is shown in Figure 25.
The loss of weight measured in weight% is plotted against the temperature. Small losses can be noted from the measurement's beginning until a temperature of approximately 400 °C. Afterwards, the sample starts to decompose. Most of the sample was decomposed at a temperature of ~495 °C. Between 500 °C and 1000 °C, the remaining 15% decomposed. The high thermal stability of emitters for OLED devices is important for a high power/brightness operation[127] and, among other things, for high processing temperatures.[128] With a stability of up to roughly 500 °C, the **PXZ-DTT** is suitable for application in an OLED device.

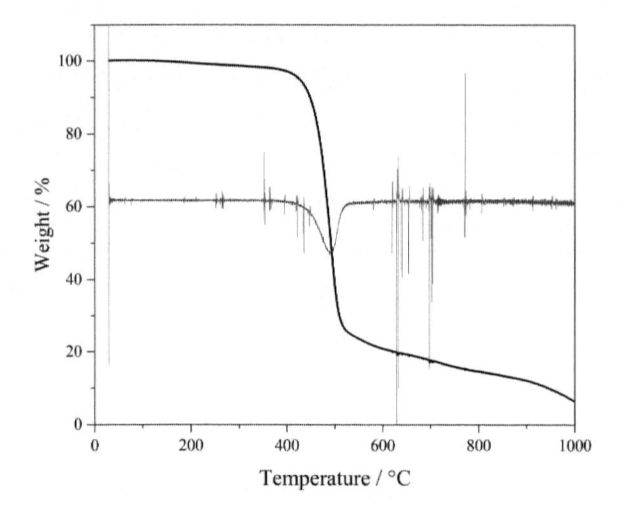

Figure 25 Thermogravimetric analysis of TADF compound **PXZ-DTT**.

3.2. Derivatization of the CzBN

The lifetime of blue TADF OLEDs was and still is an issue. Zhang *et al.* reported the CzBN series in 2016 (Figure 4).[2] They combined the novelty of efficient blue TADF emitters with a long lifetime. An efficient luminance and the modification of the peripheral substituents core seemed to be key. Bulky substituents are anticipated to enhance efficiency and stability in TADF emitters. Therefore, 3,6-di-*tert*-butyl-9*H*-carbazole donors were used to decorating the benzonitrile core, which is already an established acceptor. To design blue TADF emitters, benzonitrile is favored over phthalonitrile. Although the well-known green TADF emitter 4CzIPN, which has a phthalonitrile acceptor, is as efficient as 30% EQE, the fabrication of efficient blue TADFs is difficultly rendered by the strong electron-withdrawing abilities of the two nitrile moieties. By removing one nitrile from the benzene core, the electron-withdrawing abilities are weakened, resulting in a relatively large energy gap, hence bluer TADF emission.
Zhang *et al.* introduced four TADF emitters (**4CzBN**, **4TCzBN**, **5CzBN**, and **5TCzBN** (**9**), Figure 4) based on benzonitrile with carbazole or 3,6-di-*tert*-butyl-9*H*-carbazole donors in structures with either four or five donors attached to the core and investigated the different effects impacting the TADF properties.
They observed that the **tCz** moiety is a shield covering the luminant core, promoting photoluminescence efficiency and improving stability. Furthermore, **tCz** donors enhance solubility, which is often an issue for TADF compounds. Their devices with **tCz** achieved higher EQEs than those with the same system without *tert*-butyl groups. **4CzBN** achieved an EQE_{max} of 10.6%, whereas **4TCzBN** yielded 16.2%. The same was observed for the five-donor system yielding an EQE_{max} of 16.7% for **5CzBN** and 21.2% for **5TCzBN**. All emitters showed a sky-blue to blue emission with 442, 456, 464, and 480 nm for **4CzBN**, **4TCzBN**, **5CzBN**, and **5TCzBN**, respectively. Increasing the number of donors and/or introducing *tert*-butyl groups resulted in a red emission shift.

On top of that, they revealed that multiple donor systems facilitate sufficiently rISC, meaning k_{rISC} increases, by mixing sufficiently CT states with ^{3}LE states.[129]

In 2020, Zhang *et al.* expanded the CzBN system by introducing a phenyl bridge between the benzene core and the nitrile acceptor.[129] They intended to blue-shift the emission since blue emitters below 470 nm are still challenging. As already described, a blue shift can be achieved by reducing the number of donors. However, in correspondence, k_{rISC} decreases.[130] Two possible reasons can be attributed to this observation. First, fewer donors lead to fewer possible intermediate triplet and delocalized states. Second, fewer donors mean a less crowded structure, which can facilitate non-radiative vibrational processes.
They incorporated one or two phenylnitriles, which are weaker than benzonitrile, into their emitters. A weaker acceptor can fine-tune CT levels to reach the locally excited triplet states easier. Furthermore, these D-π-A systems improve the delocalization of excited states.[129]

The following projects were designed based on the outstanding results of the CzBN series. The aim lay on the acceptor modification. Therefore, the donor group itself was not changed during the subsequent studies. 3,6-Di-*tert*-butyl-9*H*-carbazole was chosen because the carbazole shows excellent electron-donating abilities, and the *tert*-butyl substituents have a great steric, which enhances TADF and help to make the compounds more soluble. Beneficial is also that the *tert*-butyl groups function as a probe to facilitate the analysis by NMR spectroscopy.

3.2.1. DFT Calculations

Through a collaboration with Anna Mauri and Mariana Kozlowska from the Wenzel group, KIT, the precursor **5TCzPhBN**, **5TCzPhT**, **5TCzBN** and **5TCzT** were calculated using DFT methods. It is noted that most calculations are performed under gas phase conditions, and measurements in the experiments are conducted in solution. Different DFT functionals were tried and later compared with the result of the experiments, which were conducted through collaboration with Idoia Garin. The HOMO-LUMO distributions, singlet S_1 and triplet T_1-T_n energy levels, SOC and PLQY, were calculated for **5TCzPhBN**. Due to the structural similarity of these molecules, compounds **5TCzPhT**, **5TCzBN** and **5TCzT** will not be further discussed in this work. It should be noted that according to the calculations, the HOMO is located on the donors and the LUMO on the acceptor for all four molecules, which is a good starting point for TADF characteristics.

The results of **5TCzPhBN** are discussed in the following. The absorption maximum from the experiment was measured at 343 nm. The BMK DFT functional delivered the closest result to this experiment. The PLQY for the neat sample was calculated to be 30% with an emission maximum of 465 nm. Further, the molecular orbitals were calculated with a HOMO value of -5.47 eV and a LUMO value of -2.06 eV (Figure 26 and Table 16). A high charge transfer rate is seen, which is favorable for TADF. After performing excited state optimization with TD-DFT, ΔE_{ST} was calculated to 0.32 eV. This ΔE_{ST} is on the upper limit for sufficient TADF but can be nonetheless further investigated. A high SOC from S_1 of 0.9 cm^{-1} between S_1 and T_1 was calculated. The higher the SOC, the more probable ISC is between the states.

Table 16 Calculation of the HOMO-LUMO energy levels, S_1, T_1, and ΔE_{ST} for **5TCzPhBN**.

Entry	Compound	#	HOMO [eV]	LUMO [eV]	S_1 [eV]	T_1 [eV]	ΔE_{ST} [eV]
1	**5TCzPhBN**	**37**	-5.47	-2.06	3.18	2.86	0.32

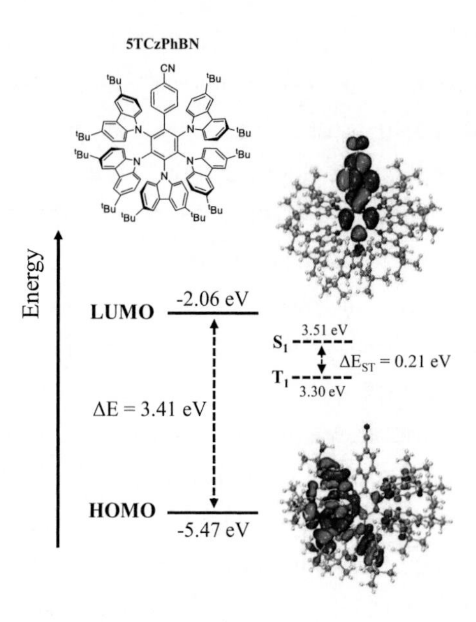

Figure 26 Visualization of the HOMO and LUMO distribution of the molecular orbitals and excited state energy levels of **5TCzPhBN**. Results provided by Anna Mauri.

3.2.2. Tetrazole Derivatives of the CzBN Class

In this project, the literature-known **5TCzBN** and a new system to this class, **5TCzPhBN**, were used as a starting point for acceptor modifications. As discussed in the previous chapter, a phenyl bridge between the acceptor and the benzene core results in a weaker acceptor, hence potential better TADF. The nitrile acceptor moiety was transformed into a tetrazole and then derivatized with various functional groups. The synthesis was designed for a modulative approach such that the acceptor modifications can be done via a single step. The aim was to investigate the impact of altering the acceptor on the overall system.

3.2.2.1. Syntheses

Syntheses of the Precursors

The **5TCzBN** and **5TCzPhBN** scaffolds were synthesized within two and three steps, respectively. As shown in Scheme 13, 2',3',4',5',6'-pentafluoro-4-carbonitrile (**32**) and 3,6-di-*tert*-butyl-9*H*-carbazole (**31**) were reacted in a nucleophilic aromatic substitution reaction giving product **5TCzBN** (**9**) in a 60% yield. The nitrile **9** was transformed into a tetrazole using sodium azide. After purification, **5TCzT** (**33**) was isolated in a 79% yield. The yield over two steps amounts to 47%.

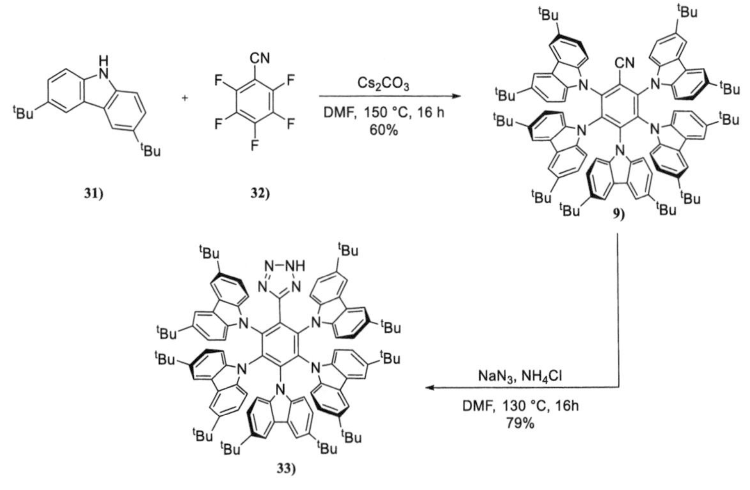

Scheme 13 Synthesis of **5TCzT** within two steps using 2',3',4',5',6'-pentafluoro-4-carbonitrile (**32**) and 3,6-di-*tert*-butyl-9*H*-carbazole (**31**). The overall yield amounts to 47%.

In the first step of the **5TCzPhBN** synthesis, the nitrile-phenyl-bridge was connected to the fluorinated benzene (Scheme 14). Therefore, (4-cyanophenyl)boronic acid (**34**) and 1-bromo-2,3,4,5,6-pentafluorobenzene (**35**) were reacted with potassium phosphate under Pd catalysis in a SUZUKI cross-coupling reaction obtaining 2',3',4',5',6'-pentafluoro-[1,1'-biphenyl]-4-carbonitrile (**36**) with a 79% yield. The **tCz** donors were attached via a nucleophilic aromatic substitution reaction in the second step. Thus, **36** was reacted with **tCz** and cesium carbonate to obtain **5TCzPhBN** (**37**) in a 74% yield. Lastly, a 1,3-dipolar cycloaddition using sodium azide gave the precursor **5TCzPhT** (**38**) in an excellent yield of 98%. The overall yield over three steps amounts to 57%.

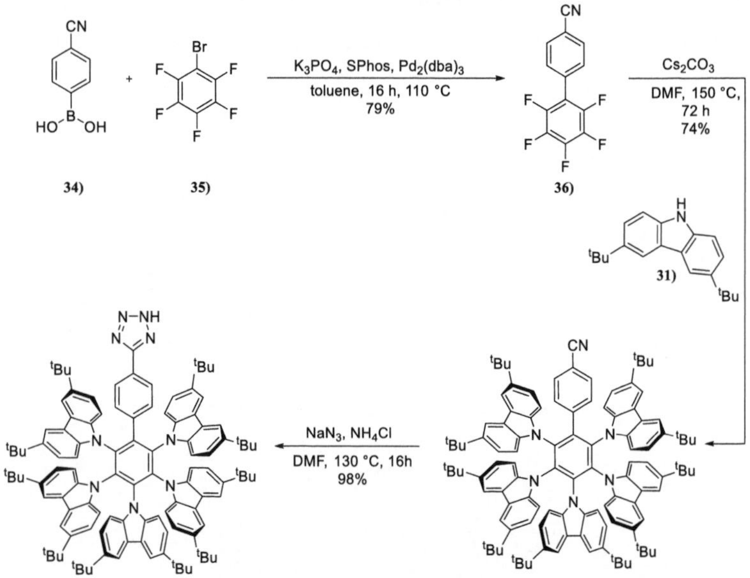

Scheme 14 Synthesis of **5TCzPhT** within three steps with an overall yield of 57%.

Synthesis of Tetrazole Derivatives

The rationale for the choice of substituents at the acceptor had three aims. The effect on the molecules' photophysical characteristics was studied by implementing electron-donating and -withdrawing groups. Further, the feasibility of implementing sterically hindered groups into the already highly crowded core structure was investigated. And lastly, the tolerance of functional groups was examined.

Tetrazole **5TCzPhT** was derivatized with aromatic systems such as phenyl-, *tert*-butyl-phenyl, anisole, trifluoromethyl-phenyl, benzonitrile-, indole-, and nitrobenzene groups and aliphatic groups like propyl, propenyl, propynyl, and *neo*-pentyl.

Aromatic substituents were attached with the corresponding boronic acid using potassium carbonate under copper catalysis with oxygen. Five products **5TCzPhT-PhtBu**, **5TCzPhT-PhOMe**, **5TCzPhT-PhtCz**, **5TCzPhT-PhCF3**, **5TCzPhT-PhIn** (**39a-e**) were successfully synthesized. It was observed that electron-rich aromatic systems with *tert*-butyl (**39a**) or methoxy groups (**39b**) facilitated the reaction. The compounds were isolated in very good yields ranging from 74% for **39a** and 71% for **39b**. A sixth donor directly attached to the tetrazole, as in **39c**, led to an average yield of 43%. The boronic acid precursor was synthesized with 9-(4-bromophenyl)-3,6-di-*tert*-butyl-9*H*-carbazole, *n*-butyllithium, and trimethyl borate. Electron-withdrawing groups like trifluoromethyl and indole showed lower yields, with 37% and 54% for **39d** and **39e**, respectively. The attachment of nitrobenzene was not successful, and no conversion of the respective compound was observed. Benzonitrile could not be attached via the original route. Therefore it was tried under different conditions using 4-fluorobenzonitrile and potassium carbonate according to a literature procedure.[131] However, no conversion was observed.

Aliphatic substituents were attached using reaction condition B with the associated alkyl bromide and triethyl amine as a base. Tetrazoles with propyl- (**5TCzPhT-PhPayl**, **39f**), propenyl- (**5TCzPhT-PhPeyl**, **39g**), and propynyl- (**5TCzPhT-PhPiyl**, **39h**) groups were successfully synthesized with yields of 95%, 81%, and 53%, respectively. The reaction with *neo*-pentyl did not result in any conversion. In a second-order nucleophilic substitution (S_N2), the nucleophile attacks the backside of the substrate. The *tert*-butyl groups in *neo*-pentyl are too sterically demanding, which makes a backside attack impossible.

Overall, the aliphatic-attributed reactions performed better with higher yields than the aromatic ones.

A) R-B(OH)$_2$, Cu(OH)(TMEDA))$_2$Cl$_2$, K$_2$CO$_3$, O$_2$ DCM, rt, 16 h
B) R-Br, NEt$_3$ MeCN, 60 °C, 24 h

38) → **39a-h)**

R =

39a) **5TCzPhT-PhtBu** 74% A)

39b) **5TCzPhT-PhOMe** 71% A)

39c) **5TCzPhT-PhtCz** 43% A)

39d) **5TCzPhT-PhCF3** 37% A)

39e) **5TCzPhT-PhIn** 54% A)

39f) **5TCzPhT-PhPayl** 95% B)

39g) **5TCzPhT-PhPeyl** 81% B)

39h) **5TCzPhT-PhPiyl** 53% B)

Scheme 15 Syntheses of tetrazole derivatives **39a-h**. Aromatic substituents were reacted under condition A), whereas aliphatic substituents were reacted under condition B).

Next, tetrazoles of **5TCzT** were derivatized (Scheme 16). 1,3-Di-*tert*-butyl-benzene, phenyl, benzonitrile, tri-methoxy benzene, *tert*-butyl, propyl, propynyl, and *neo*-pentyl were screened. Only two aliphatic tetrazoles were successfully synthesized with yields of 29% for **5TCzT-ayl** (**40a**) and 57% for **5TCzT-iyl** (**40b**) (Scheme 16). Aromatic groups could neither be attached via the original route under condition A) nor via another route with the associated fluorine and potassium carbonate.[131] *Neo*-pentyl could not be attached for the same reason already evaluated for compound **5TCzPhT**.

R-Br, NEt_3
MeCN, 60 °C, 24 h

33) **40a-b)**

R =

40a)
5TCzT-ayl
29%

40b)
5TCzT-iyl
57%

Scheme 16 Derivatization of **5TCzT** with propyl and propynyl using the associated alkyl bromide and triethylamine.

3.2.2.2. Photophysical Characterization

First, photoluminescence spectra of the precursors **5TCzPhBN**, **5TCzPhT**, **5TCzBN**, and **5TCzT** have been measured. **5TCzPhT** gave the bluest prompt fluorescent emission in this series with 444 nm. A very weak delayed component was detected; however, the noise of the spectra is too high to determine it accurately. **5TCzPhBN** showed an emission maximum of 453 nm and a very small delayed component with a small decay. Because of these results, the scaffolds of **5TCzPhBN** and **5TCzPhT** were not considered to give appropriate TADF properties. Precursor **5TCzBN**, in contrast, showed an emission maximum at 485 nm and a delayed emission at 472 nm. The intensity of the delayed emission is stronger than the prompt one (Figure 27). The tetrazole precursor **5TCzT** showed a blue-shifted prompt emission at 452 nm and a similarly delayed emission at 487 nm, compared to the associated nitrile acceptor. The prompt emission is more intense than the delayed one, unfavorable for efficient TADF. Comparing nitrile **5TCzBN** and tetrazole **5TCzT** precursors without the phenylene spacer, the tetrazole shows a higher emission intensity; however, the nitrile shows a stronger delayed component than its prompt one. The full width at half maximum (FWHM) value was measured to be smaller for the nitrile compound **5TCzBN** with 75 (prompt) and 73 nm (delayed) than for the tetrazole compound **5TCzT** with 95 nm.

The measurement process consumes a significant amount of time, which is why the derivatization of both tetrazole precursors was started without the knowledge of the experiment. After measuring, it became clear that the tetrazole compound with a phenylene spacer is unsuitable for TADF. Nevertheless, the research was focused on the synthesis and the development of a proof of principle to show the success of these types of reactions and the tolerance of varied functional groups as well as coping with steric hindrance. Ten novel structures based on the carbazole-tetrazole type were successfully synthesized, and eight were photophysically characterized.

Table 17 Photophysical data of the precursors **5TCzPhBN**, **5TCzBN**, and **5TCzT** measured in toluene solution.

Entry	Compound	#	$\lambda_{max, PF}$ [nm]	$FWHM_{PF}$ [nm]	$\lambda_{max, DF}$ [nm]	$FWHM_{DF}$ [nm]	lifetime τ_{PF} [ns]	lifetime τ_{DF} [ms]
1	**5TCzPhBN**	**37**	453	-	-	-	-	-
2	**5TCzPhT**	**38**	444	-	-	-	-	-
3	**5TCzBN**	**9**	485	75	472	73	5.6	14-19
4	**5TCzT**	**33**	452	95	487	95	5.2	5.5

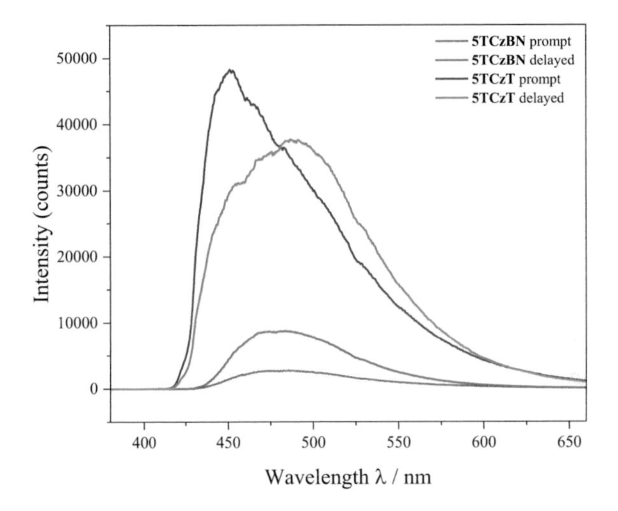

Figure 27 Emission spectra with prompt and delayed emission of **5TCzBN** and **5TCzT** measured in toluene solution. Results provided by Idoia Garin.

Compounds **39a-d**, **39f**, **39h**, and **40a-b** were studied photophysically. The basic photophysical data and the photoluminescence spectra, such as emission maxima and investigation of a delayed component, are shown in Table 18 and Figure 28. The other molecules could not be studied because a great amount of the compounds is necessary for the photophysical experiments. The photoluminescence spectra in toluene showed that all compounds emit in the blue part of the spectrum. Emission maxima λ_{max} of 454 nm, 445 nm, 450 nm, 444 nm, 449 nm, 445 nm, 450 nm, and 450 nm were measured for **39a**, **39b**, **39c**, **39f**, **39d**, **39h**, **40a**, and **40b**, respectively. The emission maxima were close and a trend, if electron-donating or-withdrawing groups influence the color cannot be undoubtedly seen. All spectra show similar broad emissions with FWHM values ranging from 45-93 nm.

Experiments to measure the excited state lifetimes have not been finished yet and are therefore missing. The tetrazoles with a phenylene spacer showed only a prompt emission and no delayed component. Only compounds **40a-b**, those tetrazoles without a spacer, showed a delayed emission, although very weak.

Table 18 Photophysical data of the tetrazole compounds **39a-d**, **-f**, **-h**, and **40a-b** measured in toluene solution.

Entry	Compound	#	$\lambda_{max, PL}$ [nm]	FWHM [nm]
1	**5TCzPhT-PhtBu**	**39a**	454	72
2	**5TCzPhT-PhOMe**	**39b**	445	86
3	**5TCzPhT-PhtCz**	**39c**	450	93
4	**5TCzPhT-PhCF3**	**39d**	449	62
5	**5TCzPhT-PhPayl**	**39f**	444	45
6	**5TCzPhT-PhPiyl**	**39h**	445	52
7	**5TCzT-ayl**	**40a**	450	71
8	**5TCzT-iyl**	**40b**	450	75

The phenylene spacer in the **39**-tetrazole series probably separates the frontier orbitals too extensively, so an interaction between HOMO and LUMO is impossible. Without a spacer like tetrazoles **40a-b**, the HOMO and LUMO can overlap sufficiently, making TADF possible. The Photoluminescence spectra of **40a-b** are shown in Figure 29. The prompt emission is more intense than the delayed one, assuming both compounds are rather poor TADF emitters. Furthermore, the emission is very broad, which is unsuitable for a TADF application.

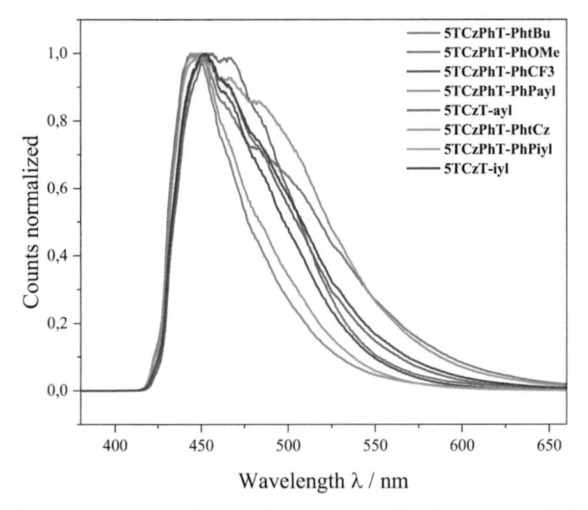

Figure 28 Photoluminescence spectra of the tetrazole series measured in toluene solution displaying compounds **39a-d**, **-f**, **-h**, and **40a-b**. The Intensity counts are normalized to 1 for a better overview. Only the prompt fluorescence component is shown. Results provided by Idoia Garin.

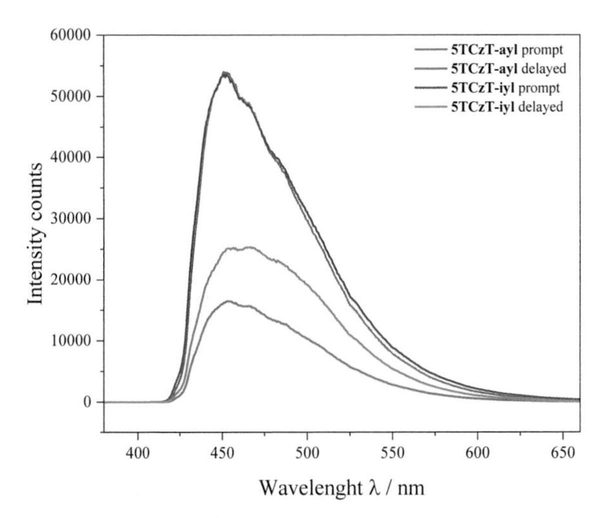

Figure 29 Photoluminescence spectra of **5TCzT-ayl** and **5TCzT-iyl** with prompt and delayed emission, measured in toluene solution. Results provided by Idoia Garin.

3.2.3. Oxadiazole Derivatives of the CzBN Class

To investigate more acceptor derivatives, **5TCzT** and **5TCzPhT** were further reacted, giving oxadiazoles, another acceptor class for TADF emitters. In 2018, Zhang *et al.* reported the TADF emitter **5CzOXD**, a phenyl substituted oxadiazole acceptor linked directly to a benzene ring with five carbazole groups. The emitter showed a light-blue emission at 496 nm, and, when doped in a PMMA film, an EQE of 9.3%.[132]

Oxadiazoles are easily accessible due to the work of Huisgen and coworkers in 1960. They reacted 5-phenyl-tetrazole with carboxylic acid chlorides in pyridine. Their proposed mechanism is shown in Scheme 17.[133] The ring opening and closing perform under relatively mild conditions, despite the tetrazole's aromatic stabilization. In the first step, the acetylated intermediate **II** is built. Carbonamide mesomery and aromatic tetrazole resonance compete with the lone electron pair of the 2-position. **II** rearranges at temperatures between 60 and 100 °C. The ring opens to give intermediate **III**. The additional electron pair is stabilized through the carbonyl group. Mesomeric conformers of **III** compensate for the lost aromaticity of the original tetrazole ring. The cleavage of nitrogen gives **IV**, followed by cyclization to the final 1.3.4-oxadiazole **V**.

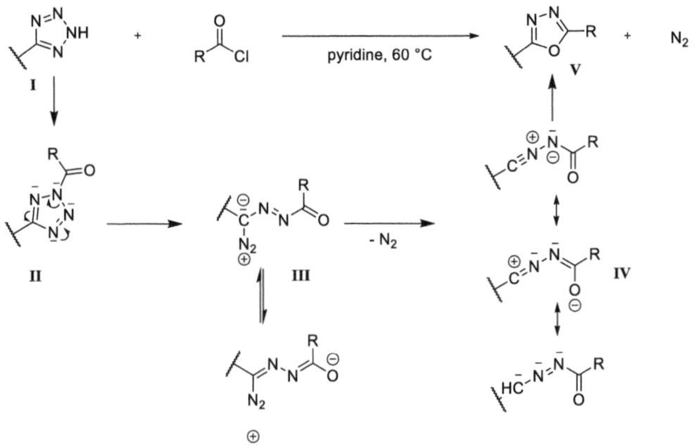

Scheme 17 Huisgen mechanism for synthesizing oxadiazoles using a tetrazole and carboxylic acid chloride.

The reaction conditions in this work were adjusted according to the scope of previous work within our group.[134] As solvent chloroform was used and the reaction was carried out at 100 °C to enhance the yield.

Eight and four oxadiazole derivatives of **5TCzT** and **5TCzPhT** were successfully synthesized (Scheme 18). Functional group tolerance and the effects of electron-withdrawing and -donating groups were in focus. Electron-withdrawing moieties like benzonitrile, pentafluorophenyl, and a perfluoro chain were selected. Tri-*tert*-butyl phenyl, adamantyl, methoxy, cyclohexyl, *tert*-butyl, and *n*-butyl were picked as electron-donating groups. The last four moieties were only reacted with **5TCzT**, because the derivatives of **5TCzPhT** did not show TADF in the later caried out photophysical measurements. This will be discussed in the photophysical part.

In the case of **5TCzPhT**, only the tri-*tert*-butyl phenyl compound **5TCzPhOx-Me** (**41a**) was successfully attached through an oxadiazole with a yield of 94%. The reaction with 1-adamantanecarbonyl chloride did not give any conversion. Good yields were obtained for the electron-withdrawing groups attached to the oxadiazole group with 71%, 81% and 82% for **5TCzPhOx-CN** (**41b**), **5TCzPhOx-ArF** (**41c**) and **5TCzPhOx-AlF** (**41d**), respectively.

The oxadiazole compounds **5TCzOx-Me**, **5TCzOx-ArF**, **5TCzOx-AlF**, **5TCzOx-OMe**, **5TCzOx-2OMe**, **5TCzOx-Hex**, **5TCzOx-tBu**, and **5TCzOx-Bu** (**42a-h**) with precursor **5TCzT** were isolated in good to excellent yields ranging from 56% to quantitative for **42h** and **42g**. The yields do not seem to follow a pattern electronically or sterically.

A) Carboxylic acid chloride
$CHCl_3$, 100 °C, 16h

38) → 41a-d)

B) Carboxylic acid chloride
$CHCl_3$, 100 °C, 16h

33) → 42a-h)

R =

41a) **5TCzPhOx-Me** 94% A)
42a) **5TCzOx-Me** 97% B)

41b) **5TCzPhOx-CN** 71% A)

41c) **5TCzPhOx-ArF** 81% A)
42b) **5TCzOx-ArF** 93% B)

41d) **5TCzPhOx-AlF** 82% A)
42c) **5TCzOx-AlF** 68% B)

42d) **5TCzOx-OMe** 85% B)

42e) **5TCzOx-2OMe** 76% B)

42f) **5TCzOx-Hex** 63% B)

42g) **5TCzOx-tBu** quant. B)

42h) **5TCzOx-Bu** 56% B)

Scheme 18 Oxadiazole syntheses with precursor **5TCzT** and **5TCzPhT**. Twelve derivatives of the oxadiazole core have been synthesized employing electron-donating-, electron-withdrawing groups, and bulky substituents.

3.2.3.1. DFT Calculations

All DFT calculations were carried out through a collaboration with the Wenzel group at the KIT.

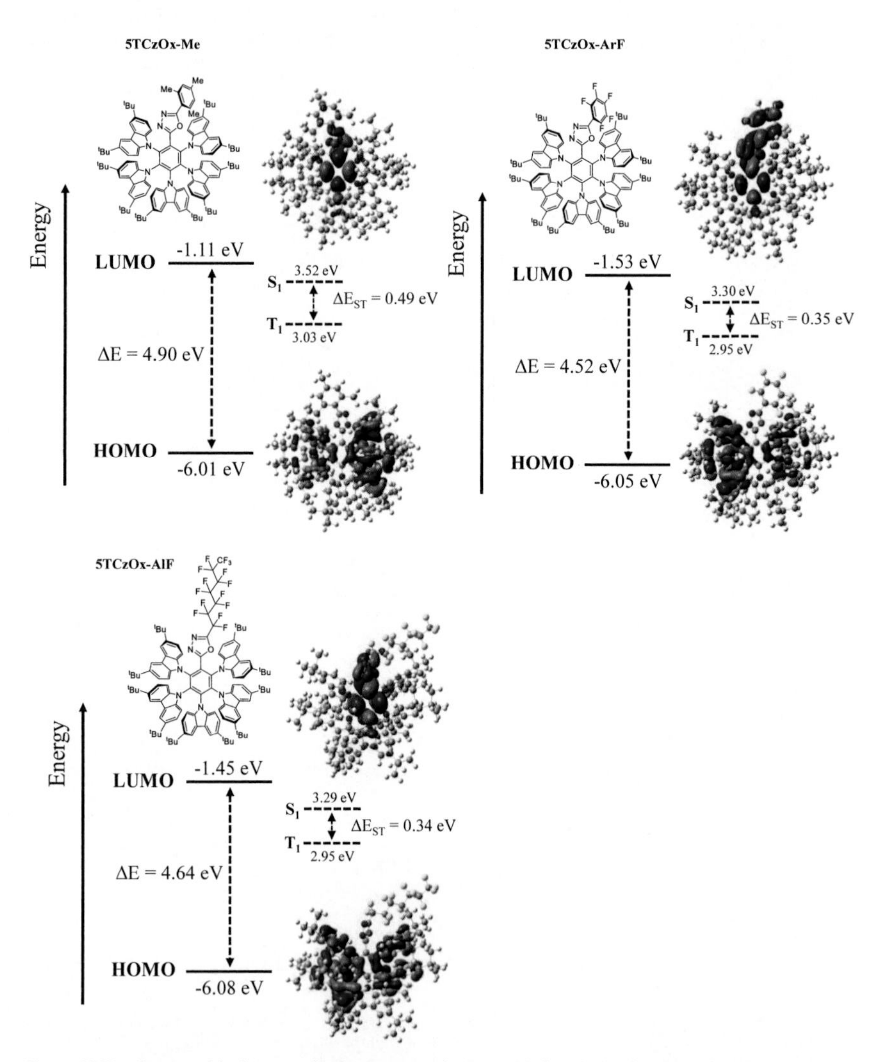

Figure 30 Visualization of the frontier orbitals of compounds **42a-c** with the calculated HOMO and LUMO energy levels and the values for S_1, T_1, and ΔE_{ST}. Results provided by Anna Mauri.

From the results of the tetrazole series, it was expected that the oxadiazole system with a phenylene spacer might not show TADF. Since DFT calculations are time-consuming, only some of the oxadiazole molecules without the phenylene spacer were calculated and compared with the photophysical experiments, following in the next Chapter.

Calculations for compounds **42a-c** have been run. The distribution of the frontier orbitals is shown in Figure 30. The HOMO is located on the *tert*-butyl-carbazole donors for all three compounds. The LUMO of **42a** is more located on the benzyl- and oxadiazole ring rather than on the trimethylphenyl attached to the acceptor. The same is observed in compound **42c**, where the LUMO is rather located on the oxadiazole than on the fluorine chain. In this series, compound **42b** is the only one where the LUMO outskirts further to the pentafluorophenyl substituent on the oxadiazole.

The singlet and triplet energy levels were calculated with the BMK/def2-TZVP basis set. With the singlet state S_1 and triplet state T_1, ΔE_{ST} was determined. All values are visualized in Figure 30. The S_1 values of **42a-c** were determined to be 3.52, 3.30, and 3.29 eV, respectively, which equals an emission in the ultraviolet region. For compound **42a**, the highest ΔE_{ST} was determined with 0.49 eV, followed by 0.35 eV for **42b** and 0.34 eV for **42c**. Expecting such high energy emissions, these emitters may not be suitable for applications in OLEDs. Nevertheless, for a final evaluation, photophysical experiments must be carried out.

3.2.3.2. Photophysical Characterization

Emission spectra of all oxadiazole emitters were measured in a toluene solution. The oxadiazole compounds with a phenylene spacer showed all similar emission maxima, irrelevant of the substituent on the oxadiazole acceptor. Emission in the blue spectra of 451 nm for **5TCzPhOx-Me**, **5TCzPhOx-Me**, and **5TCzPhOx-ArF** and 454 nm for **5TCzPhOx-AlF** was measured. The structures with the phenylene spacer did not or only showed a very weak delayed component, hence no

TADF. In contrast, the compounds without the phenylene spacer showed a delayed component, hence TADF.
Comparing the above-discussed DFT results, the experiments do not seem to match. Emitters **42a-c** showed emission wavelengths in the blue to the green region and not in the ultraviolet spectrum. With 470 nm, compound **42a** showed the bluest emission. The more electron-withdrawing character the substituent has attached to the acceptor, the more red-shifted the emission. The emission of **42b** and **42c** shifted bathochromically with a greenish emission of 505 nm and 513 nm, respectively. The FWHMs were in a similar range of 74 – 89 nm. Time-resolved spectroscopy was measured for compounds **42a-c**. The prompt fluorescence lifetimes for **42a-c** are small within the ns range. The delayed fluorescence lifetimes are rather long within the ms range, which can mitigate the TADF properties. All oxadiazole emitters with electron-donating groups at the oxadiazoles showed an emission below 500 nm, hence a sky-blue/blue emission. Emitters **42d-h** have not been fully photophysically characterized yet; the experiments are still ongoing.

Table 19 Photophysical data of the oxadiazole series measured in a toluene solution.

Entry	Compound	#	$\lambda_{max, PF}$ [nm]	$FWHM_{PF}$ [nm]	$\lambda_{max, DF}$ [nm]	$FWHM_{DF}$ [nm]	lifetime τ_{PF} [ns]	lifetime τ_{PF} [ms]
1	**5TCzPhOx-Me**	**41a**	451	-	-	-	-	-
2	**5TCzPhOx-Me**	**41b**	451	-	-	-	-	-
3	**5TCzPhOx-ArF**	**41c**	451	-	-	-	-	-
4	**5TCzPhOx-AlF**	**41d**	454	-	-	-	-	-
5	**5TCzOx-Me**	**42a**	470	80	470	74	3.75	8
6	**5TCzOx-ArF**	**42b**	505	86	508	83	4.43	6
7	**5TCzOx-AlF**	**42c**	513	89	516	89	6.31	1.5

8	**5TCzOx-OMe**	**42d**	491	-	492	-	-	-
9	**5TCzOx-2OMe**	**42e**	485	-	487	-	-	-
10	**5TCzOx-Hex**	**42f**	445, 471	-	487	-	-	-
11	**5TCzOx-tBu**	**42g**	490	-	487	-	-	-
12	**5TCzOx-Bu**	**42h**	450	-	471	-	-	-

Emitter **42c** was embedded in films with the host mCP and PMMA. After screening which host concentration gives the best performance, photoluminescence and time-resolved spectra were measured. The results are shown in Table 20. A bluer delayed emission with 495 nm of the emitter doped with mCP was achieved compared to 504 nm in PMMA. The proportion of delayed emission is higher, with 84.2% compared to 73% (PMMA). Furthermore, the FWHM values for the emissions were smaller, with 87 nm prompt and 84 nm delayed compared to 93 nm prompt and 91 nm delayed (PMMA). The prompt and delayed fluorescence lifetimes are longer, with 3.85 ns and 1.41 ms, compared to 3.45 ns and 0.87 ms in PMMA, respectively. The film with PMMA achieved a PLQY of 42%. The PLQY for the film with mCP could not yet be accurately determined because the host absorption was a problem.

Table 20 Emission maxima and fluorescence lifetime results of **42c** measured in film with mCP and PMMA. The proportion of prompt and delayed fluorescence is written in brackets.

Entry	Compound / host	#	PLQY [%]	$\lambda_{max, PF}$ [nm]	$FWHM_{PF}$ [nm]	$\lambda_{max, DF}$ [nm]	$FWHM_{DF}$ [nm]	lifetime τ_{PF} [ns]	lifetime τ_{PF} [ms]
1	**5TCzOx-AlF** / mCP (5wt%)	**42c**	-	496 (15.8%)	87	495 (84.2%)	84	3.85	1.41
2	**5TCzOx-AlF** / PMMA (5wt%)	**42c**	42	495 (27%)	93	504 (73%)	91	3.45	0.87

Temperature dependency measurements were carried out for **42c** with the host mCP. The temperature-dependent time-resolved decay curves were measured at temperatures of 75 K, 100 K, 140 K, 180 K, 220 K, 260 K, and ambient temperature. Increased intensity of the delayed emission by increasing the temperature can be observed, which is typical for TADF emitters. With these measurements, a very small ΔE_{ST} of 0.018 eV was determined.

3.2.4. Oxadiazole Dimers – D-A-A-D Skeleton

Usually, TADF emitters are designed in a D-A or D-A-D frame. The emitters of the previous project were expanded by reconstructing the oxadiazole acceptors from a D-A to a D-A-A-D scaffold. According to the proposed design strategy, two acceptor groups were bonded to enhance their acceptor abilities. Furthermore, different bridging units were investigated. Aromatic rings like phenyl and pyridine and an aliphatic chain, such as ethyl were selected. Figure 31 shows the modulative construction of the D-A-A-D scaffold to visualize what will be named a bridge and a spacer. The green marked part is attributed to the bridging unit between the two acceptors, whereas the orange marked brackets indicate the spacer between the donor and acceptor.

3.2.4.1. D-A-A-D with five and four Donors

During the first run of this project, the same donor system was used as in the previous chapter 3.2.3.

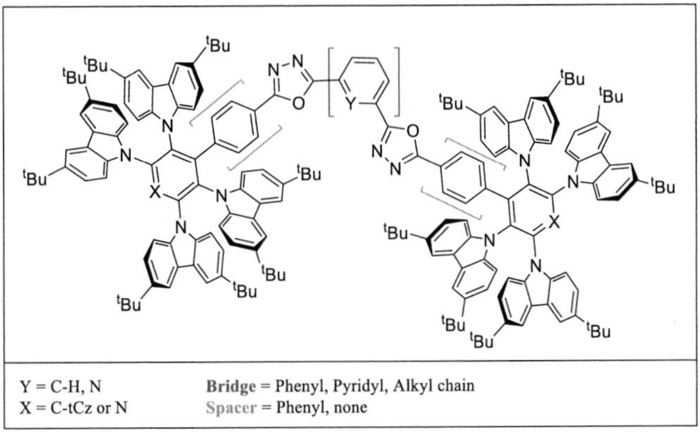

Y = C-H, N
X = C-tCz or N

Bridge = Phenyl, Pyridyl, Alkyl chain
Spacer = Phenyl, none

Figure 31 Modulative approach to the D-A-A-D scaffold with four and five **tCz** donors each side. The marked green region is named bridge, whereas the orange marked part is named spacer.

In addition, the **5TCzPhT** was slightly altered such that four **tCz** donors and pyridine were combined to build the **4TCzPyPhT** (**45**). **4TCzPyPhT** was synthesized within three steps starting with 4-bromo-2,3,5,6-tetrafluoropyridine reacting with (4-cyanophenyl)boronic acid in a SUZUKI cross-coupling reaction. In the second step **tCz** was attached through a nucleophilic aromatic substitution reaction to 4-(perfluoropyridin-4-yl)benzonitrile (**43**), which was the product from step one. At last 4-((2r,3s,5s,6s)-2,3,5,6-tetrakis(3,6-di-*tert*-butyl-9*H*-carbazol-9-yl)pyridin-4-yl)benzonitrile (**4TCzPyPhBN**, **44**) was reacted using sodium azide to obtain **4TCzPyPhT** with an overall yield of 24%. The detailed procedure and structures can be found in the experimental section.

For this type of reactions, two equivalents of the associated tetrazole precursor were used to react with one equivalent of the dicarboxylic acid chloride, which bears the selected bridging group.

All desired products were successfully synthesized, despite the highly crowded structure due to the five or four **tCz** groups. The yields varied dependent on the donor structure and the bridging unit. The lowest yield of 12% was obtained for compound **5TCzPhOx-PhDim** (**49a**), which has two **5TCzPh** groups and a phenyl bridge between the two oxadiazole acceptors. The yields of compounds containing the **4TCzPh** donors, **4TCzPhOx-PhDim** and **4TCzPhOx-PyDim** (**50a-b**), were higher with 69% and 82% as those for the **5TCzPh** compounds, **49a** and **5TCzPhOx-PyDim** (**49b**) with 12% and 30%. Furthermore, the reactions for **49b** and **50b**, involving the pyridine bridge, gave better yields than the compounds with phenyl as a bridging unit such as compounds **49a** and **50a**. This trend is reasonable since pyridine, as opposed to a phenyl bridge, is an electron-poor aromatic ring and thus facilitates the reaction with the tetrazole nucleophile. The introduction of an aliphatic bridge was attempted for the **5TCzPh** system. Succinyl dichloride (**48**) was reacted with **5TCzPhT** to obtain **5TCzPhOx-EtDim** (**49c**) with a yield of 45%.

Dimer **5TCzOx-PyDim** (**51**) containing two **5TCz** and a pyridine bridge was isolated with a low yield of 16%. Compared to the **49** and **50** series, the low yields might result from the steric nature of the direct connection of the donor systems with the acceptors. Because of this result, no different dimers with **5TCz** were pursued.

No DFT calculations were carried out because of the largeness and heaviness of these dimers. Photoluminescence spectra were measured for two of the six dimers (Table 21).

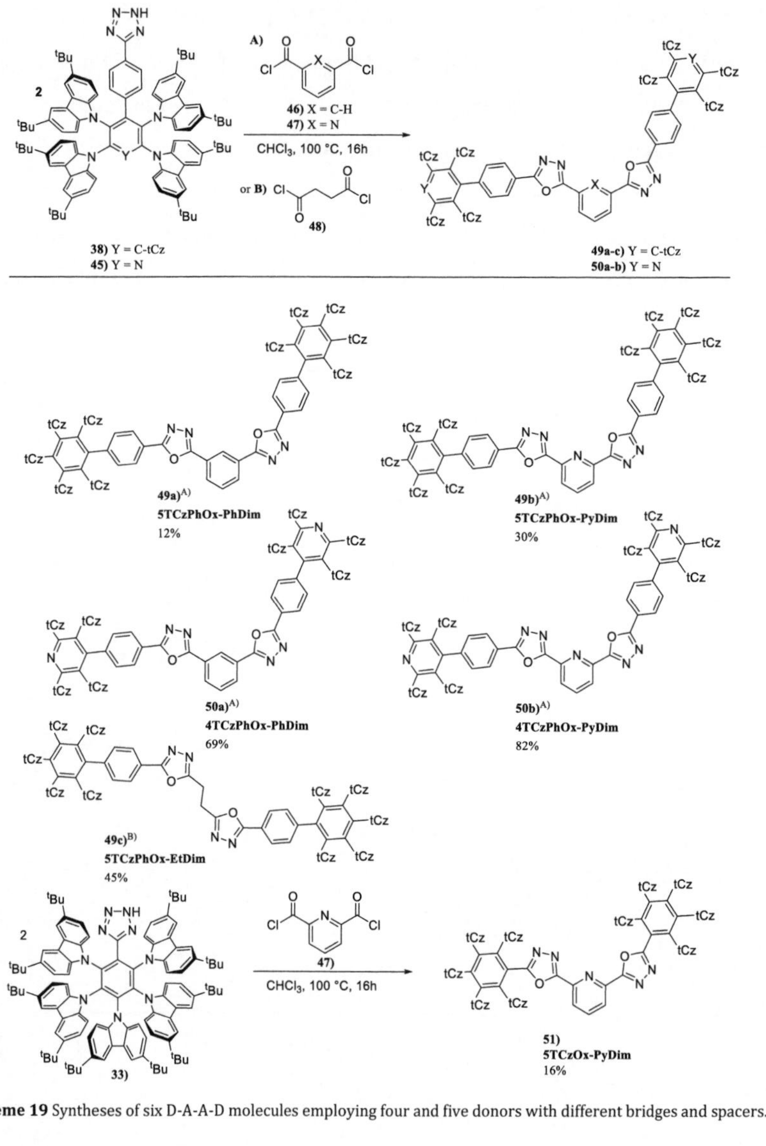

38) Y = C-tCz
45) Y = N

A) 46) X = C-H
47) X = N
CHCl₃, 100 °C, 16h
or B) 48)

49a-c) Y = C-tCz
50a-b) Y = N

49a)[A)]
5TCzPhOx-PhDim
12%

49b)[A)]
5TCzPhOx-PyDim
30%

50a)[A)]
4TCzPhOx-PhDim
69%

50b)[A)]
4TCzPhOx-PyDim
82%

49c)[B)]
5TCzPhOx-EtDim
45%

33)

47)
CHCl₃, 100 °C, 16h

51)
5TCzOx-PyDim
16%

Scheme 19 Syntheses of six D-A-A-D molecules employing four and five donors with different bridges and spacers.

Table 21 Data of the photoluminescence spectra of **47b** and **48b**.

Entry	Compound	#	$\lambda_{max, PL}$ [nm]
1	**5TCzPhOx-PyDim**	**49b**	450
2	**4TCzPhOx-PyDim**	**50b**	484

Compounds **47b** and **48b** showed a blue emission with 450 nm and 484 nm. Both did not show a delayed component as discovered in their time-resolved spectroscopy measurements; therefore, the dimer series with five and four donors was not further pursued. This project was rather designed to show proof of principle that the established oxadiazole acceptor syntheses apply to larger dimer structures.

The next project focused on mitigating the dimer structure to get a better inside into these D-A-A-D molecules and to apply theoretical back up calculations. This was done by reducing the donor count from five or four **tCz** to only two on each side.

3.2.4.2. D-A-A-D with two Donors

DFT Calculations

The initial idea was to install a simplified TADF structure consisting of an acceptor like a benzonitrile or tetrazole connected via a phenylene spacer with just two **tCz** donors. The 3',5'-bis(3,6-di-*tert*-butyl-9*H*-carbazol-9-yl)-[1,1'-biphenyl]-4-carbonitrile (**2tCzCN**) and 9,9'-(4'-(2*H*-tetrazol-5-yl)-[1,1'-biphenyl]-3,5-diyl)bis(3,6-di-*tert*-butyl-9*H*-carbazole) (**2tCzT**) were calculated with DFT methods through the collaboration with the Wenzel group, KIT. The frontier orbital distribution and the HOMO-LUMO energy levels were calculated using the ground state. The excited states were calculated with the optimized structures using TD-DFT. The results are illustrated in Figure 32. The LUMO is in both samples located on the acceptor and the HOMO on the donors. The frontier orbitals outskirts to the phenylene spacer in both compounds. The S_1 energy levels were determined to be 3.28 eV and 3.50 eV for **2tCzCN** and **2tCzT**, respectively. The energy gap between S_1 and T_1, ΔE_{ST}, accounted for 0.41 eV for **2tCzCN** and 0.65 eV for **2tCzT.** The two molecules are not just too blue to be suitable for any application; their ΔE_{ST} also appears to be too large for sufficient TADF.

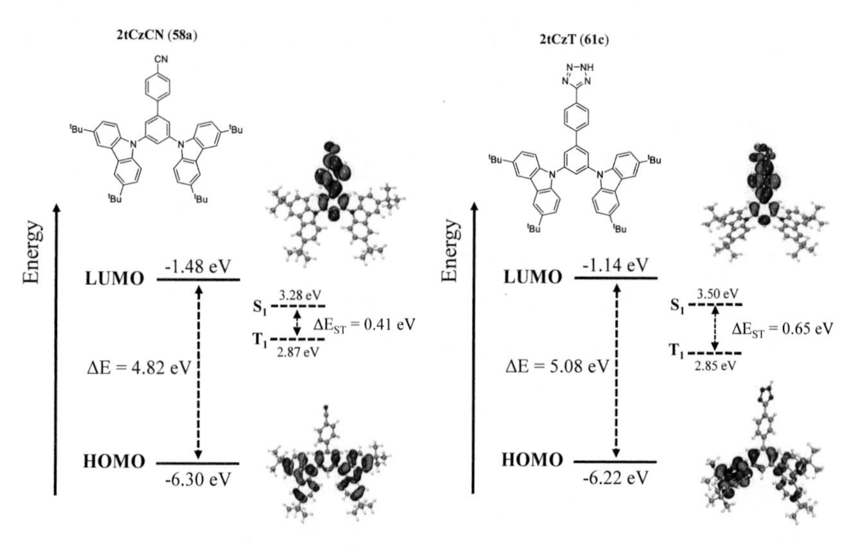

Figure 32 Illustration of the frontier orbital distribution and calculation of the HOMO-LUMO energy levels and the excited states of **2tCzCN** and **2tCzT**. Results provided by Anna Mauri.

Examples in the literature showed that D-A-A-D molecules could red shift the TADF emission compared to their associated D-A or D-A-D compounds.[135] With this in mind, the **2tCzCN** and **2tCzT** were chosen to be reconstructed into the D-A-A-D scaffold. The D-A-A-D scaffold was investigated and combined with different bridges and spacers to learn about the structural and electronical properties on the photophysical characteristics (Figure 33).

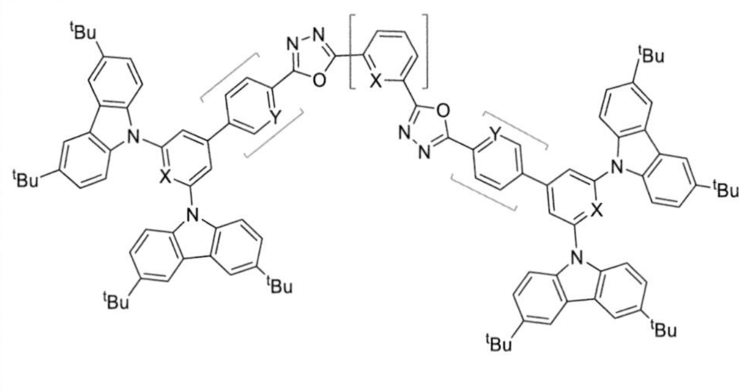

Y = C-H, N	Bridge = Phenyl (ortho, meta, para), Pyridil, Paracyclophan, Alkyl chain
X = C-H, N	Spacer = Phenyl, Pyridiyl, none

Figure 33 Modulative construction of the D-A-A-D scaffold using the **2tCzT** precursor. The marked green region is named bridge, whereas the orange marked part is named spacer.

Synthesis of the Nitrile Precursors

Four tetrazole precursors were synthesized to get variations and combinations of different donor configurations with spacers and bridges. First, the different spacers were connected to the fluorinated benzonitrile precursors (Scheme 20). Nitrile **PhFBN (53a)** and **PhFPyCN (53b)** were synthesized under SUZUKI cross-coupling conditions using potassium phosphate, SPhos, and $Pd_2(dba)_3$. **53a** was isolated in a quantitative yield and **53b** in an excellent yield of 92%. **PyFBN (53c)** was obtained with different SUZUKI conditions using potassium phosphate, $Pd_2(dba)_3°CHCl_3$, and tricyclohexylphosphine in a mixture of 1,4-dioxane and water giving full conversion. The fourth nitrile precursor, 3,5-difluorobenzonitrile (**59**), was commercially available. All four nitrile precursors were reacted with **tCz** using potassium phosphate in a nucleophilic aromatic substitution reaction to attach the donor moieties (Scheme 21). The Nitrile precursor **2TCzPhBN (58a)** with the phenyl rings in the spacer and the connecting part of the donors was obtained in a quantitative yield. **2TCzPhPyCN (58b)** containing a pyridine spacer and a phenyl ring as the connecting donor part was obtained with 92%.

Compound **2TCzPyBN** (**58c**), which has a phenylene spacer and a pyridine connecting donor part, was obtained with an 89% yield. The nitrile **2TCzBN** (**60**) without a spacer was isolated with an 85% yield.

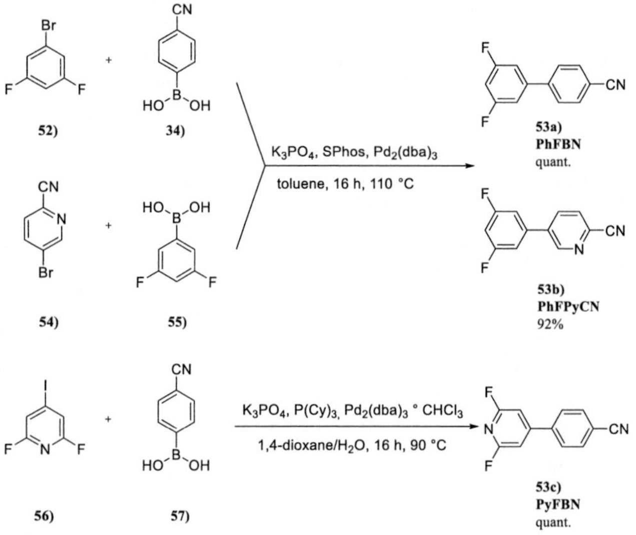

Scheme 20 Syntheses of the nitrile precursors for the installation into the D-A-A-D scaffold.

Synthesis of the Tetrazole Precursors

The four nitrile precursors **58a-c** and **60** were transformed into tetrazoles using sodium azide (Scheme 21). All four products **2TCzPhT, 2TCzPhPyT, 2TCzPyBN** (**61a-c**), and **2TCzT** (**62**) were obtained with full conversion, hence quantitative yields.

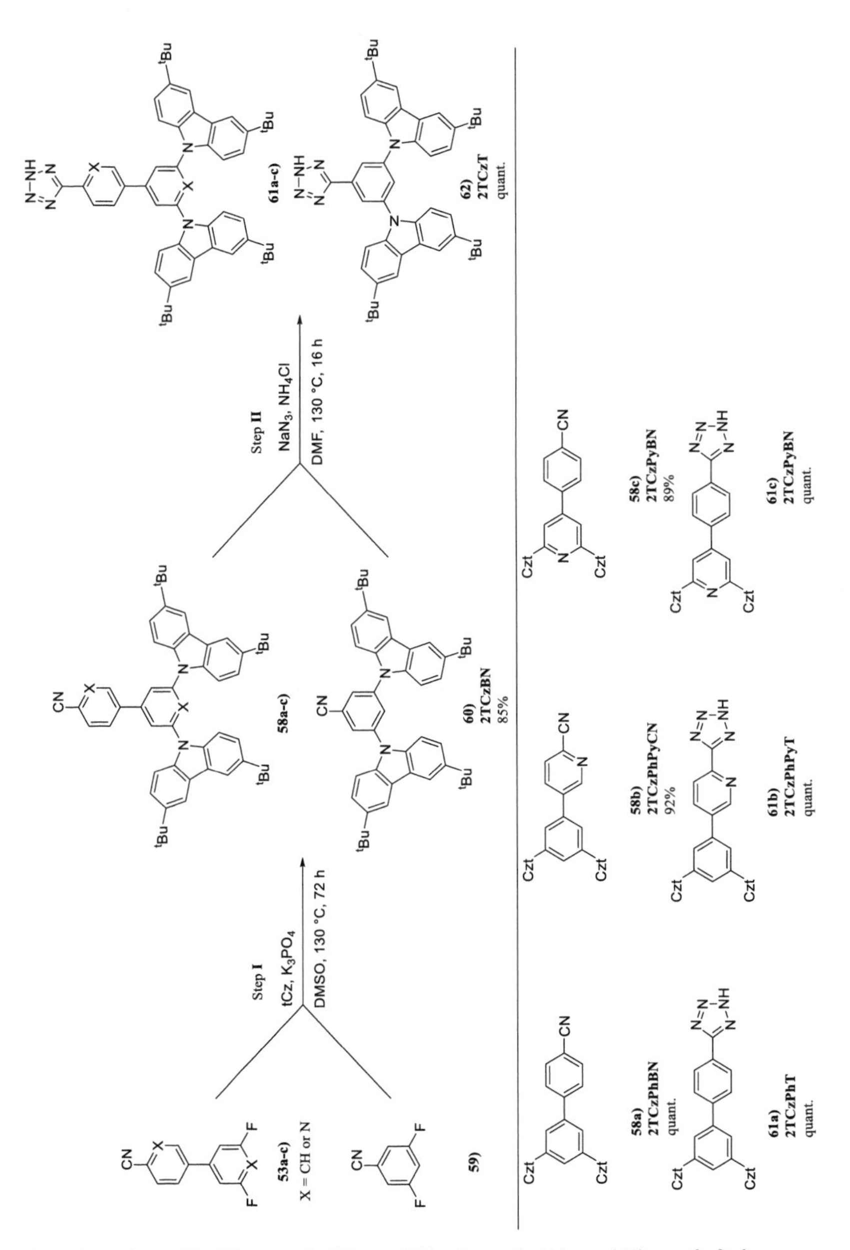

Scheme 21 Syntheses of the different nitriles (**58a-c** and **60**) and tetrazoles (**61a-c** and **62**) to get the final precursors for the D-A-A-D syntheses.

This project aimed to investigate different donor-spacer-bridge patterns within the D-A-A-D scaffold and to explore their impact on the emitting properties. A library of 15 D-A-A-D combinations was synthesized. Different bridging units such as phenylene, pyridyl, paracyclophane, alkyl chains, and adamantyl groups or no bridge were varied with different donor pattern combinations. The syntheses were adopted from the former dimer structure protocol of the previous project using one equivalent carboxylic acid chloride and two equivalents of the selected tetrazole precursor in chloroform at 100 °C. The compounds are sorted by their bridging group between the two oxadiazoles.

D-A-A-D Dimers with a Pyridine Bridge

The final D-A-A-D compounds with a pyridine bridge **2TCzPhOx-PyDim, 2TCzPyOx-PyDim, 2TCzPyPhOx-PyDim** and **2TCzOx-PyDim** (**63a-d**) are displayed in Scheme 22. The reactions to synthesizing the three compounds with a spacer **63a-c**, whether a phenylene or pyridyl, behaved almost the same, with a yield of ~77% for all three. The associated compound with a phenylene spacer **63a** was obtained with a 77% yield. The combination of a phenylene spacer with a pyridine connecting both donors (**63c**) gave a yield of 76%. Compound **63b** containing a pyridyl spacer was isolated with a 77% yield. No spacer between the donor and acceptor units resulted in a poor yield, compared to the ones with a spacer. No spacer compound **63d**, with a pyridine bridge, gave a yield of 18%. A sterical challenge might be the reason for the low yield because the demanding **tCz** groups block each other for the nucleophile to attack freely, hence the diminished reactivity.

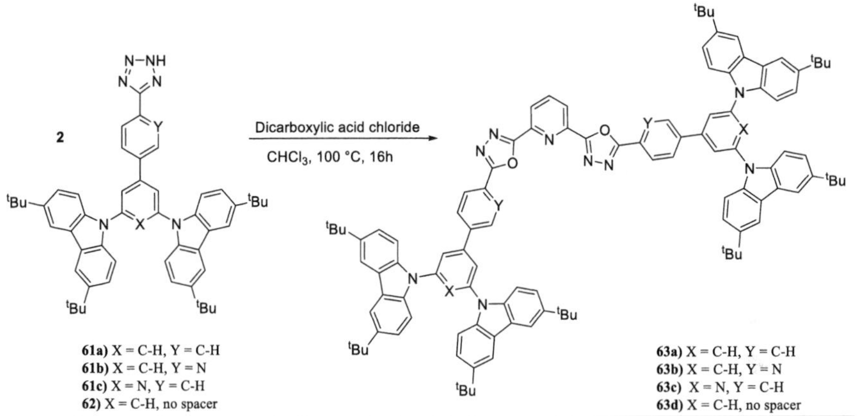

2

Dicarboxylic acid chloride

$CHCl_3$, 100 °C, 16h

61a) X = C-H, Y = C-H
61b) X = C-H, Y = N
61c) X = N, Y = C-H
62) X = C-H, no spacer

63a) X = C-H, Y = C-H
63b) X = C-H, Y = N
63c) X = N, Y = C-H
63d) X = C-H, no spacer

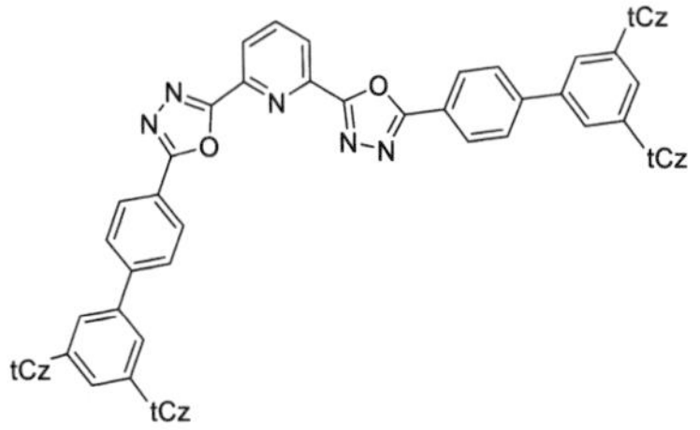

63a)
2TCzPhOx-PyDim
77%

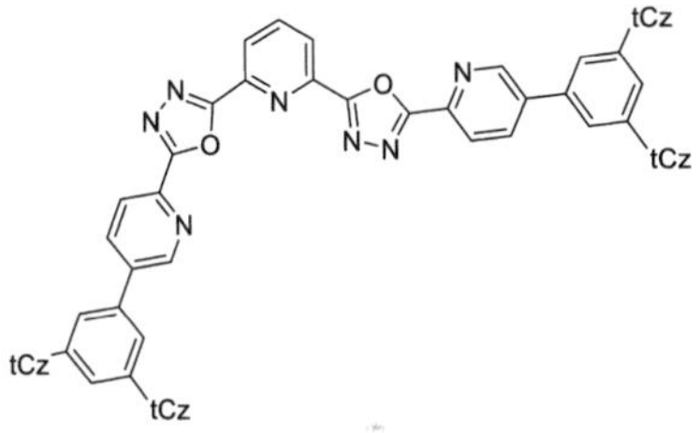

63b)
2TCzPyOx-PyDim
77%

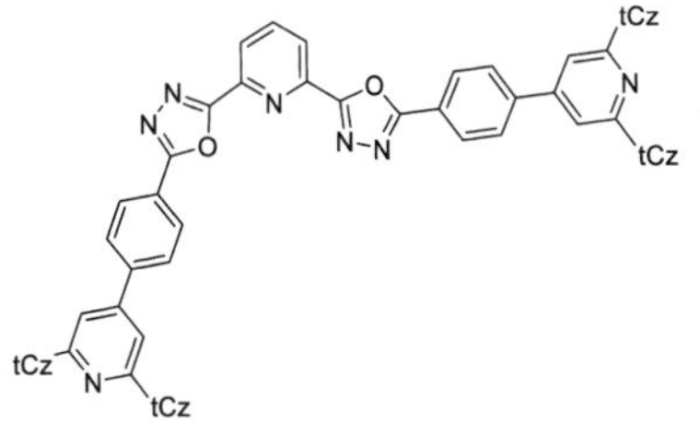

63c)
2TCzPyPhOx-PyDim
76%

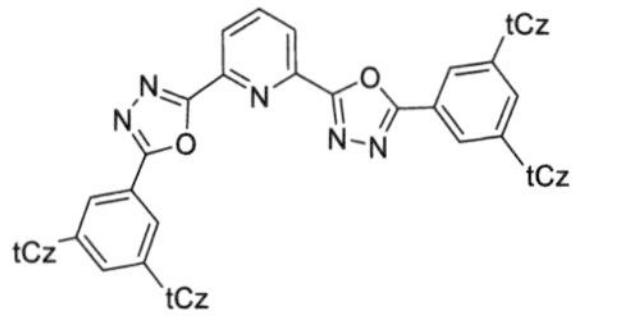

63d)
2TCzOx-PyDim
18%

Scheme 22 Syntheses of four D-A-A-D molecules with a pyridine bridge cored between the acceptors. Varied spacers like phenylene, pyridyl, or none were investigated.

D-A-A-D Dimers with a Phenylene Bridge

The D-A-A-D dimers with a phenylene bridge are shown in Scheme 23. Tetrazole precursor **61c** was reacted with isophthaloyl dichloride, phtaloyl dichloride, and terephthaloyl dichloride to connect the donor unit at the phenylene spacer in the *ortho-*, *meta-* and *para*-position, respectively. A change in the position of the acceptor led to significantly different yields. The *ortho*-product **2TCzPyPhOx-*o*PhDim** (**64b**) was isolated with the highest yield of 85%. The first attached oxadiazole group has a strong electron-withdrawing effect and could impact therefore the direct neighboring carbonyl group. This situation facilitates the second attack of the nucleophile, which targets the second carbonyl. Furthermore, the stabilization of the intermediate is better for the *ortho-* and *para*-product, which would explain the higher yield of **2TCzPyPhOx-*p*PhDim** (**64d**) with 51% compared to the *meta*-product **2TCzPyPhOx-*m*PhDim** (**64c**) with 37%. Compound **2TCzPhOx-PhDim** (**64a**), which bears the donor unit **61a**, was connected to a phenylene spacer with phtaloyl dichloride in a 55% yield.
Analyzing these compounds using ^{1}H NMR was challenging because the signals were broadened and ambiguous. Figure 34 shows a cutout of the ^{1}H NMR spectrum of the pyridine bridge compound **63a** (top) compared to the associated compound **64a** (bottom) with a phenyl bridge.

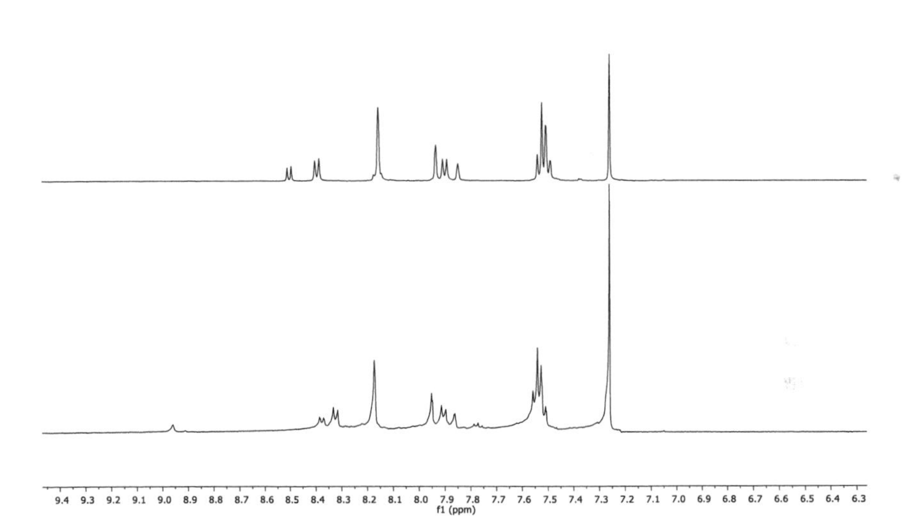

Figure 34 Comparison of the ^{1}H NMR spectra of **63a** (top) and **64a** (bottom). A cutout of the chemical shift spectra in the range of 6.0 ppm and 9.4 ppm is shown.

The NMR spectrum of **63a** shows sharp signals that can be assigned surely. In contrast, the spectrum of **64a** shows broadened signals, which makes the assignment of the integrals to the protons challenging. This signal broadens is due to the rotation latitude that originates from the phenyl bridge, resulting in different rotamers. In contrast, the donor groups attached via the oxadiazoles to the pyridine bridge cannot rotate because the nitrogen of the pyridine is in resonance with the oxygens of the oxadiazoles. The nitrogen in the pyridine bridge and the nitrogen in the oxadiazoles repulse each other. Therefore, the donor groups in compounds with a pyridine bridge are locked and cannot rotate. Compounds containing the phenyl bridge can show different rotamers, visible in their ^{1}H NMR spectra. These rotamers, like in **64c**, **64a** or **50a**, make it difficult to analyze the spectra and assign the protons correctly. Mass spectrometry analysis was therefore especially important to confirm the formation of these molecules. Phenyl

bridge compound **64b** does not show rotamers according to ^{1}H NMR. The proximity of the donor units owing to the *ortho*-position might hinder a rotation in the C – C axis from the donor-acceptor system to the bridge.
To show the rotamer effect, the molecules have been drawn in the conformation where the oxygen of the oxadiazole is in neighboring proximity to the pyridine nitrogen.
Comparing **64a** and **64c**, it is notable that the yield for **64c**, with pyridine in the connecting donor part, is lower than that for **64a**, with phenyl between the donors. A reason might be product loss during the second cycle of flash column chromatography because no pure product was isolated in the first attempt.

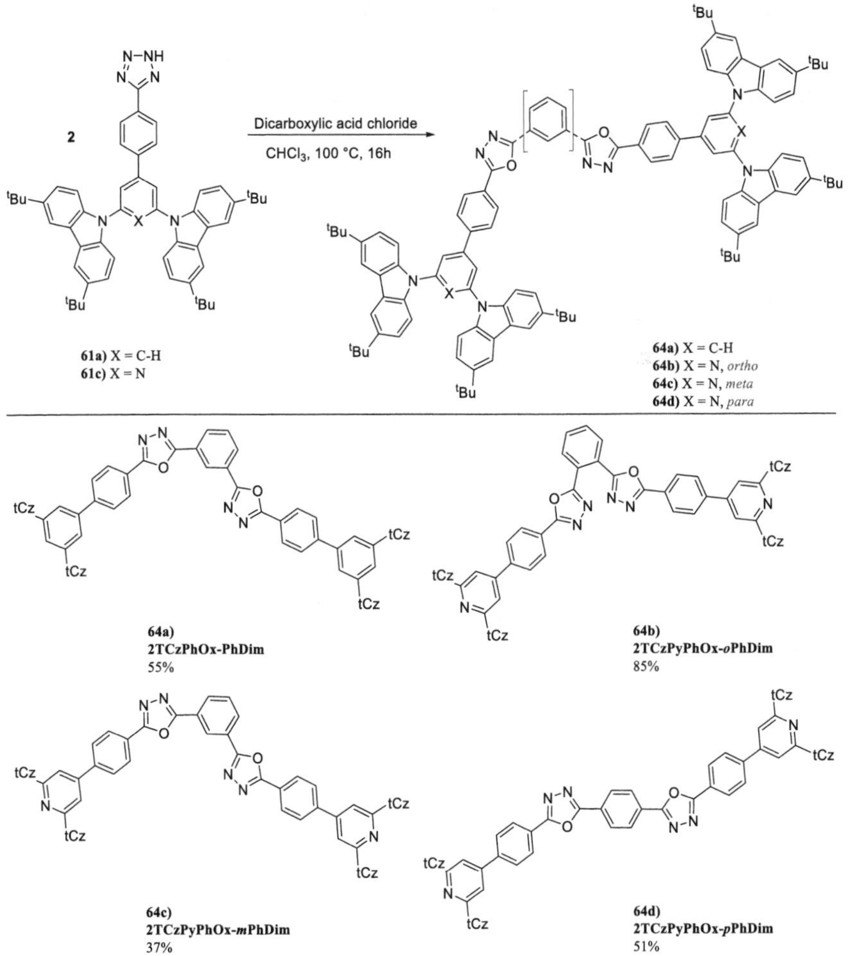

Dicarboxylic acid chloride

$CHCl_3$, 100 °C, 16h

61a) X = C-H
61c) X = N

64a) X = C-H
64b) X = N, *ortho*
64c) X = N, *meta*
64d) X = N, *para*

64a)
2TCzPhOx-PhDim
55%

64b)
2TCzPyPhOx-*o*PhDim
85%

64c)
2TCzPyPhOx-*m*PhDim
37%

64d)
2TCzPyPhOx-*p*PhDim
51%

Scheme 23 Syntheses of four D-A-A-D molecules with a phenylene bridge cored between the acceptors. The donors were installed in multiple positions to the acceptor (*ortho, meta,* and *para*).

D-A-A-D Dimers without a Bridge

Two compounds, **2TCzPhOx-Dim** and **2TCzPyPhOx-Dim (65a-b)** were designed to have no bridge between the acceptors (Scheme 24). The acceptors are directly linked together using oxalyl chloride. For both, a phenylene spacer was used. The connection between the donors was varied by phenylene and pyridine. **65a** and **65b** were obtained with 52% and 57% yields, respectively. A comparison of the dimers, which were all synthesized with tetrazole **61c**, shows that compound **63c** has the highest yield at 76%, followed by **65b** with 57% and **64c** with the lowest yield of 37%. A similar pattern is seen by the cross-comparison of compounds **63a**, **65a** and **64a**, all using tetrazole precursor **61a**, with yields of 77%, 52%, and 55%, respectively.

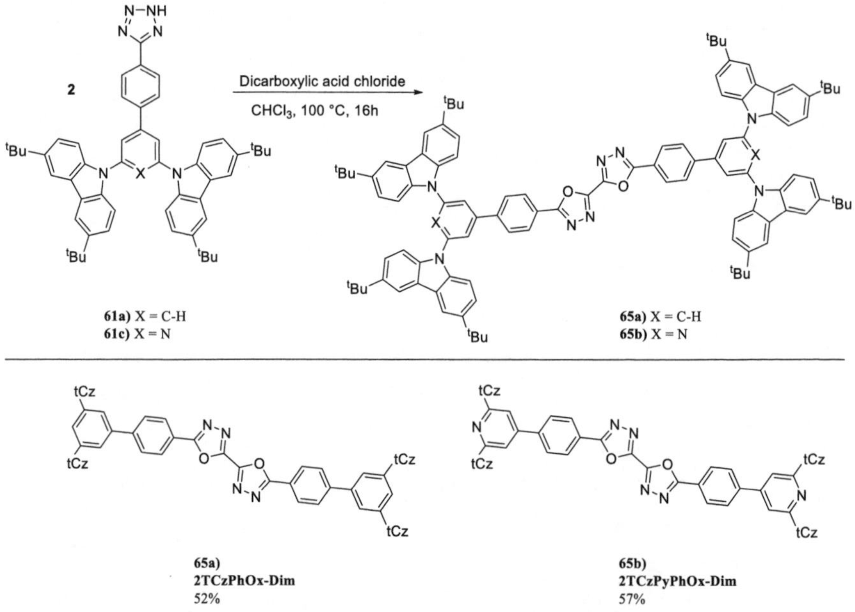

Dicarboxylic acid chloride
$CHCl_3$, 100 °C, 16h

61a) X = C-H
61c) X = N

65a) X = C-H
65b) X = N

65a)
2TCzPhOx-Dim
52%

65b)
2TCzPyPhOx-Dim
57%

Scheme 24 Syntheses of the two D-A-A-D molecules, **65a-b**, without a bridge between the acceptors.

D-A-A-D Dimers with aliphatic Bridges

All five compounds **2TCzPhOx-EtDim**, **2TCzPyPhOx-AdDim**, **2TCzPhOx-AdDim**, **2TCzPyPhOx-PcPDim**, and **2TCzPhOx-PcPDim** (**66a-e**) are displayed in Scheme 25. Adopted from compound **49c**, an ethyl bridge was tried for the two-donor system. Compound **66a** was obtained with a yield of 39%. Another alkyl bridge was incorporated using an adamantyl group. Compound **66b** with phenylene spacer and pyridine between the **tCz**'s and **66c** with phenylene as the spacer and connecting part between the donors were isolated in a 65% and 64% yield, respectively.

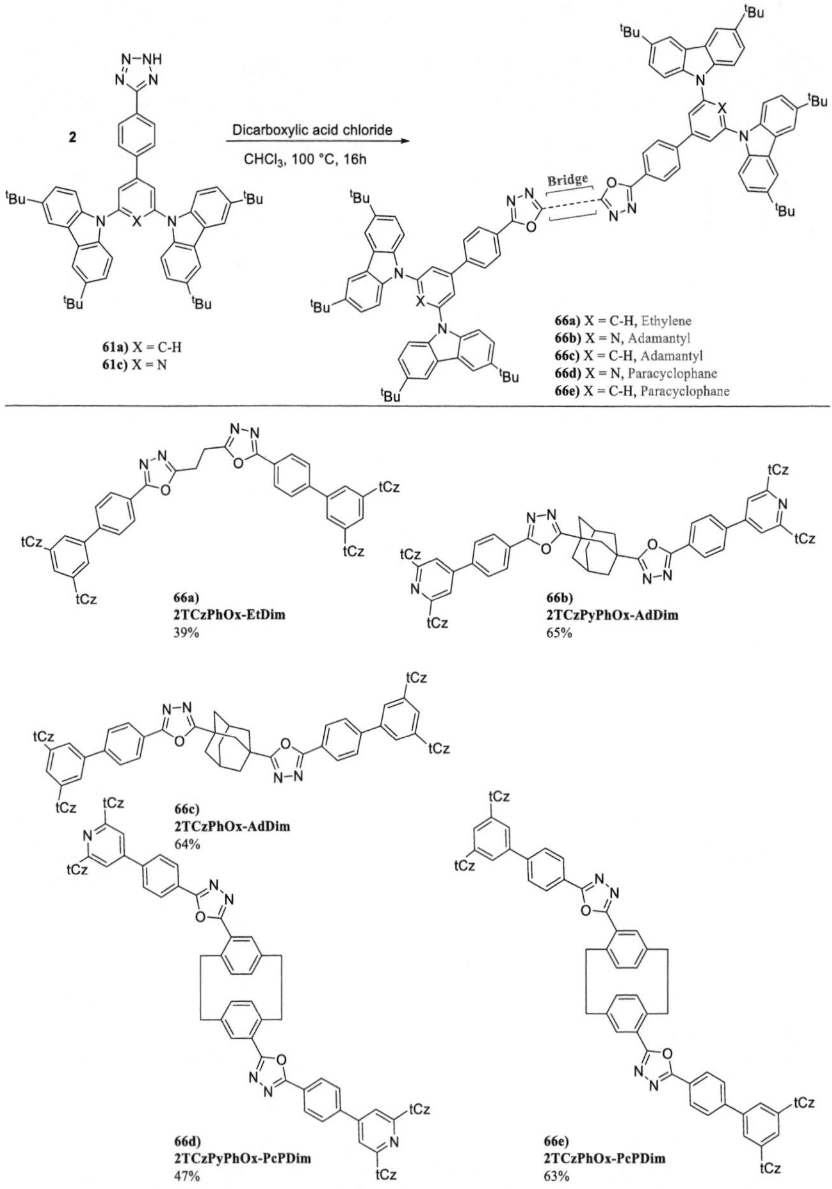

2

Dicarboxylic acid chloride

$CHCl_3$, 100 °C, 16h

Bridge

61a) X = C-H
61c) X = N

66a) X = C-H, Ethylene
66b) X = N, Adamantyl
66c) X = C-H, Adamantyl
66d) X = N, Paracyclophane
66e) X = C-H, Paracyclophane

66a)
2TCzPhOx-EtDim
39%

66b)
2TCzPyPhOx-AdDim
65%

66c)
2TCzPhOx-AdDim
64%

66d)
2TCzPyPhOx-PcPDim
47%

66e)
2TCzPhOx-PcPDim
63%

Scheme 25 Syntheses of five D-A-A-D molecules with aliphatic bridges like ethylene, adamantyl and paracyclophane cored between the acceptors.

Lastly, an interesting chemical structure, the [2.2]paracyclophane was incorporated between the acceptors. Both compounds **66e** and **66d** have a phenylene spacer with phenylene and pyridine between the donors. The yields of the isolated compounds were good, with 63% (**66e**) and 47% (**66d**).

Photophysical Characterization

The photophysical properties were characterized of a selection of the D-A-A-D series (Table 22). Not all compounds can be measured because great amounts are necessary. At least one sample was characterized from bridging compounds with phenylene, pyridyl and no bridge compounds. Photoluminescence spectra were measured and illustrated in Figure 35.

Table 22 Photospectroscopic data of **63a**, **64a-b**, and **65a-b** measured in toluene solution.

Entry	Compound	#	$\lambda_{max, PL}$ [nm]	FWHM [nm]
1	**2TCzPyPhOx-PyDim**	**63c**	456	105
2	**2TCzPhOx-PhDim**	**64a**	444	37
3	**2TCzPyPhOx-oPhDim**	**64b**	469	86
4	**2TCzPhOx-Dim**	**65a**	444	44
5	**2TCzPyPhOx-Dim**	**65b**	488	84

The emission spectra showed λ_{max} in the blue region with 444 nm, 456 nm, 469 nm, and 488 nm for compounds **64a** and **65a**, **63c**, **64b**, and **65b**, respectively. The FWHM values were small with 37 nm for **64a** and 44 nm for **65a**. **64b** and **65b** showed similar FWHM values with 86 and 84 nm, respectively. Compound **63c** showed the largest FWHM with 105 nm, which is unsuitable for an OLED application.

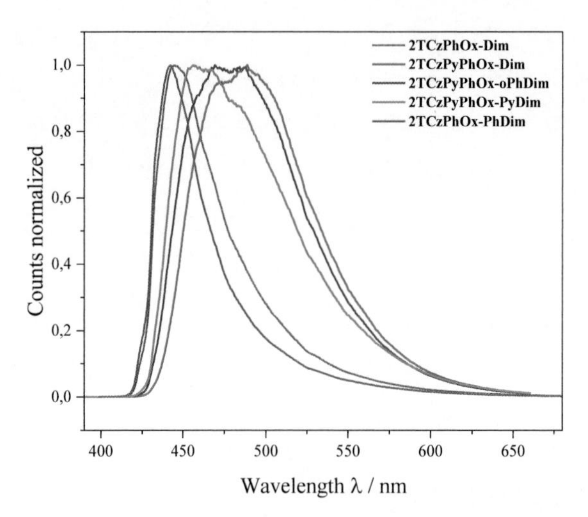

Figure 35 Emission spectra of compounds **2TCzPyPhOx-PyDim**, **2TCzPhOx-PhDim**, **2TCzPyPhOx-oPhDim**, **2TCzPhOx-Dim**, and **2TCzPyPhOx-Dim**. The Intensity counts are normalized to 1. Results provided by Idoia Garin.

No D-A-A-D compound showed a delayed fluorescence. A problem might be too little rigidity in the structure and, thus, the unlimited rotatability of the donor units through the oxadiazoles, which might lead to quenching processes. Furthermore, the overlap of the frontier orbitals could be too small for efficient charge transfer.

4 Summary and Outlook

In summary, two main projects were conducted to discover new blue TADF emitters. These projects focused on synthetic approaches and the finding of new molecules. All successfully synthesized materials with promising TADF properties were thoroughly characterized to assess their suitability for application in OLEDs.

4.1. Mono[1,2,4]-triazolo[1,3,5]-triazine and Di[1,2,4]-triazolo[1,3,5]-triazine

Six new ditriazolotriazine and seven new monotriazolotriazine-based TADF emitters were successfully synthesized. The yields for the DTT emitters were fair to good ranging from 33% to 94%, while the yields for the MTT emitters were relatively low, ranging from 7% to 76%. Addressed can be the lower reactivity of the triazine precursor for the MTT syntheses. Different acridine donors, such as **PXZ**, **DMAC**, and **DPAC**, were tested through the DTT and MTT projects to find the best donor-acceptor pair regarding their TADF properties. The original MTT and DTT series were calculated with DFT methods to reinforce the later on carried-out measurements. Photophysical characterization of the original MTT and DTT was conducted through collaboration with the Zysman-Colman group. Since these are very time-consuming measurements, not all the following MTTs and DTTs were photophysically characterized to the full extent. Experiments were conducted to measure the excitation and emission wavelengths, fluorescence lifetimes, host screenings, photoluminescence quantum yields, and HOMO & LUMO energy levels.

From the original DTT and MTT series, the **DMAC-MTT** and **-DTT** performed best, embedded in the host mCP, considering a blue emission and short fluorescence lifetime (prompt and delayed). The emission range for all original MTT and DTT compounds doped with the host mCP ranged from 483, 488, 489, 530 (twice) to 540-580 nm for **DMAC-DTT** (3wt% mCP), **DPAC-DTT** (3wt% mCP), **DPAC-MTT** (15wt% mCP), **DMAC-MTT** (10wt% mCP), **PXZ-DTT** (3wt% mCP) and **PXZ-MTT** (5wt% mCP), respectively. To get the full picture of the TADF properties of these emitters, the analysis of ΔE_{ST} and the rates of ISC and rISC is a task for future works. Moreover, the implementation of selected emitters of this series into an OLED device is of interest.

The original DTT and MTT series were expanded to study the importance of the dihedral angle. Therefore, methyl and trifluoromethyl groups were installed in the spacer unit between the donor and acceptor. Photoluminescence spectra of the trifluoromethylated phenyl spacer compounds were carried out to compare the results with the original MTT and DTT emitters. A blue shift of the emission and a decrease in the fluorescence lifetime were observed compared to those MTT and DTT with an unsubstituted phenyl spacer. Whereas **PXZ-DTT** and **DMAC-DTT** showed emission maxima of 602 nm and 538 nm, the **CF3-PXZ-DTT** and **CF3-DMAC-DTT** showed maxima at 439 nm and 527 nm, respectively. With emission maxima of 586 nm and 527 nm, **CF3-PXZ-MTT** and **CF3-DMAC-MTT** are hypsochromically shifted compared to **PXZ-MTT** and **DMAC-MTT** with 596 nm and 538 nm, respectively.

Single crystal X-ray diffraction of **PXZ-MTT** and **PXZ-MTT-Me** revealed that the structure does not conform with the TTT and DTT. The L-isomer is formed in the MTT structure.

The studies of this work on materials for third-generation OLEDs resulted in the discovery of interesting new structures for blue TADF emitters. Since 2019, the fourth generation of OLEDs based on hyperfluorescence has emerged. This generation was proposed to be the solution for the troublesome blue OLEDs.[136-137] Blue OLEDs are still issued with broad emissions within the efficient second and third generations. Furthermore, pure blue OLEDs are not as stable as they need to be. Their poor stability is the generation of highly reactive hot triplet states. These states are created through exciton annihilations of long-lived triplets. They can be understood as high-energy particles that move around the molecules, leading to the degradation of the OLED device.[138] The key is an emitter that exhibits a short triplet lifetime and an OLED design that enables fast triplet consumption.

Hyperfluorescence is capable of meeting the requirements mentioned above.[139] A TADF emitter is used as an assistant dopant to a fluorescent molecule. The main exciton creation is located on the TADF assistant. This can be regulated with the doping concentration ratio of TADF and fluorescent molecule. Through the TADF pathway, the created triplet excitons are upconverted into the excited singlet state, still within the TADF assistant. FRET transfers the excited singlet energy to the excited singlet state of the fluorescent emitter. The excited singlet state decays, resulting in delayed fluorescence.[140] To summarize, hyperfluorescence combines the narrow emission of fluorescent emitters with the high efficiency of TADF materials. Embedding some of the here synthesized blue TADF emitters, e.g., **DMAC-DTT** or **DMAC-MTT**, together with a good fluorescent emitter, could be interesting to enter the fourth generation of OLEDs.

4.2. Derivatization of the CzBN Class

The known, high-performing TADF emitter **5TCzBN** was focused on during the second project. The approach encompassed the transformation of the nitrile into a tetrazole or oxadiazole to create novel TADF emitters. These new functional acceptor groups were derivatized with aromatic substituents such as benzonitrile, mesityl, *tert*-butyl-phenyl, anisole, trifluoromethyl-phenyl, indole, pentafluorophenyl or aliphatic substituents such as cyclohexyl, *tert*-butyl, *n*-butyl, fluorinated alkyl chains, propyl, propenyl, and propynyl to study its functional group tolerance and learn their TADF behavior. Installing functional groups like alkynyls allows click reactions with azides to access other functionalities easily.

It was learned that the tetrazole derivatives **40a-b** showed only weak TADF with an emission maximum of roughly 450 nm (Figure 36). In contrast, the oxadiazoles **42a-h** did show good TADF properties. Their emission spectra with the prompt and delayed emissions are displayed in Figure 36. Emitters with electron-donating groups like **42a, -d-h** showed a blue-shifted emission compared to those with electron-withdrawing groups like **42b-c**. The emitters **42a-c** were the only ones experimentally characterized to a larger extent yet. The TADF properties, such as fluorescence lifetimes, were similar. Considering a desired blue emission, emitter **42a** showed the best performance in a toluene solution. The experiments of the remaining emitters **42d-h** are ongoing. Furthermore, host screenings in mCP for all emitters that passed the first tests to confirm their TADF character were started and will be finished in future works.
The next step in the investigation of this emitter class was the incorporation of an additional phenylene spacer. It was found that this extension of the frontier orbital separation shuts off TADF for both tetrazole and oxadiazole derivatives **39a-h** and **41a-d**, respectively. All tetrazole compounds showed a blue emission in the range around 450 nm. No matter which functional groups were used, there was no discernible pattern in the change of emission color. In a further approach, a system with two acceptors and two donor systems was tested. The D-A-A-D molecules with five, four, or two donors on each side of the donor system were successfully synthesized with different bridging combinations like phenyl, pyridine, ethyl, paracyclophane, or adamantyl. In photophysical measurements of a selection of these D-A-A-D compounds, it was learned that no TADF occurs through the system. One possible explanation might be that the separation between the HOMO and LUMO is too large, coupled with too little rigidity in this framework. Theoretical support for this project was not provided because the emitters showed a lack of TADF in the photophysical experiments, and therefore, performing DFT calculations would not be reasonable, given the complexity of these molecules and the time it would take to complete them.

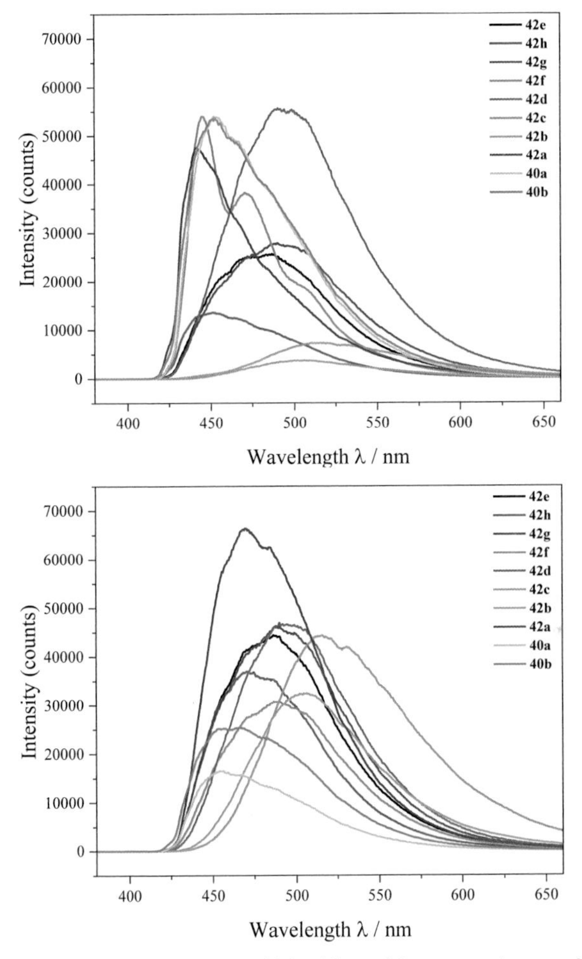

Figure 36 Emission spectra with prompt (top) and delayed (bottom) fluorescence of compounds **40a-b** and **42a-h**. Results provided by Idoia Garin.

5 Experimental Section

5.1. General Remarks – Materials and Methods

Nuclear Magnetic Resonance Spectroscopy (NMR)

The NMR spectra of the compounds described herein were recorded on a Bruker Avance 400 NMR instrument at 400 MHz for ^{1}H NMR and 101 MHz for ^{13}C NMR, and a Bruker Avance 500 NMR instrument at 500 MHz for ^{1}H NMR and 126 MHz for ^{13}C NMR. The NMR spectra were recorded at room temperature in deuterated solvents acquired from Eurisotop. The chemical shift δ is displayed in parts per million [ppm] and the references used were the ^{1}H and ^{13}C peaks of the solvents themselves: d_1-chloroform ($CDCl_3$): 7.26 ppm for ^{1}H and 77.0 ppm for ^{13}C, d_6-dimethyl sulfoxide (DMSO-d_6): 2.50 ppm for ^{1}H and 39.4 ppm for ^{13}C, d_8-tetrahydrofuran (THF-d_8): 3.58 ppm and 1.72 ppm for ^{1}H and 67.21 ppm and 25.53 ppm for ^{13}C. For the characterization of centrosymmetric signals, the signal's median point was chosen for multiplets in the signal range. The following abbreviations were used to describe the proton splitting patterns: d = doublet, t = triplet, m = multiplet, dd = doublet of doublet, ddd = doublet of doublet of doublet, dt = doublet of triplet, qd = quartet of doublets. Absolute values of the coupling constants "J" are given in Hertz [Hz] in absolute value and decreasing order.

The assignment of the signals *via* ^{13}C NMR spectra was based on the chemical shift and the multiplicity obtained via edited HSQC experiment and are described as follows: + = primary or tertiary C-atom (positive HSQC-signal), – = secondary C-atom (negative signal) and C_q = quaternary C-atom (no signal). Due to symmetry and signal overlap, signals are missing in the evaluation of the ^{13}C NMR spectra. Common solvent and solvent impurity signals as followed were not explicitly listed: in chloroform: ^{1}H NMR 8.02 (dimethylformamide), 7.25 (toluene), 7.17 (toluene), 5.30 (dichloromethane), 2.96 (dimethylformamide), 2.88 (dimethylformamide), 2.36 (toluene), 1.55 (H_2O), 1.43 (cyclohexane), 1.25 (H grease), 0.84

- 0.87 (H grease), 0.07 (silicon grease) ppm; ^{13}C NMR 162.62 (dimethylformamide), 137.89 (toluene), 129.07 (toluene), 128.26 (toluene), 125.33 (toluene), 53.52 (dichloromethane), 36.50 (dimethylformamide), 31.45 (dimethylformamide), 29.7 (H grease), 26.94 (cyclohexane), 21.46 (toluene), 1.20 (silicon grease) ppm; in dichloromethane: ^{1}H NMR 1.52 (H_2O), 1.29 (H grease), 0.84 - 0.90 (H grease), 0.09 (silicon grease) ppm; ^{13}C NMR 30.1 (H grease), 1.2 (silicon grease) ppm; in dimethyl sulfoxide: ^{1}H NMR 5.76 (dichloromethane), 3.33 (H_2O), 1.40 (cyclohexane), 1.24 (H grease), 0.82 – 0.88 (H grease), –0.06 (silicon grease) ppm; ^{13}C NMR 54.84 (dichloromethane), 29.2 (H grease), 26.33 (cyclohexane); in tetrahydrofuran: ^{1}H NMR 10.84 (THF-d_8 impurity), 5.51 (dichloromethane), 3.26 (THF-d_8 impurity), 2.50 (THF-d_8 impurity), 2.49 (H_2O), 1.44 (cyclohexane), 1.29 (H grease), 0.85 – 0.91 (H grease), 0.11 (silicon grease) ppm; ^{13}C NMR 29.9 ppm (H grease) 1.2 (silicon grease) ppm. In the case of low resolution, the signals were listed based on the ^{1}H-broadband-decoupled ^{13}C NMR spectra without phases.

Mass Spectrometry (MS)

Fast atom bombardment (FAB) experiments were conducted using a Finnigan, MAT 95 instrument, with 3-nitrobenzyl alcohol (3-NBA) as matrix and reference for high resolution. For the interpretation of the spectra, molecular peaks $[M]^+$, peaks of protonated molecules $[M+H]^+$, and characteristic fragment peaks are indicated with their mass-to-charge ratio (m/z), and in the case of EI, their intensity in percent, relative to the base peak (100%) is given. In the case of high-resolution measurements, the tolerated error is 0.0005 m/z. ESI experiments were recorded on a Q-Exactive (Orbitrap) mass spectrometer (Thermo Fisher Scientific, San Jose, CA, USA) equipped with a HESI II probe to record high resolution. The tolerated error is 5 ppm of the molecular mass. Again, the spectra were interpreted by molecular peaks $[M]^+$, peaks of protonated molecules $[M+H]^+$, and characteristic fragment peaks and indicated with their mass-to-charge ratio (m/z).

MALDI-ToF-MS (Matrix-Assisted Laser Desorption Ionisation Time of Flight Mass Spectrometry) was conducted using an Axima Confidence (type: TO-6071R00), equipped with the software Shimadzu Biotech Launchpad™ (version 2.9.3.20110624) from Shimadzu Biotech. A nitrogen laser with a wavelength of λ = 337 nm was used for the desorption and ionization of the compounds. A 384-Spots Shimadzu Kratos Analytical Standard probe plate (DE1580TA) from Shimadzu Biotech was used as a target. The compounds were dissolved in tetrahydrofuran beforehand and then applied to the target.

The matrix was made of a saturated 1:1 mixture of 2,5-dihydroxybenzoic acid and α-cyano-4-hydroxycinnamic acid (Universal-MALDI-Matrix, Sigma-Aldrich®) dissolved in a 1:1 mixture of acetonitrile and distilled water. The co-crystallization of the matrix and sample was carried out under atmospheric conditions. The sample was bombarded approximately 100 times with a frequency of 50 Hz for the measurements. The result was averaged. The protonated molecule peak $[M+H]^+$, as well as if present, the pseudo-molecule peaks with sodium $[M+Na]^+$ and potassium $[M+K]^+$ were detected.

Infrared Spectroscopy (IR)

The infrared spectra were recorded with a Bruker IFS 88 instrument. Solids were measured by the attenuated total reflection (ATR) method. The positions of the respective transmittance bands are given in wave numbers $\tilde{\nu}$ [cm^{-1}] and were measured in the range from 3600 cm^{-1} to 500 cm^{-1}. The characterization of the transmittance bands was done in a sequence of transmission strength T with the following abbreviations: vs (very strong, 0–9% T), s (strong, 10–39% T), m (medium, 40–69% T), w (weak, 70–89% T), vw (very weak, 90–100% T) and br (broad).

Elemental Analysis (EA)

Elemental analysis was done on an Elementar vario MICRO instrument. The weight scale used was a Sartorius M2P. Calculated (calc.) and found percentage by mass values for carbon, hydrogen, nitrogen, and sulfur are indicated in fractions of 100%.

Thin Layer Chromatography (TLC)

For the analytical thin layer chromatography, TLC silica plates coated with fluorescence indicator from Merck (silica gel 60 F254, thickness 0.2 mm) were used. UV-active compounds were detected at 254 nm and 365 nm excitation wavelengths with a Heraeus UV-lamp model Fluotest.

Weight Scale

For weightings of solids and liquids, a Sartorius model LC 620 S was used.

Solvents and Chemicals

Solvents of p.a. quality (per analysis) were commercially acquired from Sigma Aldrich, Carl Roth, or Acros Fisher Scientific and, unless stated otherwise, used without further purification. Dry solvents were obtained from a mbraun solvent purification system or purchased from Carl Roth, Acros, or Sigma Aldrich (< 50 ppm H_2O over molecular sieves). All used reagents were commercially acquired from abcr, Acros, Alfa Aesar, Sigma Aldrich, TCI, Chempur, Carbolution or Synchemie or were available in the group. Unless stated otherwise, all chemicals were used without further purification.

Experimental Procedure

Air- and moisture-sensitive reactions were carried out under an argon atmosphere in previously baked-out glassware using standard Schlenk techniques. Liquid reagents and solvents were injected with plastic syringes and stainless-steel cannulas of different sizes, unless specified otherwise. Reactions at low temperatures were cooled using shallow vacuum flasks produced by Isotherm, Karlsruhe, filled with a water/ice mixture for 0 °C, water/ice/sodium chloride for –20 °C or isopropanol/dry ice mixture for –78 °C. The reaction flask was equipped with a reflux condenser and connected to the argon line for reactions at high temperatures. Solvents were evaporated under reduced pressure at 40 °C using a rotary evaporator. Unless stated otherwise, solutions of inorganic salts are saturated aqueous solutions.

Reaction Monitoring

The progress of reactions in the liquid phase was monitored by TLC. UV active compounds were detected with a UV lamp at 254 nm and 365 nm excitation wavelength.

Product Purification

Unless stated otherwise, the crude compounds were purified by column chromatography. For the stationary phase of the column, silica gel, produced by Merck (silica gel 60, 0.040 × 0.063 mm, 260 – 400 mesh ASTM), and sea sand by Riedel de-Haën (baked out and washed with hydrochloric acid) were used. Solvents used were commercially acquired in HPLC-grade and individually measured volumetrically before mixing.

Repository

All reaction procedures and analytical data are available online and can be viewed under: https://www.chemotion-repository.net/home/welcome

5.2. Reaction Procedures and Analytical Data

5.2.1. Mono[1,2,4]-triazolo[1,3,5]-triazine and Di[1,2,4]-triazolo[1,3,5]-triazine

5.2.1.1. Synthesis of 9,9-diphenyl-9,10-dihydroacridine (**DPAC**)

Methyl 2-(phenylamino)benzoate (21)

In a Schlenk flask, cesium carbonate (6.82 g, 20.9 mmol, 1.50 equiv.), 1,1′-ferrocenediyl-bis(diphenylphosphine) (464 mg, 837 µmol, 0.06 equiv.) and $Pd(dba)_2$ (321 mg, 558 µmol, 0.04 equiv.) were evacuated and backfilled with argon three times. Dry toluene (180 mL), methyl 2-bromobenzoate (3.00 g, 14.0 mmol, 1.00 equiv.) and aniline (1.60 g, 16.7 mmol, 1.20 equiv.) were added, and the resulting mixture was stirred at 110 °C for 24 h. After cooling to room temperature, the reaction mixture was poured into an excess of water (100 mL) and extracted with dichloromethane (3 × 50 mL). The combined organic layers were washed with brine (50 mL), dried over sodium sulfate, and the solvents were removed under reduced pressure. The crude product was purified by flash column chromatography over silica gel (*n*-pentane/CH_2Cl_2, 1:0 to 1:1) to yield 2.85 g (12.5 mmol, 90%) of the title compound as a yellowish oil. The analytical data agree with the literature.[114]

R_f (*n*-pentane/CH_2Cl_2, 3:1) = 0.38.

^1H NMR (400 MHz, $CDCl_3$, ppm) δ = 9.46 (s, 1H, NH), 7.97 (dd, *J* = 8.0, 1.2 Hz, 1H, H_{ar}), 7.39–7.28 (m, 3H, H_{ar}), 7.24 (m, 3H, H_{ar}), 7.09 (t, *J* = 7.3 Hz, 1H, H_{ar}), 6.73 (ddd, *J* = 8.1, 6.9, 1.3 Hz, 1H, H_{ar}), 3.91 (s, 3H, CH_3).

Diphenyl(2-(phenylamino)phenyl)methanol (23)

In a Schlenk flask, methyl 2-(phenylamino)benzoate (**21**) (2.07 g, 9.11 mmol, 1.00 equiv.) was dissolved in dry tetrahydrofuran (70 mL) under argon and cooled to 0 °C. Under stirring, phenyl magnesium chloride (2 M in tetrahydrofuran, 22.8 mL, 45.5 mmol, 5.00 equiv.) was added dropwise to the reaction mixture. After 1 h, the mixture was gradually warmed to room temperature and stirred for 24 h until completion. Initially, the reaction mixture was quenched with the addition of water, then poured into an excess of water (100 mL) and extracted with dichloromethane (3 × 80 mL). The combined organic layers were washed with brine, dried over sodium sulfate, and reduced in a vacuum to yield a brownish oil. The crude product was purified by flash column chromatography over silica gel (*n*-pentane/CH_2Cl_2, 3:1 to 1:1) to yield 3.14 g (8.94 mmol, 98%) of the title compound as a yellowish oil. The analytical data agree with the literature.[115]

R_f (*n*-pentane/CH_2Cl_2, 1:1) = 0.35.

^{1}H NMR (400 MHz, $CDCl_3$, ppm) δ = 7.36 – 7.26 (m, 11H, H_{ar}), 7.23 – 7.13 (m, 3H, H_{ar}), 6.91 – 6.82 (m, 2H, H_{ar}), 6.77 (dd, *J* = 8.5, 1.1 Hz, 2H, H_{ar}), 6.62 (dd, *J* = 7.8, 1.5 Hz, 1H, H_{ar}) 5.31 (s, 1H, OH). Missing signal 1H (NH).

9,9-Diphenyl-9,10-dihydroacridine (16)

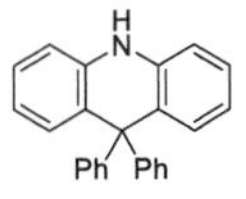

In a two-neck flask, the crude product diphenyl(2-(phenylamino)phenyl)methanol (**23**) (3.14 g. 8.93 mmol, 1.00 equiv.) was dissolved in acetic acid (40 mL), and the resulting mixture was heated at 70 °C. Concentrated hydrochloric acid (2.3 mL) was added dropwise under stirring. The resulting mixture was stirred at 100 °C for 16 h. After cooling to room temperature, the reaction mixture was poured into an excess of water (70 mL) and extracted with dichloromethane (3 × 50 mL). The combined organic layers were washed with brine (50 mL), dried over sodium sulfate, and reduced in a vacuum. The crude product was purified by flash column chromatography over silica gel (CH_2Cl_2) to yield 2.69 g (8.07 mmol, 90%) of the title compound as an off-white solid.

R_f (*n*-pentane/CH_2Cl_2, 2:1) = 0.33.

^{1}H NMR (500 MHz, $CDCl_3$, ppm) δ = 7.26 – 7.20 (m, 6H, H_{ar}), 7.20 – 7.14 (m, 2H, H_{ar}), 7.00 – 6.93 (m, 4H, H_{ar}), 6.92 – 6.74 (m, 6H, H_{ar}), 6.27 (s, 1H, NH).

^{13}C NMR (126 MHz, $CDCl_3$, ppm) δ = 146.2 (C_q,), 139.9 (C_q,), 130.4 (C_q,), 128.2 (+, CH), 127.8 (+, CH), 127.4 (+, CH), 126.5 (+, CH), 120.5 (+, CH), 113.8 (+, CH), 56.9 (C_q).

IR (ATR, cm^{-1}) ṽ = 3391 (w), 3381 (m), 3048 (w), 3021 (w), 1602 (m), 1577 (w), 1472 (vs), 1453 (s), 1442 (s), 1409 (m), 1315 (s), 1293 (w), 1273 (w), 1252 (w), 1248 (w), 1180 (w), 1149 (w), 1123 (w), 1078 (w), 1035 (w), 1000 (w), 901 (m), 868 (w), 756 (vs), 742 (vs), 696 (vs), 660 (s), 637 (vs), 630 (s), 586 (w), 569 (w), 538 (m), 516 (s), 475 (s), 462 (s), 414 (s), 391 (w), 380 (w).

MS (FAB, 3-NBA): m/z (%) = 334 (17) $[M+H]^+$, 333 (22) $[M]^+$. HRMS (FAB, $C_{25}H_{19}N$): calcd 333.1517, found 333.1517.

EA ($C_{25}H_{19}N$) calcd C: 90.06, H: 5.74, N: 4.20; found C: 89.26, H: 5.56, N: 4.16.

5.2.1.2. Syntheses of the Nitrile Precursors

4-(10*H*-Phenoxazin-10-yl)benzonitrile (25a)

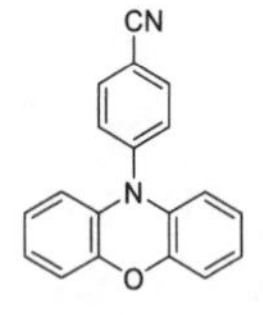

In a sealable vial, 4-fluorobenzonitrile (1.16 g, 9.55 mmol, 1.00 equiv.), 10*H*-phenoxazine (1.75 g, 9.55 mmol, 1.00 equiv.) and tripotassium phosphate (4.06 g, 19.1 mmol, 2.00 equiv.) were evacuated and backfilled with argon three times. Dry dimethyl sulfoxide (40 mL) was added, and the resulting mixture was heated at 110 °C for 16 h. After cooling to room temperature, the reaction mixture was poured into an excess of water (70 mL) and extracted with dichloromethane (3 × 50 mL). The combined organic layers were washed with brine (50 mL), dried over sodium sulfate, and the solvents were removed under reduced pressure. The crude product was purified by flash column chromatography over silica gel (*n*-pentane/CH_2Cl_2, 1:1) to yield 2.26 g of the title compound (7.91 mmol, 83%) as a yellow solid.

R_f (*n*-pentane/CH_2Cl_2, 2:1) = 0.35.

^{1}H NMR (500 MHz, $CDCl_3$, ppm) δ = 7.90 (d, *J* = 8.6 Hz, 2H, H_{ar}), 7.51 (d, *J* = 8.5 Hz, 2H, H_{ar}), 6.79–6.68 (m, 4H, H_{ar}), 6.63 (ddd, *J* = 8.0, 7.1, 1.9 Hz, 2H, H_{ar}), 5.93 (dd, *J* = 8.0, 1.3 Hz, 2H, H_{ar}).

^{13}C NMR (126 MHz, $CDCl_3$, ppm) δ = 144.2 (C_q), 144.0 (C_q), 135.1 (+, CH), 133.4 (C_q), 131.9 (+, CH), 123.5 (+, CH), 122.5 (+, CH), 118.2 (C_q), 116.1 (+, CH), 113.6 (+, CH), 112.4 (C_q).

IR (ATR, cm^{-1}) ṽ = 3063 (w), 3050 (w), 2609 (w), 2224 (w), 1629 (vw), 1596 (m), 1483 (vs), 1460 (vs), 1402 (w), 1329 (vs), 1290 (s), 1272 (vs), 1203 (m), 1184 (w), 1169 (w), 1150 (w), 1118 (w), 1101 (w), 1044 (m), 1021 (w), 926 (m), 870 (m), 856 (w), 836 (w), 822 (w), 802 (w), 756 (vs), 739 (vs), 681 (w), 647 (w), 615 (m), 585 (vs), 555 (w), 543 (s), 509 (m), 473 (w), 458 (w), 438 (w), 426 (w), 404 (w), 375 (w).

MS (FAB, 3-NBA): m/z (%) = 285 (40) $[M+H]^+$, 284 (100) $[M]^+$. HRMS (FAB, $C_{19}H_{12}N_2O$): calcd 284.0950, found 284.0950.

EA ($C_{19}H_{12}N_2O$) calcd C: 80.27, H: 4.25, N: 9.85; found C: 80.00, H: 4.13, N: 9.80.

Mp 158 °C.

4-(9,9-Dimethylacridin-10(9*H*)-yl)benzonitrile (25b)

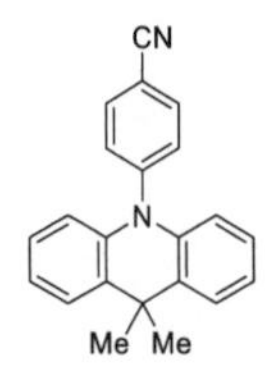

In a sealable vial, 4-fluorobenzonitrile (1.21 g, 10.0 mmol, 1.00 equiv.), 9,9-dimethyl-9,10-dihydroacridine (2.09 g, 10.0 mmol, 1.00 equiv.) and tripotassium phosphate (5.31 g, 25.0 mmol, 2.50 equiv.) were evacuated and backfilled with argon three times. Dry dimethyl sulfoxide (40 mL) was added, and the resulting mixture was heated at 110 °C for 48 h. After cooling to room temperature, the reaction mixture was poured into an excess of water (100 mL) and extracted with dichloromethane (3 × 100 mL). The combined organic layers were washed with brine (100 mL), dried over sodium sulfate, and reduced in a vacuum. The crude product was purified by flash column chromatography over silica gel (*n*-pentane/CH_2Cl_2, 5:2 to 1:1) to yield 1.46 g of the title compound (4.70 mmol, 47%) as an off-white solid.

R_f (*n*-pentane/CH_2Cl_2, 1:1) = 0.49.

^{1}H NMR (500 MHz, $CDCl_3$, ppm) δ = 7.92–7.89 (m, 2H, H_{ar}), 7.51–7.47 (m, 4H, H_{ar}), 7.04–6.97 (m, 4H, H_{ar}), 6.30–6.28 (m, 2H, H_{ar}), 1.67 (s, 6H, CH_3).

^{13}C NMR (126 MHz, $CDCl_3$, ppm) δ = 146.3 (C_q), 140.3 (C_q), 134.9 (+, CH), 131.7 (C_q), 131.2 (+, CH), 126.6 (+, CH), 125.6 (+, CH), 121.8 (+, CH), 118.5 (C_q), 114.9 (+, CH), 111.5 (C_q), 36.3 (C_q), 31.0 (+, CH_3, 2C).

IR (ATR, cm^{-1}) $\tilde{\nu}$ = 3064 (w), 3053 (w), 3033 (w), 2992 (w), 2973 (w), 2915 (w), 2908 (w), 2857 (w), 2230 (w), 1587 (s), 1502 (m), 1470 (vs), 1445 (vs), 1402 (w), 1390 (w), 1356 (w), 1320 (vs), 1266 (vs), 1186 (w), 1170 (w), 1125 (w), 1109 (w), 1101 (w), 1084 (w), 1044 (m), 1016 (w), 980 (w), 966 (w), 942 (w), 924 (m), 894 (w), 849 (w), 837 (w), 796 (w), 744 (vs), 713 (m), 667 (w), 650 (w), 622 (s), 598 (m), 582 (w), 557 (vs), 531 (s), 487 (w), 473 (w), 432 (m), 408 (w), 394 (w), 382 (w).

MS (FAB, 3-NBA): m/z (%) = 311 (53) $[M+H]^+$, 310 (55) $[M]^+$. HRMS (FAB, $C_{22}H_{18}N_2$): calcd 310.1470, found 310.1465.

EA ($C_{22}H_{18}N_2$) calcd C: 85.13, H: 5.85, N: 9.03; found C: 84.79, H: 5.72, N: 8.99.

Mp 148 °C.

4-(9,9-Diphenylacridin-10(9*H*)-yl)benzonitrile (25c)

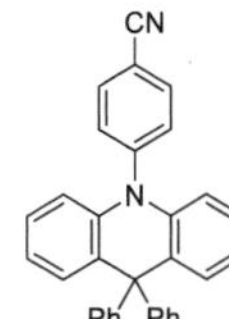

In a sealable vial, 4-fluorobenzonitrile (472 mg, 3.90 mmol, 1.00 equiv.), 9,9-diphenyl-9,10-dihydroacridine (1.30 g, 3.90 mmol, 1.00 equiv.) and tripotassium phosphate (2.07 g, 9.75 mmol, 2.50 equiv.) were evacuated and backfilled with argon three times. Dry dimethyl sulfoxide (16 mL) was added, and the resulting mixture was heated at 110 °C for 16 h. After cooling to room temperature, the reaction mixture was poured into an excess of water (50 mL) and extracted with dichloromethane (3 × 50 mL). The combined organic layers were washed with brine (50 mL), dried over sodium sulfate, and the solvents were removed under reduced pressure. The crude product was purified by flash column chromatography over silica gel (*n*-pentane/CH_2Cl_2, 3:1 to 1:2) to yield 1.29 g of the title compound (2.98 mmol, 76%) as an off-white solid.

R_f (*n*-pentane/CH_2Cl_2, 1:1) = 0.32.

^{1}H NMR (500 MHz, DMSO-d$_6$, ppm) δ = 8.08 – 8.01 (m, 2H, H_{ar}), 7.36 – 7.20 (m, 8H, H_{ar}), 7.12 (td, *J* = 8.4, 7.9, 1.5 Hz, 2H, H_{ar}), 6.96 (td, *J* = 7.8, 1.1 Hz, 2H, H_{ar}), 6.88 (dd, *J* = 7.0, 1.5 Hz, 4H, H_{ar}), 6.79 (dd, *J* = 7.8, 1.4 Hz, 2H, H_{ar}), 6.40 (dd, *J* = 8.2, 0.9 Hz, 2H, H_{ar}).

^{13}C NMR (126 MHz, DMSO-d$_6$, ppm) δ = 145.7 (C_q), 144.8 (C_q), 140.9 (C_q), 135.0 (+, CH), 131.0 (C_q), 130.1 (+, CH), 129.7 (+, CH), 129.6 (+, CH), 127.9 (+, CH), 127.3 (+, CH), 126.6 (+, CH), 121.1 (+, CH), 118.4 (C_q, 1C, CN), 114.7 (+, CH), 110.8 (C_q), 56.3 (C_q).

IR (ATR, cm^{-1}) ṽ = 3058 (vw), 3030 (vw), 2230 (w), 1591 (m), 1504 (w), 1472 (s), 1448 (s), 1319 (vs), 1262 (s), 1184 (w), 1164 (w), 1101 (w), 1081 (w), 1061 (w),

1033 (w), 1017 (w), 922 (w), 840 (w), 751 (vs), 735 (vs), 700 (vs), 671 (m), 633 (s), 622 (m), 558 (s), 510 (w), 504 (w), 476 (w), 432 (m).

MS (FAB, 3-NBA): m/z (%) = 434 (9) $[M]^{+}$. HRMS (FAB, $C_{32}H_{22}N_2$): calcd 434.1783, found 434.1783.

EA ($C_{32}H_{22}N_2$) calcd C: 88.45, H: 5.10, N: 6.45; found C: 85.93, H: 4.89, N: 6.22.

Mp 248 °C.

3,5-Dimethyl-4-(10*H*-phenoxazin-10-yl)benzonitrile (25d)

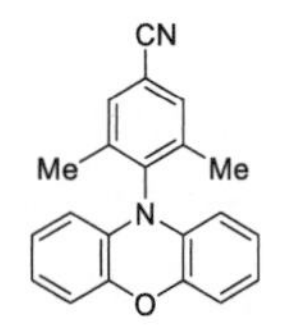

In a sealable vial, 4-fluoro-3,5-dimethylbenzonitrile (1.63 g, 10.9 mmol, 1.00 equiv.), 10*H*-phenoxazine (2.00 g, 10.9 mmol, 1.00 equiv.) and tripotassium phosphate (4.63 g, 21.8 mmol, 2.00 equiv.) were evacuated and backfilled with argon three times. Dry dimethyl sulfoxide (44 mL) was added, and the resulting mixture was heated at 110 °C for 16 h. After cooling to room temperature, the reaction mixture was poured into an excess of water (70 mL) and extracted with dichloromethane (3 × 50 mL). The combined organic layers were washed with brine (50 mL), dried over sodium sulfate, and the solvents were removed under reduced pressure. The crude product was purified by flash column chromatography over silica gel (*n*-pentane/CH_2Cl_2, 3:1 to 1:1 and afterwards toluene) to yield 2.41 g of the title compound (7.72 mmol, 71%) as a yellow solid.

R_f (*n*-pentane/CH_2Cl_2, 1:1) = 0.38.

^{1}H NMR (500 MHz, $CDCl_3$, ppm) δ = 7.57 (s, 2H, H_{ar}), 6.76–6.62 (m, 4H, H_{ar}), 6.63–6.55 (m, 2H, H_{ar}), 5.63 (d, *J* = 9.3 Hz, 2H, H_{ar}), 2.27 (s, 6H, CH_3).

^{13}C NMR (126 MHz, $CDCl_3$, ppm) δ = 143.8 (C_q), 141.3 (C_q), 139.7 (C_q), 133.3 (+, CH), 131.3 (C_q), 123.9 (+, CH), 122.0 (+, CH), 118.5 (C_q), 116.0 (+, CH), 112.8 (C_q), 111.6 (+, CH), 18.0 (+, CH_3, 2C).

IR (ATR, cm^{-1}) $\tilde{\nu}$ = 3055 (w), 3036 (vw), 2917 (vw), 2615 (vw), 2225 (w), 1628 (vw), 1591 (w), 1485 (vs), 1460 (s), 1397 (w), 1381 (w), 1339 (vs), 1327 (s), 1290 (m), 1269 (vs), 1203 (s), 1159 (w), 1152 (w), 1139 (w), 1113 (w), 1095 (w), 1037 (w), 955 (w), 926 (w), 914 (w), 899 (w), 882 (m), 860 (w), 840 (w), 833 (w), 779 (w), 732 (vs), 684 (w), 626 (s), 562 (w), 557 (w), 534 (w), 510 (w), 490 (w), 477 (w), 458 (w), 436 (w), 418 (vw), 402 (w), 388 (vw).
MS (FAB, 3-NBA): m/z (%) = 312 (100) $[M]^{+}$. HRMS (FAB, $C_{21}H_{16}N_2O$): calcd 312.1263, found 312.1420.
EA ($C_{21}H_{16}N_2O$) calcd C: 80.75, H: 5.16, N: 8.97; found C: 80.82, H: 5.11, N: 8.94.
Mp 165 °C.

4-(10*H*-Phenoxazin-10-yl)-3-(trifluoromethyl)benzonitrile (25e)

In a sealable vial, 4-fluoro-3-(trifluoromethyl)benzonitrile (1.81 g, 9.55 mmol, 1.00 equiv.), 10*H*-phenoxazine (1.75 g, 9.55 mmol, 1.00 equiv.) and tripotassium phosphate (4.06 g, 19.1 mmol, 2.00 equiv.) were evacuated and backfilled with argon three times. Dry dimethyl sulfoxide (40 mL) was added, and the resulting mixture was heated at 110 °C for 48 h. After cooling to room temperature, the reaction mixture was poured into an excess of water (100 mL) and extracted with dichloromethane (3 × 50 mL). The combined organic layers were washed with brine (100 mL), dried over sodium sulfate and reduced in a vacuum. The crude product was purified by flash column chromatography over silica gel (*n*-pentane/CH_2Cl_2, 4:1 to 1:1) to yield 2.00 g of the title compound (5.68 mmol, 59%) as a yellow solid.
R_f (*n*-pentane/CH_2Cl_2, 2:1) = 0.33.
^{1}H NMR (500 MHz, $CDCl_3$, ppm) δ = 8.24 (s, 1H, H_{ar}), 8.13 (dd, *J* = 8.1, 2.1 Hz, 1H, H_{ar}), 7.68 (d, *J* = 8.2 Hz, 1H, H_{ar}), 6.80 – 6.72 (m, 4H, H_{ar}), 6.63 (t, *J* = 7.6 Hz, 2H, H_{ar}), 5.66 (d, *J* = 8.0 Hz, 2H, H_{ar}).

^{13}C NMR (126 MHz, $CDCl_3$, ppm) δ = 143.8 (C_q), 142.9 (C_q), 138.7 (+, CH), 135.6 (+, CH), 133.4 (C_q), 132.6 (q, J = 5.1 Hz, +, CH), 123.5 (+, CH), 122.7 (+, CH), 116.8 (C_q), 116.1 (+, CH), 113.9 (+, CH).

^{19}F NMR (470 MHz, $CDCl_3$, ppm) δ = -62.2 (s, CF_3).

IR (ATR, cm^{-1}) ṽ = 2234 (vw), 1602 (w), 1592 (w), 1506 (w), 1483 (vs), 1462 (s), 1418 (w), 1329 (vs), 1312 (vs), 1290 (m), 1268 (vs), 1211 (w), 1196 (m), 1180 (m), 1166 (s), 1133 (vs), 1119 (vs), 1096 (m), 1057 (vs), 1041 (m), 979 (w), 966 (w), 928 (m), 905 (s), 866 (s), 837 (w), 785 (w), 766 (m), 747 (vs), 738 (vs), 681 (w), 673 (m), 639 (w), 615 (m), 599 (w), 585 (s), 557 (w), 548 (w), 541 (w), 521 (w), 479 (w), 473 (w), 446 (m), 416 (w), 399 (w), 388 (w).

MS (ESI): m/z (%) = 352 (100) $[M]^{+}$. HRMS (ESI, $C_{20}H_{11}N_2OF_3$): calcd 352.0823, found 352.0913.

EA ($C_{20}H_{11}N_2OF_3$) calc. C: 68.18, H: 3.15, N: 7.95; found C: 67.89, H: 3.06, N: 7.93.

Mp 214 °C.

4-(9,9-Dimethylacridin-10(9*H*)-yl)-3-(trifluoromethyl)benzonitrile (25f)

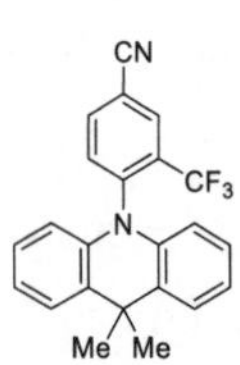

In a sealable vial, 4-fluoro-3-(trifluoromethyl)benzonitrile (662 mg, 3.50 mmol, 1.00 equiv.), 9,9-dimethyl-9,10-dihydroacridine (733 mg, 3.50 mmol, 1.00 equiv.) and tripotassium phosphate (1.86 g, 8.76 mmol, 2.50 equiv.) were evacuated and backfilled with argon three times. Dry dimethyl sulfoxide (14 mL) was added, and the resulting mixture was heated at 110 °C for 48 h. After cooling to room temperature, the reaction mixture was poured into an excess of water (100 mL) and extracted with dichloromethane (3 × 100 mL). The combined organic layers were washed with brine (100 mL), dried over sodium sulfate, and reduced in a vacuum. The crude product was purified by flash column chromatography over silica gel (*c*Hex/CH_2Cl_2, 2:1) to yield 472 mg of the title compound (1.25 mmol, 36%) as a yellow solid.

R_f (*c*Hex/CH_2Cl_2, 1:1) = 0.33.

^{1}H NMR (500 MHz, $CDCl_3$, ppm) δ = 8.30 (s, 1H, H_{ar}), 8.14 (d, *J* = 8.2 Hz, 1H, H_{ar}), 7.60 (d, *J* = 8.2 Hz, 1H, H_{ar}), 7.51 (d, *J* = 7.2 Hz, 2H, H_{ar}), 7.00 (p, *J* = 7.3 Hz, 4H, H_{ar}), 5.95 (d, *J* = 7.2 Hz, 2H, H_{ar}), 1.93 (s, 3H, CH_3), 1.42 (s, 3H, CH_3).

^{13}C NMR (126 MHz, $CDCl_3$, ppm) δ = 144.9 (C_q), 140.7 (C_q), 137.9 (C_q), 136.5 (+, CH), 134.0 (C_q), 133.7 (C_q), 133.5 (C_q), 133.3 (C_q), 132.8 (q, J = 5.3 Hz, C_q), 130.5 (C_q), 126.6 (+, CH), 125.4 (+, CH), 123.2 (C_q), 121.9 (+, CH), 121.0 (C_q), 117.1 (C_q), 113.8 (+, CH), 113.5 (C_q), 36.2 (C_q), 34.4 (+, CH_3, 1C), 27.2 (+, CH_3, 1C).

^{19}F NMR (376 MHz, $CDCl_3$, ppm) δ = -62.3 (s, CF_3).

IR (ATR, cm^{-1}) $\tilde{\nu}$ = 3070 (w), 2972 (w), 2925 (w), 2235 (w), 1594 (m), 1503 (w), 1494 (w), 1477 (vs), 1463 (m), 1448 (vs), 1415 (w), 1385 (w), 1329 (s), 1317 (vs), 1272 (vs), 1238 (w), 1196 (w), 1180 (m), 1173 (m), 1163 (vs), 1139 (vs), 1126 (vs), 1092 (m), 1055 (vs), 1006 (w), 938 (w), 926 (s), 904 (w), 853 (w), 846 (w), 761 (vs), 745 (vs), 718 (m), 674 (w), 669 (w), 639 (w), 625 (s), 605 (m), 582 (w), 557 (m), 545 (w), 534 (w), 482 (w), 456 (s), 429 (w), 385 (w).

MS (FAB, 3-NBA): m/z (%) = 379 (39) $[M+H]^+$, 378 (55) $[M]^+$. HRMS (FAB, $C_{23}H_{17}N_2F_3$): calcd 378.1338, found 378.1337.

Mp 235 °C.

5.2.1.3. Syntheses of the Tetrazole Precursors

10-(4-(2*H*-Tetrazol-5-yl)phenyl)-10*H*-phenoxazine (26a)

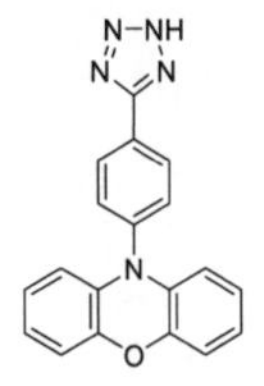

In a sealable vial, 4-(10*H*-phenoxazin-10-yl)benzonitrile (**25a**) (1.80 g. 6.33 mmol, 1.00 equiv.), sodium azide (1.24 g, 19.0 mmol, 3.00 equiv.) and ammonium chloride (1.02 g, 19.0 mmol, 3.00 equiv.) were evacuated and backfilled with argon three times. Dry dimethylformamide (36 mL) was added, and the resulting mixture was heated at 130 °C for 16 h until completion. After cooling to room temperature, the reaction mixture was poured into an excess of 1 M hydrochloric acid (100 mL) and mixed thoroughly. The yellow solid was filtered off, washed several times with water (100 mL), once with cold toluene (20 mL), and thoroughly dried to yield 2.05 g of the title compound (6.27 mmol, 99%) as a yellow solid.

^{1}H NMR (500 MHz, DMSO-d$_6$, ppm) δ = 8.38–8.27 (m, 2H, H$_{ar}$), 7.74–7.63 (m, 2H, H$_{ar}$), 6.77 (dd, *J* = 7.6, 1.8 Hz, 2H, H$_{ar}$), 6.74–6.62 (m, 4H, H$_{ar}$), 5.95 (dd, *J* = 7.6, 1.8 Hz, 2H, H$_{ar}$). Missing signal (1H) NH due to H/D exchange in DMSO-d$_6$.

^{13}C NMR (126 MHz, DMSO-d$_6$, ppm) δ = 155.2 (C$_q$), 143.2 (C$_q$), 140.9 (C$_q$), 133.5 (C$_q$), 131.7 (+, CH), 130.0 (+, CH), 124.6 (C$_q$), 123.8 (+, CH), 121.8 (+, CH), 115.5 (+, CH), 113.4 (+, CH).

IR (ATR, cm^{-1}) ṽ = 3064 (vw), 2734 (w), 2721 (w), 2696 (w), 2643 (w), 2635 (w), 2618 (w), 2606 (w), 2571 (w), 2561 (w), 2538 (w), 2534 (w), 2516 (w), 2497 (w), 2487 (w), 2466 (w), 1605 (m), 1594 (m), 1485 (vs), 1463 (s), 1438 (m), 1401 (w), 1327 (vs), 1305 (m), 1290 (m), 1271 (vs), 1204 (w), 1193 (w), 1157 (w), 1119 (w), 1095 (w), 1052 (w), 1044 (w), 1023 (w), 1011 (w), 992 (w), 963 (w), 926 (w), 895 (w), 870 (m), 844 (w), 799 (w), 751 (s), 741 (vs), 722 (m), 701 (w), 615 (w), 584 (w), 520 (w), 458 (w).

MS (FAB, 3-NBA): m/z (%) = 327 (52) [M]$^{+}$. HRMS (FAB, $C_{19}H_{13}N_5$): calcd 327.1120, found 327.1118.

EA ($C_{19}H_{13}N_5O$) calcd C: 69.71, H: 4.00, N: 21.39; found C: 69.49, H: 3.92, N: 21.19.

Mp 261 °C.

10-(4-(2*H*-Tetrazol-5-yl)phenyl)-9,9-dimethyl-9,10-dihydroacridine (26b)

In a sealable vial, 4-(9,9-dimethylacridin-10(9*H*)-yl)benzonitrile (**26b**) (1.55 g, 5.00 mmol, 1.00 equiv.), sodium azide (975 mg, 15.0 mmol, 3.00 equiv.) and ammonium chloride (802 mg, 15.0 mmol, 3.00 equiv.) were evacuated and backfilled with argon three times. Dry dimethylformamide (30 mL) was added and the resulting mixture was heated at 130 °C for 16 h until completion. After cooling to room temperature, the reaction mixture was poured into an excess of 1 M hydrochloric acid (100 mL) and mixed thoroughly. The pale-yellow solid was filtered off, washed several times with water (100 mL), once with cold toluene (20 mL), and thoroughly dried to yield 1.68 g of the title compound (4.75 mmol, 95%) as a pale-yellow solid.

^{1}H NMR (500 MHz, DMSO-d$_6$, ppm) δ = 8.35 (d, *J* = 8.5 Hz, 2H, H_{ar}), 7.64 (d, *J* = 8.4 Hz, 2H, H_{ar}), 7.52 (dd, *J* = 7.7, 1.6 Hz, 2H, H_{ar}), 7.00 (ddd, *J* = 8.3, 7.2, 1.6 Hz, 2H, H_{ar}), 6.93 (td, *J* = 7.4, 1.3 Hz, 2H, H_{ar}), 6.21 (dd, *J* = 8.2, 1.3 Hz, 2H, H_{ar}), 3.34 (s, 1H, NH), 1.64 (s, 6H, CH_3).

^{13}C NMR (126 MHz, DMSO-*d*$_6$, ppm) δ = 143.3 (C_q), 140.0 (C_q), 132.2 (+, CH), 129.9 (+, CH), 129.8 (C_q), 126.5 (+, CH), 125.5 (+, CH), 120.9 (+, CH), 113.7 (+, CH), 35.6 (C_q), 31.2 (+, CH_3, 2C).

IR (ATR, cm^{-1}) $\tilde{\nu}$ = 2965 (m), 2927 (w), 2918 (w), 2904 (w), 2856 (w), 1592 (s), 1502 (w), 1483 (s), 1438 (vs), 1432 (vs), 1332 (vs), 1319 (s), 1285 (m), 1266 (vs), 1214 (w), 1154 (m), 1111 (w), 1086 (w), 1054 (m), 1045 (s), 1023 (w), 990 (m), 925 (m), 866 (w), 850 (m), 741 (vs), 725 (vs), 701 (s), 660 (w), 622 (s), 606 (w), 533 (m), 520 (m), 469 (m), 462 (m).

MS (FAB, 3-NBA): m/z (%) = 354 (67) $[M+H]^+$, 353 (57) $[M]^+$. HRMS (FAB, $C_{22}H_{19}N_5$): calcd 353.1640, found 353.1635.
EA ($C_{22}H_{19}N_5$) calcd C: 74.77, H: 5.42, N: 19.82; found C: 74.45, H: 5.34, N: 19.37.
Mp 255 °C.

10-(4-(2*H*-Tetrazol-5-yl)phenyl)-9,9-diphenyl-9,10-dihydroacridine (26c)

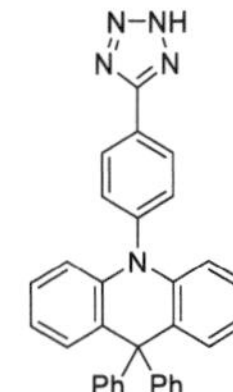

In a sealable vial, 4-(9,9-diphenylacridin-10(9*H*)-yl)benzonitrile (**25c**) (1.20 g, 2.76 mmol, 1.00 equiv.), sodium azide (539 mg, 8.29 mmol, 3.00 equiv.) and ammonium chloride (443 mg, 8.29 mmol, 3.00 equiv.) were evacuated and backfilled with argon three times. Dry dimethylformamide (17 mL) was added, and the resulting mixture was heated at 130 °C for 16 h until completion. After cooling to room temperature, the reaction mixture was poured into an excess of 1 M hydrochloric acid (75 mL) and mixed thoroughly. The white solid was filtered off, washed several times with water (100 mL), once with cold toluene (20 mL), and thoroughly dried to yield 1.25 g of the title compound (2.62 mmol, 95%) as an off-white solid.

^{1}H NMR (500 MHz, DMSO-d$_6$, ppm) δ = 8.28 (d, *J* = 8.5 Hz, 2H, H$_{ar}$), 7.38–7.24 (m, 8H, H$_{ar}$), 7.10 (ddd, *J* = 8.5, 7.3, 1.5 Hz, 2H, H$_{ar}$), 6.96–6.87 (m, 6H, H$_{ar}$), 6.79 (dd, *J* = 7.8, 1.4 Hz, 2H, H$_{ar}$), 6.40 (dd, *J* = 8.3, 0.9 Hz, 2H, H$_{ar}$), Missing signal (1H) NH due to H/D exchange in DMSO-d$_6$.

^{13}C NMR (126 MHz, DMSO-d$_6$, ppm) δ = 146.0 (C$_q$), 142.6 (C$_q$), 141.2 (C$_q$), 131.8 (+, CH), 129.7 (+, CH), 129.7 (+, CH), 129.6 (+, CH), 129.0 (C$_q$), 127.9 (+, CH), 127.3 (+, CH), 126.6 (+, CH), 120.6 (+, CH), 114.0 (+, CH), 56.1 (C$_q$).

IR (ATR, cm^{-1}) ṽ = 3067 (vw), 3026 (vw), 2723 (vw), 2609 (vw), 1592 (m), 1490 (m), 1472 (vs), 1449 (s), 1392 (vw), 1320 (s), 1266 (s), 1156 (w), 1081 (vw), 1051 (w), 1034 (w), 992 (w), 924 (w), 905 (w), 898 (w), 844 (w), 751 (vs), 735 (s), 698 (vs), 666 (w), 639 (w), 630 (w), 622 (w), 554 (w), 541 (m), 456 (w), 424 (w).

MS (FAB, 3-NBA): m/z (%) = 478 (4) [M+H]$^+$. HRMS (FAB, $C_{32}H_{24}N_5$): calcd 478.2032 [M+H]$^+$, found 478.2033 [M+H]$^+$.

EA ($C_{32}H_{23}N_5$) calcd C: 80.48, H: 4.85, N: 14.66; found C: 79.29, H: 4.77, N: 13.95.

Mp 277 °C.

10-(2,6-Dimethyl-4-(2*H*-tetrazol-5-yl)phenyl)-10*H*-phenoxazine (26d)

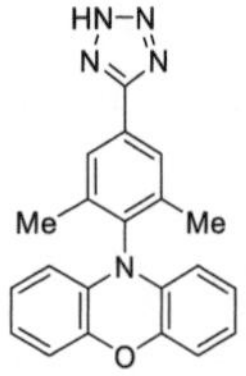

In a sealable vial, 3,5-dimethyl-4-(10*H*-phenoxazin-10-yl)benzonitrile (**25d**) (2.00 g, 6.40 mmol, 1.00 equiv.), sodium azide (1.25 g, 19.2 mmol, 3.00 equiv.) and ammonium chloride (1.03 g, 19.2 mmol, 3.00 equiv.) were evacuated and backfilled with argon three times. Dry dimethylformamide (38 mL) was added, and the resulting mixture was heated at 130 °C for 16 h until completion. After cooling to room temperature, the reaction mixture was poured into an excess of 1 M hydrochloric acid (100 mL) and mixed thoroughly. The yellow solid was filtered off, washed several times with water (100 mL), once with cold toluene (20 mL), and thoroughly dried to yield 1.87 g of the title compound (5.26 mmol, 82%) as a pale-yellow solid.

R_f (CH_2Cl_2/CH_3OH, 9:1) = 0.38.

^1H NMR (500 MHz, DMSO-d_6, ppm) δ = 8.03 (s, 2H, H_{ar}), 6.80–6.73 (m, 2H, H_{ar}), 6.73–6.63 (m, 4H, H_{ar}), 5.75–5.68 (m, 2H, H_{ar}), 2.24 (s, 6H, CH_3). Missing signal (1H) NH due to H/D exchange in DMSO-d_6.

^{13}C NMR (126 MHz, DMSO-d_6, ppm) δ = 142.9 (C_q), 139.7 (C_q), 136.7 (C_q), 131.2 (C_q), 128.9 (C_q), 128.2 (+, CH), 124.2 (+, CH), 121.7 (+, CH), 115.6 (+, CH), 111.5 (+, CH), 17.4 (+, CH_3, 2C).

IR (ATR, cm^{-1}) $\tilde{\nu}$ = 2912 (w), 2857 (w), 2847 (w), 2839 (w), 2764 (w), 2737 (w), 2707 (w), 2694 (w), 2635 (w), 2616 (w), 2567 (w), 2557 (w), 1645 (w), 1629 (w), 1612 (w), 1608 (w), 1591 (w), 1562 (w), 1507 (w), 1485 (vs), 1462 (s), 1438 (w), 1421 (w), 1414 (w), 1390 (w), 1381 (w), 1337 (vs), 1316 (m), 1290 (m), 1272 (vs), 1259 (s), 1205 (m), 1145 (w), 1112 (w), 1094 (w), 1086 (w), 1072 (w), 1043 (m), 1035 (m), 990 (w), 959 (w), 926 (m), 888 (m), 870 (w), 863 (w), 847 (m), 738 (vs), 714 (m), 694 (w), 683 (w), 673 (w), 633 (w), 620 (m), 602 (w), 591 (w), 579 (w), 571 (w), 550 (w), 531 (w), 520 (w), 509 (w), 455 (w), 441 (w), 416 (w), 398 (w), 387 (w), 377 (w).

MS (FAB, 3-NBA): m/z (%) = 356 (56) $[M+H]^+$, 355 (100) $[M]^+$. HRMS (FAB, $C_{21}H_{17}N_5O$): calcd 355.1433, found 355.1531.

Mp 222 °C.

10-(4-(2*H*-Tetrazol-5-yl)-2-(trifluoromethyl)phenyl)-10*H*-phenoxazine (26e)

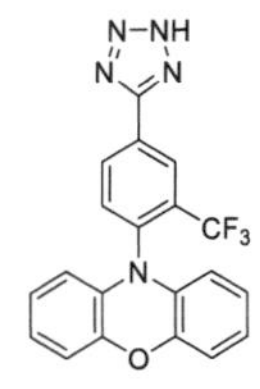

In a sealable vial, 4-(10*H*-phenoxazin-10-yl)-3-(trifluoromethyl)benzonitrile (**25e**) (50.0 mg, 142 µmol, 1.00 equiv.), sodium azide (27.7 mg, 426 µmol, 3.00 equiv.) and ammonium chloride (23.0 mg, 426 µmol, 3.00 equiv.) were evacuated and backfilled with argon three times. Dry dimethylformamide (3 mL) was added, and the resulting mixture was heated at 130 °C for 16 h until completion. After cooling to room temperature, the reaction mixture was poured into an excess of 1 M hydrochloric acid and mixed thoroughly. The yellow solid was filtered off, washed several times with water, and thoroughly dried to yield 41.3 mg of the title compound (105 µmol, 74%) as a yellow solid.

^{1}H NMR (500 MHz, DMSO-d$_6$, ppm) δ = 8.66 – 8.62 (m, 2H, H_{ar}), 7.95 (d, *J* = 8.2 Hz, 1H, H_{ar}), 6.81 (d, *J* = 7.8 Hz, 2H, H_{ar}), 6.75 (t, *J* = 7.5 Hz, 2H, H_{ar}), 6.69 (t, *J* = 7.6 Hz, 2H, H_{ar}), 5.81 (d, *J* = 7.9 Hz, 2H, H_{ar}). Missing signal (1H) NH due to H/D exchange in DMSO-d$_6$.

^{13}C NMR (126 MHz, DMSO-d$_6$, ppm) δ = 143.0 (C_q), 139.1 (C_q), 135.4 (+, CH), 134.6 (+, CH), 133.5 (C_q), 127.1 (q, J = 5.1 Hz, C_q), 126.3 (+, CH), 123.9 (C_q), 122.4 (+, CH), 115.6 (+, CH), 114.0 (+, CH).

^{19}F NMR (470 MHz, DMSO-d$_6$, ppm) δ = -60.9 (s, CF_3).

IR (ATR, cm^{-1}) $\tilde{\nu}$ = 1647 (w), 1619 (w), 1606 (w), 1594 (w), 1483 (vs), 1463 (m), 1435 (w), 1385 (w), 1366 (w), 1332 (s), 1316 (m), 1306 (s), 1290 (m), 1272 (vs), 1211 (m), 1196 (w), 1171 (m), 1153 (m), 1130 (vs), 1098 (m), 1055 (vs), 1044 (s), 1028 (s), 1000 (m), 929 (w), 916 (w), 870 (m), 841 (w), 824 (m), 805 (w),

735 (vs), 718 (s), 704 (m), 679 (m), 662 (m), 616 (m), 598 (w), 586 (m), 555 (w), 545 (w), 533 (w), 527 (w), 500 (m), 482 (m), 469 (m), 441 (m), 428 (m), 412 (w), 402 (w), 392 (m), 385 (m).

MS (FAB, 3-NBA): m/z (%) = 395 $[M]^+$. HRMS (FAB, $C_{20}H_{12}N_5OF_3$): calcd 395.0988, found 395.0990.

EA ($C_{20}H_{12}N_5OF_3$) calc. C: 60.76, H: 3.06, N: 17.71; found C: 60.27, H: 3.06, N: 17.44.

Mp 199 °C.

10-(4-(2*H*-Tetrazol-5-yl)-2-(trifluoromethyl)phenyl)-9,9-dimethyl-9,10-dihydroacridine (26f)

In a sealable vial, 4-(9,9-dimethylacridin-10(9*H*)-yl)-3-(trifluoromethyl)-benzonitrile (**25f**) (350 mg, 925 µmol, 1.00 equiv.), sodium azide (180 mg, 2.78 mmol, 3.00 equiv.) and ammonium chloride (148 mg, 2.78 mmol, 3.00 equiv.) were evacuated and backfilled with argon three times. Dry dimethylformamide (5 mL) was added and the resulting mixture was heated at 130 °C for 16 h until completion. After cooling to room temperature, the reaction mixture was poured into an excess of 1 M hydrochloric acid and mixed thoroughly. The yellow solid was filtered off, washed several times with water, and thoroughly dried to yield 390 mg of the title compound (925 µmol, quant.) as a yellow solid.

^{1}H NMR (500 MHz, DMSO-d_6, ppm) δ = 8.70 (s, 1H, H_{ar}), 8.65 (dd, *J* = 8.2, 1.5 Hz, 1H, H_{ar}), 7.81 (d, *J* = 8.2 Hz, 1H, H_{ar}), 7.54 (d, *J* = 7.5 Hz, 2H, H_{ar}), 6.98 (dt, *J* = 15.4, 7.2 Hz, 4H, H_{ar}), 6.04 (d, *J* = 7.9 Hz, 2H, H_{ar}), 1.87 (s, 3H, CH_3), 1.36 (s, 3H, CH_3). Missing signal (1H) NH due to H/D exchange in DMSO-d_6.

^{13}C NMR (126 MHz, DMSO-d_6, ppm) δ = 141.2 (C_q), 140.3 (C_q), 136.1 (+, CH), 133.8 (+, CH), 130.6 (q, J = 30.7 Hz, C_q), 129.7 (C_q), 127.0 (q, J = 4.8 Hz, C_q), 126.6 (+, CH),

126.1 (+, CH), 125.9 (+, CH), 125.4 (+, CH), 123.9 (+, CH), 121.7 (+, CH), 121.3 (+, CH), 113.7 (+, CH), 35.6 (C_q), 34.3 (+, CH_3), 27.5 (+, CH_3).

^{19}F NMR (376 MHz, DMSO-d_6, ppm) δ = -60.9 (s, CF_3).

IR (ATR, cm^{-1}) ṽ = 1642 (w), 1626 (w), 1592 (w), 1502 (w), 1492 (w), 1475 (s), 1465 (w), 1446 (s), 1383 (w), 1367 (w), 1315 (vs), 1286 (w), 1264 (s), 1163 (m), 1143 (vs), 1130 (vs), 1113 (m), 1085 (m), 1077 (w), 1054 (s), 1026 (m), 1018 (m), 972 (w), 956 (w), 925 (w), 904 (w), 895 (w), 846 (w), 813 (w), 742 (vs), 710 (w), 700 (m), 667 (w), 657 (w), 623 (m), 603 (w), 582 (w), 548 (w), 534 (w), 499 (m), 483 (w), 453 (m), 433 (w), 416 (w), 411 (w), 391 (w), 378 (w).

MS (FAB, 3-NBA): m/z (%) = 422 (90) $[M+H]^+$, 421 (89) $[M]^+$. HRMS (FAB, $C_{23}H_{18}N_5F_3$): calcd 421.1509, found 421.1509.

Mp 170 °C.

5.2.1.4. Syntheses of the Di[1,2,4]-triazolo[1,3,5]-triazine Compounds

10,10'-((5-Phenylbis([1,2,4]triazolo)[4,3-a:4',3'-c][1,3,5]triazine-3,9-diyl)bis(4,1-phenylene))bis(10*H*-phenoxazine) (28a)

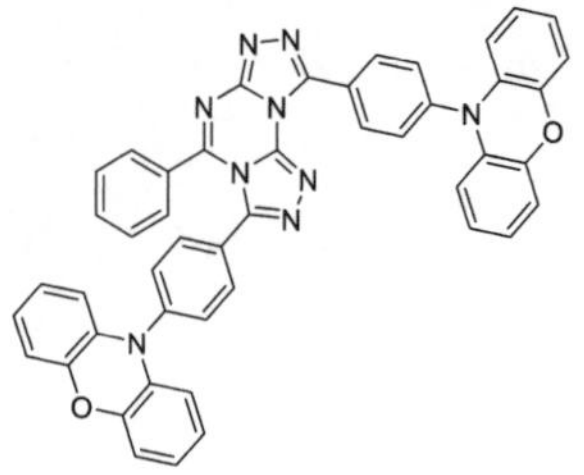

In a sealable vial, 10-(4-(2*H*-tetrazol-5-yl)phenyl)-10*H*-phenoxazine (**26a**) (524 mg, 1.60 mmol, 1.00 equiv.) and 2,4-dichloro-6-phenyl-1,3,5-triazine (163 mg, 720 µmol, 0.45 equiv.) were evacuated and backfilled with argon three times. Dry toluene (32 mL) and 2,6-lutidine (429 mg, 4.00 mmol, 2.50 equiv.) were added, and the resulting mixture was heated at 80 °C for 24 h until completion. After cooling to room temperature, the reaction mixture was poured into an excess of water (50 mL) and extracted with dichloromethane (3 × 50 mL). The combined organic layers were washed with brine (50 mL), dried over sodium sulfate, reduced in a vacuum, and purified by flash column chromatography over silica gel (CH_2Cl_2/EtOAc, 99:1 to 91:9) to yield 339 mg of the title compound (452 µmol, 63%) as a yellow solid.

R_f (CH_2Cl_2/EtOAc, 9:1) = 0.36.

^{1}H NMR (500 MHz, DMSO-d_6, ppm) δ = 8.59 (d, *J* = 8.5 Hz, 2H, H_{ar}), 7.79 (d, *J* = 8.5 Hz, 2H, H_{ar}), 7.58 (d, *J* = 7.0 Hz, 2H, H_{ar}), 7.53 (d, *J* = 8.4 Hz, 2H, H_{ar}), 7.50 (d, *J* = 7.5 Hz, 1H, H_{ar}), 7.36 (t, *J* = 7.8 Hz, 2H, H_{ar}), 7.25 (d, *J* = 8.4 Hz, 2H, H_{ar}), 6.82 – 6.71 (m, 12H, H_{ar}), 6.08 – 6.03 (m, 2H, H_{ar}), 5.74 (dd, *J* = 7.9, 1.5 Hz, 2H, H_{ar}).

^{13}C NMR (126 MHz, THF-d_8, ppm) δ = 161.8 (C_q), 153.1 (C_q), 148.8 (C_q), 147.7 (C_q), 145.2 (C_q), 145.2 (C_q), 143.1 (C_q), 142.5 (C_q), 135.2 (C_q), 135.1 (C_q), 134.0 (C_q), 133.4 (C_q), 132.6 (+, CH), 132.0 (+, CH), 131.9 (+, CH), 131.2 (+, CH), 130.3 (+, CH), 129.6 (C_q), 127.7 (C_q), 124.4 (+, CH), 124.3 (+, CH), 122.6 (+, CH), 116.5 (+, CH), 114.5 (+, CH), 114.3 (+, CH).

IR (ATR, cm^{-1}) $\tilde{\nu}$ = 3061 (w), 3037 (w), 1625 (m), 1602 (w), 1568 (w), 1483 (vs), 1462 (s), 1449 (m), 1417 (w), 1388 (m), 1323 (vs), 1292 (s), 1271 (vs), 1207 (m),

1191 (w), 1169 (w), 1154 (w), 1132 (w), 1116 (w), 1098 (w), 1041 (w), 1010 (w), 986 (w), 960 (w), 936 (w), 924 (w), 868 (m), 846 (w), 832 (w), 796 (w), 772 (w), 738 (vs), 714 (m), 703 (m), 696 (m), 676 (w), 650 (w), 616 (m), 592 (m), 558 (w), 535 (w), 524 (w), 501 (w), 486 (w), 467 (w), 438 (w), 422 (w), 414 (w), 404 (w), 384 (w).

MS (ESI): m/z (%) = 752 (54) $[M+H]^+$, 751 (100) $[M]^+$. HRMS (ESI, $C_{47}H_{29}N_9O_2$): calcd 751.2444, found 751.2420.

Mp 290 °C.

10,10'-((5-Phenylbis([1,2,4]triazolo)[4,3-a:4',3'-c][1,3,5]triazine-3,9-diyl)bis(4,1-phenylene))bis(9,9-dimethyl-9,10-dihydroacridine) (28b)

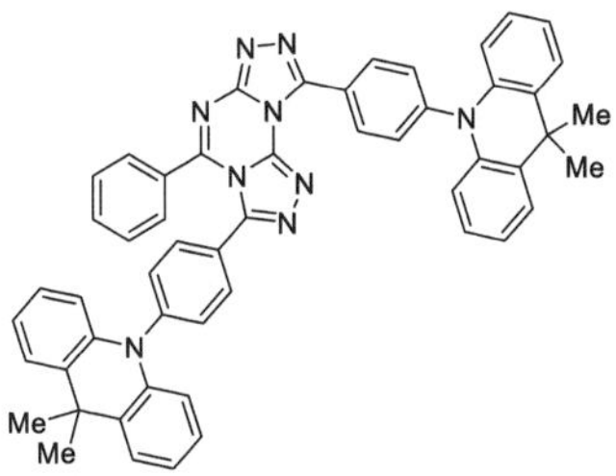

In a sealable vial, 10-(4-(2*H*-tetrazol-5-yl)phenyl)-9,9-diphenyl-9,10-dihydroacridine (**26b**) (250 mg, 707 µmol, 1.00 equiv.) and 2,4-dichloro-6-phenyl-1,3,5-triazine (72.0 mg, 318 µmol, 0.45 equiv.) were evacuated and backfilled with argon three times. Dry toluene (13 mL) and 2,6-lutidine (227 mg, 2.12 mmol, 3.00 equiv.) were added, and the resulting mixture heated at 80 °C for 24 h until completion. After cooling to room temperature, the reaction mixture was poured into an excess of water (30 mL) and extracted with dichloromethane (3 × 40 mL). The combined organic layers were washed with brine (50 mL), dried over sodium sulfate, reduced in a vacuum, and purified by flash column chromatography (toluene/EtOAc, 1:0 to 96:4) to yield 151 mg of the title compound (188 µmol, 59%) as a yellow solid.

R_f (toluene/EtOAc, 11:2) = 0.37.

^{1}H NMR (500 MHz, $CDCl_3$, ppm) δ = 8.84 (d, *J* = 7.3 Hz, 2H, H_{ar}), 8.67 (d, *J* = 8.1 Hz, 2H, H_{ar}), 8.50 (d, *J* = 8.4 Hz, 2H, H_{ar}), 7.77 (t, *J* = 7.4 Hz, 1H, H_{ar}), 7.70 (t, *J* = 7.5 Hz,

2H, H_{ar}), 7.60 (d, J = 8.0 Hz, 2H, H_{ar}), 7.51 – 7.46 (m, 6H, H_{ar}), 7.01 (td, J = 8.3, 7.8, 1.6 Hz, 2H, H_{ar}), 6.95 (ddd, J = 8.7, 5.5, 1.5 Hz, 6H, H_{ar}), 6.43 (dd, J = 8.1, 1.3 Hz, 2H, H_{ar}), 6.32 – 6.26 (m, 2H, H_{ar}), 1.69 (s, 12H, CH_3).

^{13}C NMR (126 MHz, $CDCl_3$, ppm) δ = 161.9 (C_q), 151.5 (C_q), 148.4 (C_q), 147.5 (C_q), 146.6 (C_q), 144.7 (C_q), 144.3 (C_q), 140.7 (C_q), 140.6 (C_q), 134.0 (C_q), 132.6 (+, CH), 132.1 (+, CH), 131.6 (+, CH), 131.5 (+, CH), 130.5 (+, CH), 130.3 (+, CH), 129.0 (+, CH), 128.2 (C_q), 126.6 (+, CH), 126.5 (+, CH), 125.6 (+, CH), 125.5 (+, CH), 125.2 (C_q), 121.2 (+, CH), 121.1 (+, CH), 114.3 (+, CH), 114.2 (+, CH), 36.2 (C_q), 31.4 (+, CH_3, 2C), 31.3 (+, CH_3, 2C), 29.9 (C_q).

IR (ATR, cm^{-1}) $\tilde{\nu}$ = 3060 (w), 3033 (w), 2956 (w), 2919 (w), 2868 (w), 2853 (w), 1633 (m), 1601 (w), 1591 (m), 1551 (w), 1524 (w), 1500 (w), 1472 (vs), 1448 (vs), 1424 (m), 1402 (w), 1384 (w), 1357 (w), 1327 (vs), 1292 (w), 1272 (s), 1211 (w), 1187 (w), 1169 (w), 1137 (w), 1112 (w), 1101 (w), 1082 (w), 1047 (w), 1030 (w), 1020 (w), 977 (vw), 936 (w), 925 (w), 844 (w), 744 (vs), 728 (vs), 694 (s), 686 (m), 667 (w), 650 (w), 637 (w), 623 (m), 602 (w), 592 (w), 579 (w), 560 (w), 537 (w), 526 (w), 511 (w), 465 (w), 455 (w), 445 (w), 429 (w).

MS (FAB, 3-NBA): m/z (%) = 805 (43) $[M+H]^+$, 804 (74) $[M^{13}C]^+$, 803 (20) $[M]^+$.

HRMS (FAB, $C_{53}H_{41}N_9$): calcd 804.3558, found 804.3556.

Mp 300 °C.

10,10'-((5-Phenylbis([1,2,4]triazolo)[4,3-a:4',3'-c][1,3,5]triazine-3,9-diyl)bis(4,1-phenylene))bis(9,9-diphenyl-9,10-dihydroacridine) (28c)

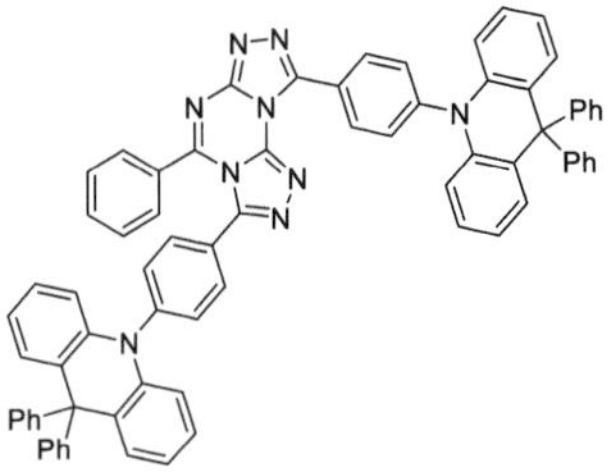

In a sealable vial, 10-(4-(2*H*-tetrazol-5-yl)phenyl) -9,9-diphenyl- 9,10-dihydroacridine (**26c**) (300 mg, 628 µmol, 1.00 equiv.) and 2,4-dichloro-6-phenyl-1,3,5-triazine (63.9 mg, 283 µmol, 0.45 equiv.) were evacuated and backfilled with argon three times. Dry toluene (13 mL) and 2,6-lutidine (202 mg, 1.89 mmol, 3.00 equiv.) were added, and the resulting mixture was heated at 80 °C for 24 h until completion. After cooling to room temperature, the reaction mixture was poured into an excess of water (30 mL) and extracted with dichloromethane (3 × 40 mL). The combined organic layers were washed with brine (50 mL), dried over sodium sulfate, reduced in a vacuum, and purified by flash column chromatography over silica gel (CH_2Cl_2/EtOAc, 1:0 to 24:1 and afterwards toluene/EtOAc, 95:5 to 93:7) to yield 99.3 mg of the title compound (94.4 µmol, 33%) as a yellow solid. (Yield calculated to 2,4-dichloro-6-phenyl-1,3,5-triazine.)

R_f (toluene/EtOAc, 47:3) = 0.30.

^{1}H NMR (500 MHz, $CDCl_3$, ppm) δ = 8.62 (d, J = 8.4 Hz, 2H, H_{ar}), 7.51 (dd, J = 23.4, 7.4 Hz, 3H, H_{ar}), 7.42 – 7.21 (m, 20H, H_{ar}), 7.15 – 7.07 (m, 4H, H_{ar}), 7.02 (dd, J = 8.0, 1.5 Hz, 4H, H_{ar}), 7.00 – 6.85 (m, 12H, H_{ar}), 6.59 (d, J = 8.1 Hz, 2H, H_{ar}), 6.25 (d, J = 8.1 Hz, 2H, H_{ar}).

^{13}C NMR (126 MHz, $CDCl_3$, ppm) δ = 148.7 (C_q), 148.6 (C_q), 148.5 (C_q), 146.2 (C_q), 146.1 (C_q), 141.7 (C_q), 141.4 (C_q), 131.9 (+, CH), 131.5 (+, CH), 131.3 (+, CH), 130.3 (+, CH), 130.2 (+, CH), 130.1 (C_q), 130.0 (+, CH), 129.8 (C_q), 129.4 (C_q), 128.9 (C_q), 128.4 (C_q), 128.1 (C_q), 127.6 (+, CH), 126.9 (C_q), 126.6 (C_q), 126.3 (+, CH), 126.2 (+, CH), 125.2 (C_q), 124.8 (C_q), 120.6 (+, CH), 120.4 (C_q), 114.2 (+, CH), 114.0 (+, CH), 56.7 (C_q), 56.6 (C_q).

IR (ATR, cm^{-1}) $\tilde{v}$ = 3054 (w), 3030 (w), 1626 (m), 1592 (m), 1551 (w), 1521 (w), 1492 (w), 1469 (vs), 1449 (vs), 1418 (m), 1381 (w), 1324 (vs), 1261 (s), 1207 (w), 1186 (w), 1159 (w), 1136 (w), 1120 (w), 1098 (w), 1079 (w), 1055 (w), 1033 (w), 1017 (w), 1001 (w), 984 (w), 958 (w), 935 (w), 922 (m), 904 (w), 839 (w), 748 (vs), 734 (vs), 696 (vs), 667 (s), 650 (m), 637 (s), 629 (s), 623 (s), 568 (m), 538 (m), 523 (m), 500 (m), 489 (m), 467 (m), 458 (m), 450 (m), 442 (m), 432 (m), 424 (m), 407 (m), 397 (m), 388 (m), 375 (m).
MS (ESI): m/z (%) = 1051 (100) $[M]^{+}$. HRMS (ESI, $C_{73}H_{49}N_9$) calcd 1051.4111, found 1051.4097.
Mp 218 °C.

10,10'-((5-Phenylbis([1,2,4]triazolo)[4,3-a:4',3'-c][1,3,5]triazine-3,9-diyl)bis(2,6-dimethyl-4,1-phenylene))bis(10*H*-phenoxazine) (28d)

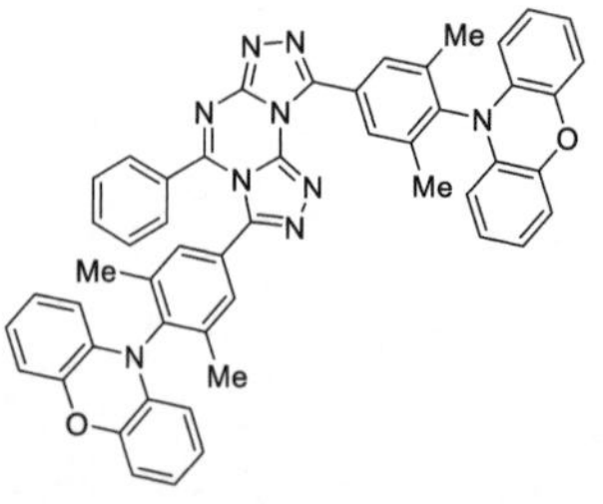

In a sealable vial, 10-(4-(2*H*-tetrazol-5-yl)phenyl)-10*H*-phenoxazine (**26d**) (355 mg, 1.00 mmol, 1.00 equiv.) and 2,4-dichloro-6-phenyl-1,3,5-triazine (102 mg, 449 µmol, 0.45 equiv.) were evacuated and backfilled with argon three times. Dry toluene (20 mL) and 2,6-lutidine (214 mg, 2.00 mmol, 2.00 equiv.) were added, and the resulting mixture was heated at 80 °C for 16 h until completion. After cooling to room temperature, the reaction mixture was poured into an excess of water (30 mL) and extracted with dichloromethane (3 × 40 mL). The combined organic layers were washed with brine (50 mL), dried over sodium sulfate, reduced in a vacuum, and purified by flash column chromatography (CH_2Cl_2/EtOAc, 99:1 to 96:4 and second column in toluene/EtOAc, 99:1 to 9:1) to yield 126 mg of the title compound (156 µmol, 35%) as an orange solid.

R_f (toluene/EtOAc, 5:1) = 0.38.

^{1}H NMR (500 MHz, $CDCl_3$, ppm) δ = 8.81 (d, J = 7.8 Hz, 2H, H_{ar}), 8.45 (s, 2H, H_{ar}), 8.14 (s, 2H, H_{ar}), 7.76 (t, J = 7.3 Hz, 1H, H_{ar}), 7.70 (t, J = 7.5 Hz, 2H, H_{ar}), 6.70 (td, J = 7.3, 1.7 Hz, 4H, H_{ar}), 6.68 – 6.54 (m, 8H, H_{ar}), 5.87 (d, J = 7.0 Hz, 2H, H_{ar}), 5.73 (d, J = 7.7 Hz, 2H, H_{ar}), 2.43 (s, 6H, CH_3), 2.30 (s, 6H, CH_3).

^{13}C NMR (126 MHz, $CDCl_3$, ppm) δ = 161.9 (C_q), 148.3 (C_q), 146.6 (C_q), 144.0 (C_q), 140.5 (C_q), 139.8 (C_q), 133.9 (+, CH), 131.8 (C_q), 131.4 (+, CH), 131.1 (C_q), 129.1 (+, CH), 129.0 (C_q), 128.8 (C_q), 128.6 (+, CH), 125.6 (C_q), 123.9 (+, CH), 123.8 (+, CH), 121.7 (+, CH), 115.8 (+, CH), 111.9 (+, CH), 111.7 (+, CH), 18.3 (+, CH_3, 2C), 18.2 (+, CH_3, 2C).

IR (ATR, cm^{-1}) $\tilde{\nu}$ = 3058 (w), 2952 (w), 2921 (w), 2851 (w), 1626 (m), 1604 (w), 1591 (w), 1578 (w), 1553 (m), 1485 (vs), 1460 (s), 1449 (s), 1419 (m), 1394 (m), 1388 (m), 1378 (w), 1337 (vs), 1327 (s), 1290 (m), 1271 (vs), 1254 (s), 1205 (s), 1190 (m), 1153 (w), 1123 (w), 1115 (w), 1095 (w), 1082 (w), 1065 (w), 1034 (w), 1014 (w), 983 (w), 953 (w), 938 (w), 924 (w), 891 (w), 858 (w), 793 (w), 778 (w), 737 (vs), 694 (m), 684 (s), 663 (w), 633 (w), 620 (m), 596 (w), 555 (w), 467 (w), 456 (w), 442 (w), 419 (w), 394 (w), 378 (w).

MS (FAB, 3-NBA): m/z (%) = 809 (49) $[M+H]^+$, 808 (100) $[M]^+$. HRMS (FAB, $C_{51}H_{37}N_9O_2$): calcd 808.3143, found 808.3146.

Mp 240 °C.

10,10'-((5-Phenylbis([1,2,4]triazolo)[4,3-a:4',3'-c][1,3,5]triazine-3,9-diyl)bis(2-(trifluoromethyl)-4,1-phenylene))bis(10*H*-phenoxazine) (28e)

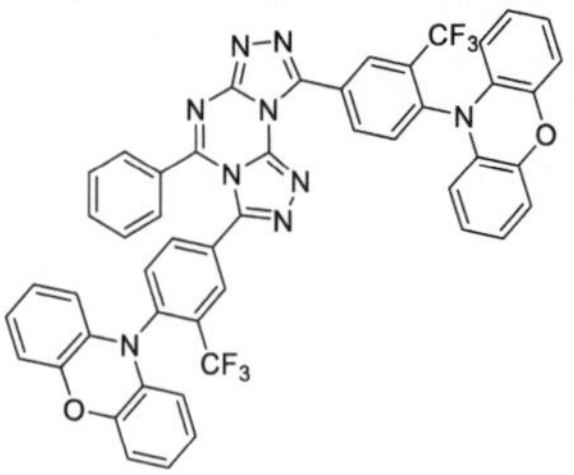

In a sealable vial, 10-(4-(2*H*-tetrazol-5-yl)-2-(trifluoromethyl)phenyl)-10*H*-phenoxazine (**26e**) (198 mg, 501 µmol, 1.00 equiv.) and 2,4-dichloro-6-phenyl-1,3,5-triazine (50.9 mg, 225 µmol, 0.45 equiv.) were evacuated and backfilled with argon three times. Dry toluene (10 mL) and 2,6-lutidine (107 mg, 1.00 mmol, 2.00 equiv.) were added, and the resulting mixture was heated at 80 °C for 16 h until completion. After cooling to room temperature, the reaction mixture was poured into an excess of water (50 mL) and extracted with dichloromethane (3 × 50 mL). The combined organic layers were washed with brine (50 mL), dried over sodium sulfate, reduced in a vacuum, and purified by flash column chromatography over silica gel (CH_2Cl_2/EtOAc, 99:1 to 97:3) to yield 188 mg of the title compound (212 µmol, 94%) as an orange solid.

R_f (CH_2Cl_2/EtOAc, 97:3) = 0.38.

^{1}H NMR (500 MHz, DMSO-d$_6$, ppm) δ = 9.29 (d, *J* = 2.1 Hz, 1H, H_{ar}), 8.86 (dd, *J* = 8.3, 2.1 Hz, 1H, H_{ar}), 8.78 (dd, *J* = 8.3, 2.0 Hz, 1H, H_{ar}), 8.67 – 8.62 (m, 3H, H_{ar}), 8.05 (d, *J* = 8.2 Hz, 1H, H_{ar}), 7.96 (d, *J* = 8.3 Hz, 1H, H_{ar}), 7.87 – 7.82 (m, 1H, H_{ar}), 7.79 (t, *J* = 7.5 Hz, 2H, H_{ar}), 6.87 – 6.65 (m, 12H, H_{ar}), 5.91 (dd, *J* = 7.8, 1.7 Hz, 2H, H_{ar}), 5.83 (dd, *J* = 8.0, 1.5 Hz, 2H, H_{ar}).

^{13}C NMR (126 MHz, CD_2Cl_2, ppm) δ = 161.0 (C_q), 152.2 (C_q), 149.2 (C_q), 147.2 (C_q), 146.7 (C_q), 144.3 (d, J = 9.1 Hz, C_q), 141.2 (C_q), 140.7 (C_q), 136.7 (+, CH), 135.3 (+, CH), 135.0 (+, CH), 134.6 (+, CH), 134.5 (+, CH), 134.4 (+, CH), 134.3 (+, CH), 133.9 (+, CH), 133.6 (+, CH), 133.2 (+, CH), 132.9 (+, CH), 131.8 (+, CH), 130.4 (q, J = 5.0 Hz, C_q), 130.1 (+, CH), 129.5 (+, CH), 129.2 (+, CH), 128.2 (q, J = 5.4, 4.8 Hz, C_q), 127.3 (+, CH), 124.6 (+, CH), 124.3 (+, CH), 123.8 (d, J = 9.0 Hz, +, CH), 122.7 (d, J = 6.4 Hz, +, CH), 122.5 (+, CH), 116.2 (+, CH), 116.1 (+, CH), 114.4 (d, J = 10.6 Hz, +, CH).

^{19}F NMR (471 MHz, CD_2Cl_2, ppm) δ = -62.1 (s, CF_3), -62.3 (s, CF_3).

IR (ATR, cm^{-1}) ṽ = 1632 (w), 1616 (w), 1592 (w), 1579 (w), 1554 (w), 1514 (w), 1483 (vs), 1463 (s), 1451 (m), 1408 (m), 1326 (s), 1315 (s), 1292 (vs), 1271 (vs), 1210 (m), 1190 (w), 1163 (s), 1129 (vs), 1118 (vs), 1054 (vs), 1044 (s), 1003 (w), 987 (w), 976 (w), 938 (m), 916 (m), 868 (m), 830 (m), 798 (w), 779 (w), 738 (vs), 713 (s), 697 (m), 684 (s), 656 (m), 635 (w), 616 (m), 599 (w), 586 (m), 541 (w), 531 (w), 524 (w), 520 (w), 503 (w), 493 (w), 480 (w), 469 (w), 458 (w), 439 (m), 422 (w), 411 (w), 382 (w).

MS (ESI): m/z (%) = 888 (56) $[M+H]^+$, 887 (100) $[M]^+$. HRMS (ESI, $C_{49}H_{27}N_9F_6O_2$): calcd 887.2192, found 887.2171.

Mp 230 °C.

10,10'-((5-Phenylbis([1,2,4]triazolo)[4,3-a:4',3'-c][1,3,5]triazine-3,9-diyl)bis(2-(trifluoromethyl)-4,1-phenylene))bis(9,9-dimethyl-9,10-dihydroacridine) (28f)

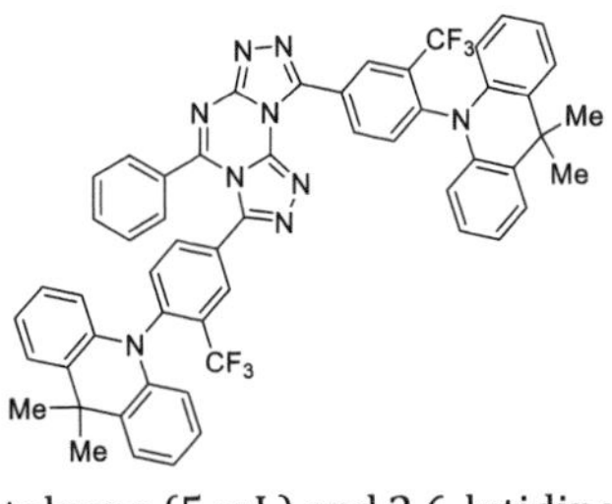

In a sealable vial, 10-(4-(2*H*-tetrazol-5-yl)-2-(trifluoromethyl)phenyl)-9,9-dimethyl-9,10-dihydro-acridine (**26f**) (70.0 mg, 166 µmol, 1.00 equiv.) and 2,4-dichloro-6-phenyl-1,3,5-triazine (16.9 mg, 75 µmol, 0.45 equiv.) were evacuated and backfilled with argon three times. Dry toluene (5 mL) and 2,6-lutidine (36.0 mg, 332 µmol, 2.00 equiv.) were added, and the resulting mixture was heated at 80 °C for 16 h until completion. After cooling to room temperature, the reaction mixture was poured into an excess of water (50 mL) and extracted with dichloromethane (3 × 50 mL). The combined organic layers were washed with brine (50 mL), dried over sodium sulfate, reduced in a vacuum, and purified by flash column chromatography over silica gel (toluene/EtOAc, 95:5 to 8:2) to yield 24.3 mg of the title compound (25.9 µmol, 35%) as a yellow solid.

R_f (toluene/EtOAc, 10:1) = 0.32.

^{1}H NMR (500 MHz, $CDCl_3$, ppm) δ = 9.49 (d, *J* = 1.6 Hz, 1H, H_{ar}), 8.94 (dd, *J* = 8.4, 1.9 Hz, 1H, H_{ar}), 8.88 – 8.86 (m, 1H, H_{ar}), 8.83 – 8.80 (m, 2H, H_{ar}), 8.73 (dd, *J* = 8.2, 1.6 Hz, 1H, H_{ar}), 7.80 (t, *J* = 7.4 Hz, 1H, H_{ar}), 7.72 (dd, *J* = 13.7, 8.1 Hz, 3H, H_{ar}), 7.59 (d, *J* = 8.2 Hz, 1H, H_{ar}), 7.50 (td, *J* = 6.6, 6.1, 1.8 Hz, 4H, H_{ar}), 7.03 – 6.94 (m, 8H, H_{ar}), 6.22 – 6.17 (m, 2H, H_{ar}), 6.11 – 6.05 (m, 2H, H_{ar}), 1.93 (d, *J* = 4.4 Hz, 6H, CH_3), 1.43 (d, *J* = 11.9 Hz, 6H, CH_3).

^{13}C NMR (126 MHz, $CDCl_3$, ppm) δ = 161.4 (C_q), 160.7 (C_q), 151.6 (C_q), 146.6 (C_q), 146.4 (C_q), 140.9 (d, J = 5.2 Hz, C_q), 135.7 (+, CH), 135.4 (+, CH), 135.3 (+, CH), 134.2 (+, CH), 131.4 (+, CH), 130.3 (d, *J* = 7.4 Hz, +, CH), 129.1 (+, CH), 128.8 (+, CH), 128.5 (+, CH), 126.4 – 126.3 (m, +, CH), 125.1 (+, CH), 121.4 (d, *J* = 8.4 Hz, +, CH), 113.9 (d, *J* = 9.8 Hz, +, CH), 36.1 (d, *J* = 3.5 Hz, C_q), 34.3 (+, CH_3), 27.0 (d, *J* = 4.4 Hz, CH_3).

^{19}F NMR (376 MHz, $CDCl_3$, ppm) δ = -61.9 (s, CF_3), -62.1 (s, CF_3).

IR (ATR, cm^{-1}) $\tilde{\nu}$ = 1636 (w), 1591 (w), 1577 (w), 1558 (w), 1514 (w), 1500 (w), 1475 (s), 1449 (s), 1412 (m), 1317 (vs), 1303 (s), 1271 (s), 1231 (w), 1193 (w), 1147 (vs), 1142 (vs), 1128 (vs), 1086 (w), 1055 (s), 1030 (w), 989 (w), 976 (w), 942 (w), 926 (w), 919 (w), 907 (w), 773 (w), 744 (vs), 727 (s), 693 (s), 667 (w), 650 (w), 625 (m), 605 (w), 581 (w), 554 (w), 541 (w), 517 (w), 477 (w), 465 (w), 453 (w), 438 (w), 384 (w).

MS (ESI): m/z (%) = 940 (65) $[M+H]^+$, 939 (100) $[M]^+$. HRMS (ESI, $C_{55}H_{39}N_9F_6$): calcd 939.3233, found 939.3213.

5.2.1.5. Syntheses of the Mono[1,2,4]-triazolo[1,3,5]-triazine Compounds

10-(4-(5,7-Diphenyl-[1,2,4]triazolo[1,5-a][1,3,5]triazin-2-yl)phenyl)-10*H*-phenoxazine (30a)

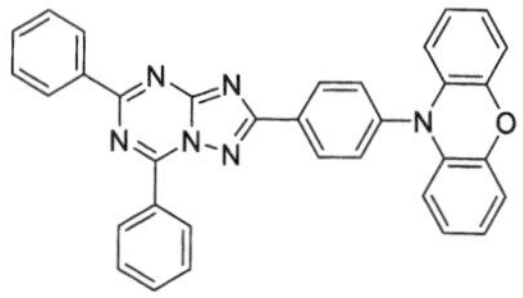

In a sealable vial, 10-(4-(2*H*-tetrazol-5-yl)phenyl)-10*H*-phenoxazine (**26a**) (150 mg, 458 µmol, 1.00 equiv.) and 2-chloro-4,6-diphenyl-1,3,5-triazine (123 mg, 458 µmol, 1.00 equiv.) were evacuated and backfilled with argon three times. Dry toluene (5 mL) and 2,6-lutidine (49.0 mg, 458 µmol, 1.00 equiv.) were added, and the resulting mixture was heated at 80 °C for 24 h until completion. After cooling to room temperature, the reaction mixture was poured into an excess of water (50 mL) and extracted with dichloromethane (3 × 50 mL). The combined organic layers were washed with brine (50 mL), dried over sodium sulfate, reduced in a vacuum, and purified by flash column chromatography over silica gel (*n*-pentane/CH_2Cl_2, 1:1 to 0:1) to yield 25.8 mg of the title compound (48.6 µmol, 11%) as a yellow solid.

R_f (toluene/EtOAc, 49:1) = 0.32.

^1H NMR (500 MHz, THF-d_8, ppm) δ = 9.21–9.11 (m, 2H, H_{ar}), 8.77 (dd, *J* = 7.9, 1.8 Hz, 2H, H_{ar}), 8.69 (d, *J* = 8.5 Hz, 2H, H_{ar}), 7.80–7.75 (m, 1H, H_{ar}), 7.75–7.69 (m, 2H, H_{ar}), 7.65–7.54 (m, 5H, H_{ar}), 6.71–6.57 (m, 6H, H_{ar}), 6.05 (dd, *J* = 7.8, 1.6 Hz, 2H, H_{ar}).

^{13}C NMR (126 MHz, THF-d_8, ppm) δ = 167.3 (C_q), 164.2 (C_q), 161.0 (C_q), 155.6 (C_q), 145.2 (C_q), 143.0 (C_q), 137.1 (C_q), 135.3 (C_q), 134.7 (+, CH), 133.2 (+, CH), 132.5 (C_q), 132.4 (C_q), 131.8 (+, CH), 131.6 (+, CH), 131.3 (+, CH), 130.2 (+, CH), 129.7 (+, CH), 124.3 (+, CH), 122.5 (+, CH), 116.4 (+, CH), 114.4 (+, CH).

IR (ATR, cm^{-1}) $\tilde{\nu}$ = 3061 (w), 1604 (w), 1592 (m), 1571 (w), 1506 (m), 1482 (vs), 1453 (m), 1442 (s), 1421 (m), 1384 (m), 1374 (m), 1332 (s), 1316 (m), 1290 (m), 1272 (vs), 1222 (m), 1207 (m), 1186 (w), 1164 (w), 1152 (w), 1137 (w), 1116 (w), 1096 (w), 1071 (w), 1043 (w), 1026 (w), 1017 (w), 1001 (w), 982 (w), 958 (w), 928 (w), 868 (w), 844 (w), 793 (w), 769 (s), 754 (m), 741 (vs), 701 (vs), 686 (vs), 669 (w), 637 (w), 628 (w), 616 (w), 609 (w), 596 (w), 584 (w), 557 (w), 524 (w), 453 (w), 433 (w), 416 (w), 407 (w), 395 (w), 381 (w).

MS (ESI): m/z (%) = 531 (49) $[M+H]^+$, 530 (100) $[M]^+$. HRMS (ESI, $C_{34}H_{22}N_6O$): calcd 530.1855, found 530.1842.

Mp 336 °C.

10-(4-(5,7-Diphenyl-[1,2,4]triazolo[1,5-a][1,3,5]triazin-2-yl)phenyl)-9,9-dimethyl-9,10-dihydroacridine (30b)

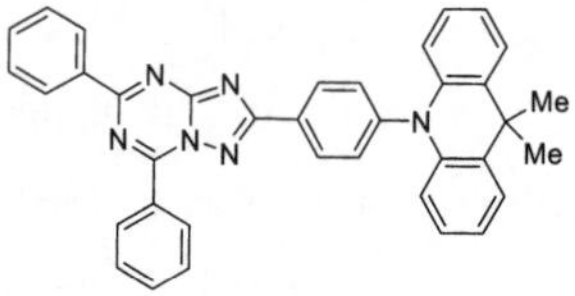

In a sealable vial, 10-(4-(2*H*-tetrazol-5-yl)phenyl)-9,9-dimethyl-9,10-dihydroacridine (**26b**) (353 mg, 1.00 mmol, 1.00 equiv.) and 2-chloro-4,6-diphenyl-1,3,5-triazine (267 mg, 1.00 mmol, 1.00 equiv.) were evacuated and backfilled with argon three times. Dry toluene (20 mL) and 2,6-lutidine (214 mg, 2.00 mmol, 2.00 equiv.) were added, and the resulting mixture was heated at 80 °C for 72 h until completion. After cooling to room temperature, the reaction mixture was poured into an excess of water (50 mL) and extracted with dichloromethane (3 × 50 mL). The combined organic layers were washed with brine (50 mL), dried over sodium sulfate, reduced in a vacuum, and purified by flash column chromatography over silica gel (*n*-pentane/CH_2Cl_2, 3:1 to 0:1) to yield 145 mg of the title compound (261 µmol, 26%) as a yellow solid.

R_f (*n*-pentane/CH_2Cl_2, 1:1) = 0.31.

^1H NMR (500 MHz, $CDCl_3$, ppm) δ = 9.14–9.12 (m, 2H, H_{ar}), 8.79–8.77 (m, 2H, H_{ar}), 8.73–8.71 (m, 2H, H_{ar}), 7.79–7.76 (m, 1H, H_{ar}), 7.75–7.71 (m, 2H, H_{ar}), 7.63–7.58 (m, 3H, H_{ar}), 7.56–7.53 (m, 2H, H_{ar}), 7.48 (dd, J = 7.6, 1.7 Hz, 2H, H_{ar}), 7.02–6.93 (m, 4H, H_{ar}), 6.38 (dd, J = 8.1, 1.4 Hz, 2H, H_{ar}), 1.72 (s, 6H, CH_3).

^{13}C NMR (126 MHz, $CDCl_3$, ppm) δ = 166.7 (C_q), 163.8 (C_q), 159.9 (C_q), 154.5 (C_q), 144.3 (C_q), 140.8 (C_q), 135.5 (+, CH), 134.3 (C_q), 132.8 (+, CH), 132.0 (+, CH), 131.7 (+, CH), 130.7 (+, CH), 130.4 (C_q), 129.9 (+, CH), 129.8 (C_q), 129.6 (C_q), 129.0 (+, CH), 128.9 (+, CH), 126.6 (+, CH), 125.5 (+, CH), 121.0 (+, CH), 114.3 (+, CH), 36.2 (C_q), 31.4 (+, CH_3, 2C).

IR (ATR, cm^{-1}) ṽ = 3070 (w), 3058 (w), 3033 (w), 2961 (w), 2952 (w), 2914 (w), 1591 (vs), 1568 (m), 1500 (m), 1475 (vs), 1439 (vs), 1421 (vs), 1374 (vs), 1319 (vs), 1269 (vs), 1221 (vs), 1188 (m), 1166 (m), 1150 (m), 1139 (w), 1111 (w), 1098 (m), 1088 (w), 1071 (w), 1044 (m), 1031 (w), 1016 (m), 1000 (w), 980 (w), 943 (w), 925 (w), 849 (w), 769 (vs), 744 (vs), 725 (w), 701 (vs), 687 (vs), 657 (m), 639 (w), 623 (m), 613 (s), 601 (w), 579 (w), 547 (w), 530 (w), 513 (w), 455 (w), 436 (w), 411 (w).

MS (ESI): m/z (%) = 557 (1) $[M+H]^+$, 556 (4) $[M]^+$. HRMS (ESI, $C_{37}H_{28}N_6$): calcd 556.2375, found 556.2372.

Mp 273 °C.

10-(4-(5,7-Diphenyl-[1,2,4]triazolo[1,5-a][1,3,5]triazin-2-yl)phenyl)-9,9-diphenyl-9,10-dihydroacridine (30c)

In a sealable vial, 10-(4-(2*H*-tetrazol-5-yl)phenyl)-9,9-diphenyl-9,10-dihydroacridine (**26c**) (189 mg, 396 µmol, 1.00 equiv.) and 2-chloro-4,6-diphenyl-1,3,5-triazine (106 mg, 396 µmol, 1.00 equiv.) were evacuated and backfilled with argon three times. Dry toluene (9 mL) and 2,6-lutidine (85.0 mg, 793 µmol, 2.00 equiv.) were added, and the resulting mixture was heated at 80 °C for 48 h until completion. After cooling to room temperature, the reaction mixture was poured into an excess of water (50 mL) and extracted with dichloromethane (3 × 50 mL). The combined organic layers were

washed with brine (50 mL), dried over sodium sulfate, reduced in a vacuum, and purified by flash column chromatography over silica gel (*n*-pentane/CH_2Cl_2, 2:1 to 0:1) to yield 18.4 mg of the title compound (27.0 µmol, 7%) as a yellow solid.
R_f (CH_2Cl_2) = 0.36.
^1H NMR (500 MHz, $CDCl_3$, ppm) δ = 9.17–9.07 (m, 2H, H_{ar}), 8.77 (dd, *J* = 8.1, 1.5 Hz, 2H, H_{ar}), 8.62 (d, *J* = 8.4 Hz, 2H, H_{ar}), 7.80–7.74 (m, 1H, H_{ar}), 7.71 (dd, *J* = 8.2, 6.8 Hz, 2H, H_{ar}), 7.65–7.54 (m, 3H, H_{ar}), 7.31–7.25 (m, 8H, H_{ar}), 7.08 (ddd, *J* = 8.5, 6.3, 2.4 Hz, 2H, H_{ar}), 7.02 (dd, *J* = 8.0, 1.6 Hz, 4H, H_{ar}), 6.96–6.86 (m, 4H, H_{ar}), 6.52 (d, *J* = 8.0 Hz, 2H, H_{ar}).
^{13}C NMR (126 MHz, $CDCl_3$, ppm) δ = 166.7 (C_q), 163.8 (C_q), 159.8 (C_q), 154.5 (C_q), 146.5 (C_q), 143.8 (C_q), 142.1 (C_q), 135.4 (C_q), 134.2 (+, CH), 132.8 (+, CH), 131.9 (+, CH), 131.7 (+, CH), 130.6 (+, CH), 130.4 (+, CH), 130.3 (+, CH), 129.9 (+, CH), 129.8 (C_q), 129.6 (C_q), 129.0 (C_q), 128.9 (+, CH), 127.8 (+, CH), 127.1 (+, CH), 126.5 (+, CH), 120.6 (+, CH), 114.3 (+, CH), 56.9 (C_q).
IR (ATR, cm^{-1}) $\tilde{\nu}$ = 3058 (w), 3030 (w), 1592 (vs), 1570 (m), 1528 (w), 1507 (m), 1483 (vs), 1470 (vs), 1451 (vs), 1441 (vs), 1421 (m), 1375 (s), 1327 (s), 1315 (m), 1275 (m), 1264 (m), 1224 (m), 1187 (w), 1167 (w), 1136 (w), 1096 (w), 1081 (w), 1055 (w), 1026 (w), 1018 (w), 1001 (w), 925 (w), 907 (w), 846 (w), 809 (w), 762 (vs), 751 (vs), 734 (s), 700 (vs), 686 (vs), 660 (m), 639 (m), 632 (m), 623 (m), 612 (m), 601 (w), 538 (m), 523 (w), 500 (w), 480 (w), 459 (w), 450 (m), 432 (w), 409 (w), 395 (w), 385 (w), 375 (w).
MS (ESI): m/z (%) = 682 (12) $[M+2H]^+$, 681 (53) $[M+H]^+$, 680 (100) $[M]^+$. HRMS (ESI, $C_{47}H_{32}N_6$): calcd 680.2688, found 680.2660.
Mp 330 °C.

10-(4-(5,7-Diphenyl-[1,2,4]triazolo[1,5-a][1,3,5]triazin-2-yl)-2,6-dimethylphenyl)-10*H*-phenoxazine (30d)

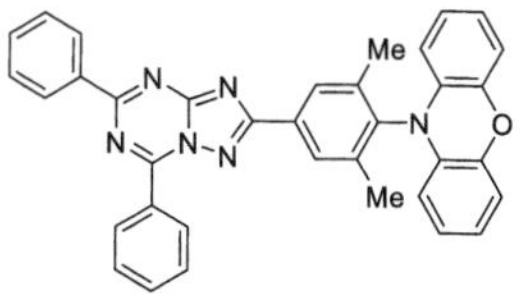

In a sealable vial, 10-(2,6-dimethyl-4-(2*H*-tetrazol-5-yl)phenyl)-10*H*-phenoxazine (**26d**) (355 mg, 1.00 mmol, 1.00 equiv.), 2-bromo-4,6-diphenyl-1,3,5-triazine (312 mg, 1.00 mmol, 1.00 equiv.), bis(dibenzylideneacetone) palladium ($Pd(dba)_2$) (23.0 mg, 40 µmol, 0.04 equiv.), 1,1-bis(diphenylphosphino)ferrocene (dppf) (33.2 mg, 60 µmol, 0.06 equiv.) and cesium carbonate (488 mg, 1.50 mmol, 1.50 equiv.) were evacuated and backfilled with argon three times. Dry toluene (20 mL) was added, and the resulting mixture was heated at 110 °C for 16 h until completion. After cooling to room temperature, the reaction mixture was poured into an excess of water (50 mL) and extracted with dichloromethane (3 × 50 mL). The combined organic layers were washed with brine (50 mL), dried over sodium sulfate, reduced in a vacuum, and purified by flash column chromatography over silica gel (*n*-pentane/CH_2Cl_2, 1:3 to 0:1) to yield 408 mg of the title compound (730 µmol, 73%) as a yellow solid.

R_f (CH_2Cl_2) = 0.48.

^{1}H NMR (500 MHz, $CDCl_3$, ppm) δ = 9.12–9.09 (m, 2H, H_{ar}), 8.78–8.75 (m, 2H, H_{ar}), 8.34 (s, 2H, H_{ar}), 7.80–7.70 (m, 3H, H_{ar}), 7.65–7.57 (m, 3H, H_{ar}), 6.75–6.56 (m, 6H, H_{ar}), 5.80 (d, *J* = 7.3 Hz, 2H, H_{ar}), 2.36 (s, 6H, CH_3).

^{13}C NMR (126 MHz, $CDCl_3$, ppm) δ = 166.8 (C_q), 163.8 (C_q), 159.8 (C_q), 154.5 (C_q), 135.5 (C_q), 134.2 (+, CH), 132.8 (+, CH), 131.7 (+, CH), 130.1 (+, CH), 129.8 (C_q), 129.6 (C_q), 129.3 (+, CH), 129.0 (+, CH), 128.9 (+, CH), 123.9 (+, CH), 115.6 (+, CH), 111.9 (+, CH), 18.2 (+, CH_3, 2C).

IR (ATR, cm^{-1}) $\tilde{\nu}$ = 1606 (w), 1594 (m), 1572 (w), 1509 (m), 1485 (vs), 1452 (m), 1443 (m), 1425 (w), 1377 (m), 1340 (s), 1313 (w), 1290 (w), 1269 (s), 1262 (s), 1235 (w), 1220 (w), 1204 (w), 1162 (w), 1009 (w), 766 (m), 738 (vs), 705 (s), 687 (s), 664 (w), 616 (w), 458 (w).

MS (FAB, 3-NBA): m/z (%) = 558 (31) [M]$^+$. HRMS (FAB, $C_{36}H_{26}N_6O$): calcd 558.2168, found 558.2162.

EA ($C_{36}H_{26}N_6O$) calcd C: 77.40, H: 4.69, N: 15.04; found C: 76.66, H: 4.46, N: 14.78. Mp 350 °C.

10-(4-(5,7-Diphenyl-[1,2,4]triazolo[1,5-a][1,3,5]triazin-2-yl)-2-(trifluoromethyl)phenyl)-10*H*-phenoxazine (30e)

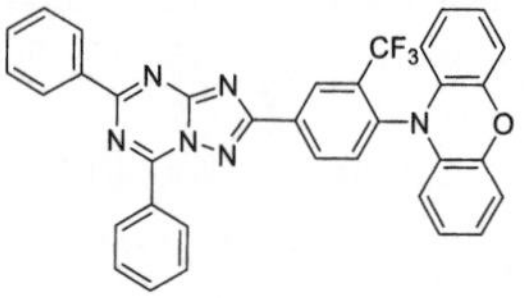

In a sealable vial, 10-(4-(2*H*-tetrazol-5-yl)-2-(trifluoromethyl)phenyl)-10*H*-phenoxazine (**26e**) (396 mg, 1.00 mmol, 1.00 equiv.) and 2-chloro-4,6-diphenyl-1,3,5-triazine (268 mg, 1.00 mmol, 1.00 equiv.) were evacuated and backfilled with argon three times. Dry toluene (20 mL) and 2,6-lutidine (215 mg, 2.00 mmol, 2.00 equiv.) were added, and the resulting mixture was heated at 80 °C for 72 h until completion. After cooling to room temperature, the reaction mixture was poured into an excess of water (100 mL) and extracted with dichloromethane (3 × 100 mL). The combined organic layers were washed with brine (100 mL), dried over sodium sulfate, reduced in a vacuum, and purified by flash column chromatography over silica gel (*n*-pentane/CH_2Cl_2, 2:1 to 1:2) to yield 159 mg of the title compound (266 µmol, 27%) as an orange solid.

R_f (*n*-pentane/CH_2Cl_2, 1:3) = 0.30.

^{1}H NMR (500 MHz, THF-d$_8$, ppm) δ = 9.14 – 9.11 (m, 2H, H_{ar}), 8.99 (d, *J* = 2.0 Hz, 1H, H_{ar}), 8.94 (dd, *J* = 8.1, 2.0 Hz, 1H, H_{ar}), 8.79 – 8.75 (m, 2H, H_{ar}), 7.85 (d, *J* = 8.1 Hz, 1H, H_{ar}), 7.83 – 7.76 (m, 1H, H_{ar}), 7.74 (dd, *J* = 8.4, 6.6 Hz, 2H, H_{ar}), 7.64 – 7.58 (m, 3H, H_{ar}), 6.73 (dd, J = 7.9, 1.7 Hz, 2H, H_{ar}), 6.69 (td, *J* = 7.6, 1.5 Hz, 2H, H_{ar}), 6.62 (td, *J* = 7.6, 1.7 Hz, 2H, H_{ar}), 5.88 (dd, *J* = 8.0, 1.5 Hz, 2H, H_{ar}).

^{13}C NMR (126 MHz, THF-d$_8$, ppm) δ = 166.0 (C_q), 164.6 (C_q), 161.1 (C_q), 156.0 (C_q), 144.9 (C_q), 136.9 (C_q), 135.8 (+, CH), 135.4 (+, CH), 135.2 (+, CH), 134.9 (+, CH),

133.4 (+, CH), 132.9 (+, CH), 132.5 (+, CH), 131.2 (+, CH), 130.2 (C_q), 129.8 (d, J = 3.8 Hz, +, CH), 124.3 (+, CH), 122.9 (+, CH), 116.4 (+, CH), 115.0 (+, CH).

^{19}F NMR (376 MHz, THF-d_8, ppm) δ = -62.5 (s, CF_3).

IR (ATR, cm^{-1}) ṽ = 3068 (w), 3051 (w), 3033 (w), 1605 (m), 1592 (s), 1571 (m), 1506 (m), 1480 (vs), 1459 (vs), 1448 (s), 1432 (m), 1418 (m), 1380 (s), 1340 (s), 1333 (s), 1305 (vs), 1290 (vs), 1272 (vs), 1232 (m), 1217 (s), 1208 (s), 1188 (m), 1166 (s), 1150 (s), 1126 (vs), 1075 (m), 1054 (s), 1043 (m), 1026 (m), 1000 (m), 982 (m), 966 (m), 931 (m), 918 (m), 868 (m), 837 (m), 807 (w), 799 (w), 769 (vs), 737 (vs), 721 (s), 705 (vs), 684 (vs), 662 (s), 632 (m), 612 (s), 586 (s), 550 (m), 533 (m), 520 (m), 507 (m), 480 (m), 462 (m), 442 (s), 415 (m), 394 (m), 385 (m), 375 (m).

MS (ESI): m/z (%) = 599 (40) $[M+H]^+$, 598 (100) $[M]^+$. HRMS (ESI, $C_{35}H_{21}N_6F_3O$): calcd 598.1729, found 598.1724.

Mp 298 °C.

10-(4-(5,7-Diphenyl-[1,2,4]triazolo[1,5-a][1,3,5]triazin-2-yl)-2-(trifluoromethyl)phenyl)-9,9-dimethyl-9,10-dihydroacridine (30f)

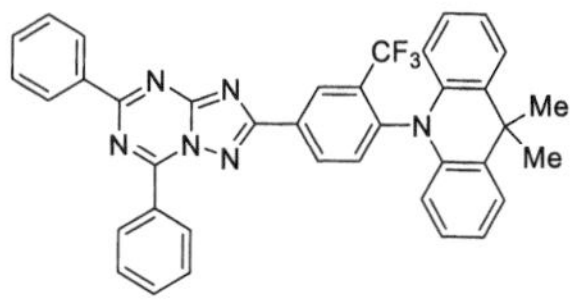

In a sealable vial, 10-(4-(2*H*-tetrazol-5-yl)-2-(trifluoromethyl)phenyl)-9,9-dimethyl-9,10-dihydroacridine (**26f**) (211 mg, 501 µmol, 1.00 equiv.) and 2-chloro-4,6-diphenyl-1,3,5-triazine (134 mg, 501 µmol, 1.00 equiv.) were evacuated and backfilled with argon three times. Dry toluene (10 mL) and 2,6-lutidine (107 mg, 1.00 mmol, 2.00 equiv.) were added, and the resulting mixture was heated at 80 °C for 72 h until completion. After cooling to room temperature, the reaction mixture was poured into an excess of water (100 mL) and extracted with dichloromethane (3 × 100 mL). The combined organic layers were washed with brine (100 mL), dried over sodium sulfate, reduced in a vacuum, and purified by flash column chromatography over silica gel (toluene/EtOAc, 1:0 to 96:4) to yield 237 mg of the title compound (379 µmol, 76%) as a yellow solid.

R_f (toluene/EtOAc, 98:2) = 0.27.

^{1}H NMR (500 MHz, $CDCl_3$, ppm) δ = 9.13 – 9.09 (m, 2H, H_{ar}), 9.07 (d, J = 2.0 Hz, 1H, H_{ar}), 8.90 (dd, J = 8.2, 2.0 Hz, 1H, H_{ar}), 8.79 (dt, J = 6.9, 1.6 Hz, 2H, H_{ar}), 7.82 – 7.78 (m, 1H, H_{ar}), 7.74 (dd, J = 8.4, 6.7 Hz, 2H, H_{ar}), 7.65 – 7.59 (m, 4H, H_{ar}), 7.54 – 7.49 (m, 2H, H_{ar}), 7.01 – 6.97 (m, 4H, H_{ar}), 6.17 – 6.14 (m, 2H, H_{ar}), 1.95 (s, 3H, CH_3), 1.45 (s, 3H, CH_3).

^{13}C NMR (126 MHz, $CDCl_3$, ppm) δ = 165.5 (C_q), 164.1 (C_q), 159.9 (C_q), 154.8 (C_q), 142.6 (C_q), 141.1 (C_q), 135.4 (+, CH), 135.2 (+, CH), 134.5 (+, CH), 133.6 (+, CH), 133.0 (+, CH), 131.7 (+, CH), 130.8 (+, CH), 130.3 (+, CH), 129.7 (+, CH), 129.5 (+, CH), 129.1 (d, J = 11.5 Hz, +, CH), 128.5 (+, CH), 128.4 (+, CH), 126.5 (+, CH), 125.2 (+, CH), 124.1 (+, CH), 121.4 (+, CH), 114.1 (+, CH), 36.2 (C_q), 34.5 (+, CH_3), 29.9 (C_q), 27.2 (+, CH_3).

^{19}F NMR (376 MHz, $CDCl_3$, ppm) δ = -61.9 (s, CF_3).

IR (ATR, cm^{-1}) $\tilde{\nu}$ = 2955 (w), 2919 (m), 2871 (w), 2851 (w), 1606 (s), 1595 (vs), 1570 (s), 1502 (m), 1483 (vs), 1462 (vs), 1445 (vs), 1418 (vs), 1383 (s), 1370 (vs), 1326 (s), 1306 (vs), 1272 (vs), 1225 (s), 1190 (m), 1143 (vs), 1128 (vs), 1106 (s), 1098 (m), 1069 (m), 1052 (vs), 1030 (m), 1000 (w), 983 (m), 967 (m), 942 (w), 926 (w), 915 (m), 867 (w), 844 (w), 769 (vs), 741 (vs), 705 (vs), 686 (vs), 667 (m), 654 (m), 619 (s), 603 (m), 581 (w), 550 (m), 540 (w), 528 (w), 507 (w), 482 (w), 456 (m), 448 (m), 431 (w), 419 (w), 402 (w), 385 (w).

MS (ESI): m/z (%) = 625 (42) $[M+H]^+$, 624 (100) $[M]^+$. HRMS (ESI, $C_{38}H_{27}N_6F_3$): calcd 624.2249, found 624.2239.

Mp 300 °C.

10-(4-(5,7-Di-p-tolyl-[1,2,4]triazolo[1,5-a][1,3,5]triazin-2-yl)phenyl)-10*H*-phenoxazine (30g)

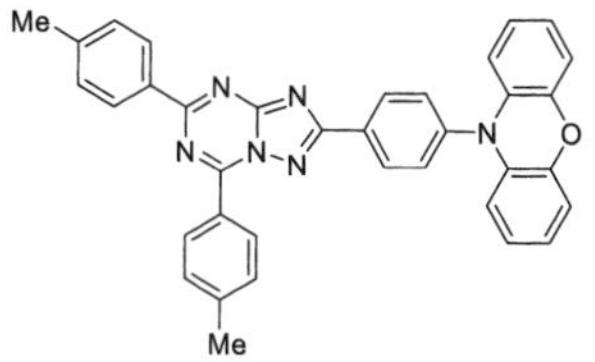

In a sealable vial, 10-(4-(2*H*-tetrazol-5-yl)phenyl)-10*H*-phenoxazine (**26a**) (82.0 mg, 250 μmol, 1.00 equiv.) and 2-chloro-4,6-di-p-tolyl-1,3,5-triazine (74.1 mg, 250 μmol, 1.00 equiv.) were evacuated and backfilled with argon three times. Dry toluene (5 mL) and 2,6-lutidine (27.0 mg, 250 μmol, 1.00 equiv.) were added, and the resulting mixture was heated at 80 °C for 16 h until completion. After cooling to room temperature, the reaction mixture was poured into an excess of water (100 mL) and extracted with dichloromethane (3 × 100 mL). The combined organic layers were washed with brine (100 mL), dried over sodium sulfate, reduced in a vacuum, and purified by flash column chromatography over silica gel (*n*-pentane/CH_2Cl_2, 1:1 to 0:1) to yield 40.5 mg of the title compound (72.5 μmol, 29%) as an orange solid.

R_f (*n*-pentane/CH_2Cl_2, 1:1) = 0.31.

^1H NMR (500 MHz, $CDCl_3$, ppm) δ = 9.02 (dd, *J* = 8.4, 2.0 Hz, 2H, H_{ar}), 8.66 (td, *J* = 9.0, 8.2, 4.5 Hz, 4H, H_{ar}), 7.52 (dd, *J* = 14.6, 8.0 Hz, 4H, H_{ar}), 7.39 (d, J = 7.9 Hz, 2H, H_{ar}), 6.76 – 6.59 (m, 6H, H_{ar}), 6.04 (d, *J* = 7.9 Hz, 2H, H_{ar}), 2.55 (d, *J* = 1.9 Hz, 3H, CH_3), 2.49 (d, *J* = 2.0 Hz, 3H, CH_3).

^{13}C NMR (126 MHz, $CDCl_3$, ppm) δ = 166.1 (C_q), 163.8 (C_q), 159.8 (C_q), 154.3 (C_q), 154.3 (C_q), 145.3 (C_q), 143.3 (C_q), 132.8 (C_q), 131.6 (+, CH), 131.2 (+, CH), 130.7 (+, CH), 130.2 (+, CH), 129.6 (+, CH), 129.5 (+, CH), 126.9 (C_q), 123.3 (C_q), 113.4 (+, CH), 29.7 (+, CH_3), 22.0 (+, CH_3), 21.8 (+, CH_3).

IR (ATR, cm^{-1}) $\tilde{v}$ = 3061 (w), 3044 (w), 3036 (w), 2952 (w), 2919 (m), 2851 (w), 1596 (s), 1579 (m), 1565 (w), 1517 (w), 1483 (vs), 1446 (vs), 1428 (s), 1411 (s), 1383 (s), 1375 (s), 1329 (vs), 1309 (s), 1290 (s), 1268 (vs), 1225 (s), 1207 (m), 1190 (s), 1171 (s), 1153 (s), 1142 (m), 1115 (m), 1096 (m), 1043 (m), 1018 (m), 980 (w), 959 (w), 926 (w), 914 (w), 870 (m), 841 (w), 829 (m), 796 (w), 785 (vs),

771 (w), 739 (vs), 710 (vs), 696 (s), 673 (m), 660 (w), 643 (w), 635 (w), 615 (m), 582 (m), 558 (m), 524 (w), 499 (m), 492 (m), 475 (m), 460 (w), 449 (m), 428 (m), 411 (w), 399 (w), 390 (w).

MS (ESI): m/z (%) = 559 (13) $[M+H]^+$, 558 (43) $[M]^+$. HRMS (ESI, $C_{36}H_{26}N_6O$): calcd 558.2169, found 558.2162.

Mp 292 °C.

5.2.2. Derivatization of the CzBN Class

5.2.2.1. Benzonitrile Precursors

3',5'-Difluoro-[1,1'-biphenyl]-4-carbonitrile (53a)

CN

F F

In a sealable vial, 1-bromo-3,5-difluorobenzene (1.16 g, 6.00 mmol, 1.00 equiv.), (4-cyanophenyl)boronic acid (1.32 g, 9.00 mmol, 1.50 equiv.), tripotassium phosphate (3.82 g, 18.0 mmol, 3.00 equiv.), $Pd_2(dba)_3$ (121 mg, 132 µmol, 0.02 equiv.) and SPhos (99.0 mg, 240 µmol, 0.04 equiv.) were evacuated and backfilled with argon three times. Toluene (20 mL) and water (2 mL) were added, and the resulting mixture was degassed with argon for 10-15 min. Subsequently, the mixture was heated at 110 °C for 16 h. After cooling to room temperature, an excess of water (100 mL) was added, and the mixture was extracted with dichloromethane (3 × 100 mL). The combined organic layers were washed with brine (100 mL), dried over sodium sulfate, and reduced in a vacuum. The crude product was purified by flash column chromatography over silica gel (*c*Hex/CH_2Cl_2, 4:1 to 0:1) to yield 1.29 g of the title compound (5.99 mmol, quant.) as a colorless solid.

R_f (*c*Hex/CH_2Cl_2, 1:1) = 0.28.

^{1}H NMR (500 MHz, $CDCl_3$, ppm) δ = 7.76 (d, J = 8.0 Hz, 2H, H_{ar}), 7.65 (d, J = 8.1 Hz, 2H, H_{ar}), 7.11 (d, J = 6.0 Hz, 2H, H_{ar}), 6.87 (tt, J = 8.8, 2.4 Hz, 1H, H_{ar}).

^{13}C NMR (126 MHz, $CDCl_3$, ppm) δ = 164.6 (d, J = 12.9 Hz, C_q), 162.6 (d, J = 13.0 Hz, C_q), 143.4 (t, J = 2.6 Hz, C_q), 142.5 (t, J = 9.5 Hz, C_q), 133.0 (+, CH), 127.9 (+, CH), 118.6 (C_q), 112.4 (C_q), 110.5 (d, J = 6.4 Hz, +, CH), 110.3 (d, J = 6.4 Hz, +, CH), 104.1 (t, J = 25.4 Hz, +, CH).

^{19}F NMR (471 MHz, $CDCl_3$, ppm) δ = -108.47 – -108.54 (m).

IR (ATR, cm^{-1}) ṽ = 3091 (w), 3064 (w), 3051 (w), 2959 (w), 2952 (w), 2928 (w), 2919 (w), 2854 (w), 2221 (w), 2198 (w), 1619 (m), 1595 (vs), 1560 (m), 1511 (w), 1497 (w), 1449 (s), 1434 (m), 1397 (s), 1337 (s), 1300 (w), 1271 (w), 1210 (w), 1197 (m), 1181 (w), 1111 (vs), 1067 (w), 989 (vs), 956 (w), 870 (s), 839 (vs),

823 (vs), 759 (m), 730 (w), 701 (w), 677 (s), 647 (w), 598 (m), 571 (m), 547 (s), 537 (s), 510 (s).

MS (ESI): m/z (%) = 216 (100) $[M+H]^+$. HRMS (ESI, $C_{13}H_8NF_2$): calc. 216.0625 $[M+H]^+$, found 216.0618 $[M+H]^+$.

EA ($C_{13}H_7NF_2$) calc. C: 72.56, H: 3.28, N: 6.51; found C: 72.36, H: 3.15, N: 6.43.

Mp 125 °C

5-(3,5-Difluorophenyl)picolinonitrile (53b)

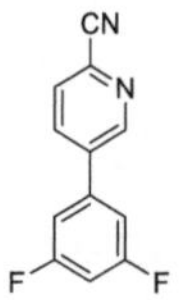

In a sealable vial, 5-bromopicolinonitrile (968 mg, 5.29 mmol, 1.00 equiv.), (3,5-difluorophenyl)boronic acid (1.25 g, 7.93 mmol, 1.50 equiv.), tripotassium phosphate (3.37 g, 15.9 mmol, 3.00 equiv.), $Pd_2(dba)_3$ (107 mg, 116 µmol, 0.02 equiv.) and SPhos (86.9 mg, 212 µmol, 0.04 equiv.) were evacuated and backfilled with argon three times. Toluene (35 mL) and water (3.5 mL) were added, and the resulting mixture was degassed with argon for 10-15 min. Subsequently, the mixture was heated at 110 °C for 16 h. After cooling to room temperature, an excess of water (100 mL) was added, and the mixture was extracted with dichloromethane (3 × 100 mL). The combined organic layers were washed with brine (100 mL), dried over sodium sulfate, and reduced in a vacuum. The crude product was purified by flash column chromatography over silica gel (*n*-pentane/CH_2Cl_2, 1:1 to 0:1) to yield 1.05 g of the title compound (4.86 mmol, 92%) as an off-white solid.

R_f (CH_2Cl_2/CH_3OH, 99.5:0.5) = 0.27.

^1H NMR (500 MHz, $CDCl_3$, ppm) δ = 8.91 (d, *J* = 2.3 Hz, 1H, H_{ar}), 7.98 (dd, *J* = 8.1, 2.3 Hz, 1H, H_{ar}), 7.80 (d, *J* = 8.1 Hz, 1H, H_{ar}), 7.13 (dt, *J* = 6.2, 2.1 Hz, 2H, H_{ar}), 6.94 (tt, *J* = 8.7, 2.3 Hz, 1H, H_{ar}).

^{13}C NMR (126 MHz, $CDCl_3$, ppm) δ = 164.8 (d, *J* = 12.9 Hz, C_q), 162.8 (d, *J* = 12.8 Hz, C_q), 149.5 (+, CH), 139.2 (t, *J* = 9.5 Hz, C_q), 137.7 (t, *J* = 2.6 Hz, C_q), 135.2 (+, CH),

133.6 (C_q), 128.7 (+, CH), 117.1 (C_q) 110.7 (d, J = 6.5 Hz, +, CH), 110.5 (d, J = 6.5 Hz, +, CH), 105.0 (t, J = 25.2 Hz, +, CH).

^{19}F NMR (376 MHz, $CDCl_3$, ppm) δ = -107.4 (s).

IR (ATR, cm^{-1}) ṽ = 3092 (w), 3057 (w), 2240 (vw), 1622 (m), 1598 (vs), 1561 (w), 1453 (s), 1438 (m), 1388 (w), 1364 (s), 1343 (vs), 1305 (w), 1265 (vw), 1221 (w), 1186 (vw), 1119 (vs), 1074 (w), 1026 (w), 990 (vs), 946 (w), 873 (vs), 847 (vs), 837 (s), 773 (w), 756 (w), 684 (s), 652 (m), 605 (w), 579 (w), 557 (w), 538 (s), 509 (m), 438 (w), 397 (w).

MS (ESI): m/z (%) = 217 (100) $[M+H]^+$. HRMS (ESI, $C_{12}H_7N_2F_2$): calc. 217.0577 $[M+H]^+$, found 217.0570 $[M+H]^+$.

Mp 173 °C.

4-(2,6-Difluoropyridin-4-yl)benzonitrile (53c)

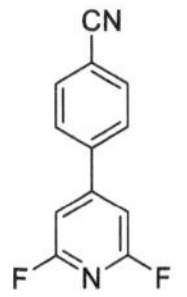

In a sealable vial, 2,6-difluoro-4-iodopyridine (100 mg, 415 µmol, 1.00 equiv.), (4-cyanophenyl)boronic acid (73.2 mg, 498 µmol, 1.20 equiv.), tripotassium phosphate (176 mg, 830 µmol, 2.00 equiv.), $Pd_2(dba)_3$°$CHCl_3$ (21.0 mg, 21.0 µmol, 0.05 equiv.) and PCy_3 (11.6 mg, 41.0 µmol, 0.10 equiv.) were evacuated and backfilled with argon three times. 1,4-Dioxane (2 mL) and water (0.85 mL) were added, and the resulting mixture was degassed with argon for 10-15 min. Subsequently, the mixture was heated at 90 °C for 16 h. After cooling to room temperature, an excess of water (50 mL) was added, and the mixture was extracted with dichloromethane (3 × 100 mL). The combined organic layers were washed with brine (100 mL), dried over sodium sulfate, and reduced in a vacuum. The crude product was purified by flash column chromatography over silica gel (cHex/CH_2Cl_2, 4:1 to 0:1) to yield 89.7 mg of the title compound (415 µmol, quant.) as an off-white solid.

R_f (n-pentane/CH_2Cl_2, 1:3) = 0.29.

^{1}H NMR (500 MHz, $CDCl_3$, ppm) δ = 7.85 (d, *J* = 8.1 Hz, 2H, H_{ar}), 7.75 (d, *J* = 8.1 Hz, 2H, H_{ar}), 7.06 (s, 2H, H_{ar}).

^{13}C NMR (126 MHz, $CDCl_3$, ppm) δ = 163.5 (d, *J* = 16.1 Hz, C_q), 161.5 (d, *J* = 16.1 Hz, C_q), 156.3 (t, *J* = 8.0 Hz, C_q), 140.6 (t, *J* = 3.2 Hz, C_q), 133.3 (+, CH, 2C), 128.0 (+, CH, 2C), 118.1 (C_q), 114.2 (C_q), 104.7 - 104.2 (m, +, CH, 2C).

^{19}F NMR (376 MHz, $CDCl_3$, ppm) δ = -66.9 (s).

IR (ATR, cm^{-1}) ṽ = 3097 (w), 3065 (w), 2230 (m), 1623 (vs), 1601 (s), 1577 (m), 1550 (s), 1513 (m), 1462 (w), 1426 (s), 1398 (s), 1364 (s), 1320 (w), 1264 (m), 1217 (m), 1201 (s), 1132 (w), 1116 (w), 1067 (w), 1030 (s), 999 (m), 975 (w), 962 (w), 868 (s), 829 (vs), 762 (m), 737 (s), 704 (m), 653 (w), 630 (w), 605 (w), 571 (w), 555 (vs), 545 (vs), 537 (vs), 514 (m), 494 (w), 486 (w), 480 (w), 467 (w), 438 (w), 424 (w), 405 (w), 397 (w), 385 (w).

MS (ESI): m/z (%) = 217 (100) $[M+H]^+$. HRMS (ESI, $C_{12}H_7N_2F_2$): calc. 217.0577 $[M+H]^+$, found 217.0571 $[M+H]^+$.

Mp 139 °C.

4-(2,3,4,5,6-Pentafluorophenyl)benzonitrile (36)

CN

F F F F F

In a sealable vial, 1-bromo-2,3,4,5,6-penta-fluorobenzene (2.02 g, 8.19 mmol, 1.00 equiv.), (4-cyanophenyl)boronic acid (1.81 g, 12.3 mmol, 1.50 equiv.), tripotassium phosphate (5.22 g, 24.6 mmol, 3.00 equiv.), tris(dibenzylideneacetone)dipalladium(0) (165 mg, 180 µmol, 0.02 equiv.) and 2-dicyclohexylphosphino-2′,6′-dimethoxybiphenyl (135 mg, 328 µmol, 0.04 equiv.) were evacuated and backfilled with argon three times. Toluene (40 mL) and water (4 mL) were added, and the resulting mixture was degassed with argon for 10-15 min. Subsequently, the mixture was heated at 110 °C for 16 h. After cooling to room temperature, an excess of water was added (100 mL), and the mixture was extracted with dichloromethane

(3 × 50 mL). The combined organic layers were washed with brine (100 mL), dried over sodium sulfate, and reduced in a vacuum. The crude product was purified by flash column chromatography over silica gel (*c*Hex/CH_2Cl_2, 3:1 to 1:1) to yield 1.74 g of the title compound (6.46 mmol, 79%) as an off-white solid.

R_f (*c*Hex/CH_2Cl_2, 2:1) = 0.33.

^{1}H NMR (500 MHz, $CDCl_3$, ppm) δ = 7.82–7.78 (m, 2H, H_{ar}), 7.59–7.54 (m, 2H, H_{ar}).

^{13}C NMR (126 MHz, $CDCl_3$, ppm) δ = 132.6 (+, CH), 131.3 (C_q), 131.1 (+, CH), 118.3 (C_q), 113.5 (C_q).

^{19}F NMR (376 MHz, $CDCl_3$, ppm) δ = -142.7 – -142.8 (m), -152.8 (t), -160.8 – -161.0 (m).

IR (ATR, cm^{-1}) $\tilde{\nu}$ = 3081 (w), 3058 (w), 2235 (m), 1650 (w), 1523 (m), 1504 (s), 1482 (vs), 1431 (m), 1408 (s), 1392 (m), 1373 (w), 1364 (w), 1334 (w), 1322 (w), 1186 (w), 1062 (s), 984 (vs), 962 (s), 858 (m), 844 (vs), 779 (m), 756 (m), 742 (w), 659 (w), 552 (vs), 472 (w).

MS (FAB, 3-NBA): m/z (%) = 270 (12) $[M+H]^+$, 269 (100) $[M]^+$. HRMS (FAB, $C_{13}H_4NF_5$): calcd 269.0258, found 269.0260.

Mp 128 °C.

4-(Perfluoropyridin-4-yl)benzonitrile (43)

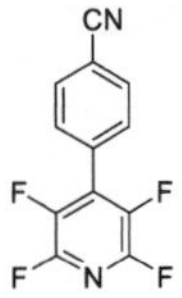

In a sealable vial, 4-bromo-2,3,5,6-tetrafluoropyridine (680 mg, 3.00 mmol, 1.00 equiv.), (4-cyanophenyl)boronic acid (652 mg, 4.44 mmol, 1.50 equiv.), tripotassium phosphate (1.88 g, 8.87 mmol, 3.00 equiv.), $Pd_2(dba)_3$ (60.0 mg, 65 µmol, 0.02 equiv.) and SPhos (48.6 mg, 118 µmol, 0.04 equiv.) were evacuated and backfilled with argon three times. Toluene (25 mL) and water (2.5 mL) were added, and the resulting mixture was degassed with argon for 10-15 min. Subsequently, the mixture was heated at 110 °C for 16 h. After cooling to room temperature, an excess of water (100 mL) was added, and the mixture was extracted with dichloromethane (3 × 100 mL). The combined organic layers were washed with brine (100 mL), dried over sodium sulfate, and reduced in a vacuum. The crude product was purified by flash column chromatography over silica gel (*c*Hex/CH_2Cl_2, 4:1 to 2:1) to yield 220 mg of the title compound (872 µmol, 30%) as a colorless solid.

R_f (*c*Hex/CH_2Cl_2, 1:1) = 0.18.

^{1}H NMR (500 MHz, $CDCl_3$, ppm) δ = 7.86 (d, *J* = 8.2 Hz, 2H, H_{ar}), 7.67 (d, *J* = 7.9 Hz, 2H, H_{ar}).

^{13}C NMR (126 MHz, $CDCl_3$, ppm) δ = 132.8 (+, CH), 130.7 (+, CH), 130.4 (C_q), 117.9 (C_q), 114.8 (C_q).

^{19}F NMR (376 MHz, $CDCl_3$, ppm) δ = -89.0 – -89.2 (m), -144.3 – -144.5 (m).

IR (ATR, cm^{-1}) $\tilde{\nu}$ = 3496 (w), 3075 (w), 2952 (w), 2921 (w), 2873 (w), 2853 (w), 2225 (w), 1649 (m), 1609 (w), 1592 (w), 1557 (w), 1509 (w), 1462 (vs), 1434 (s), 1417 (m), 1402 (vs), 1317 (w), 1290 (m), 1276 (m), 1272 (m), 1248 (w), 1210 (w), 1194 (w), 1183 (w), 1146 (vs), 1125 (w), 1060 (w), 1023 (w), 1011 (w), 1006 (w), 963 (vs), 950 (vs), 877 (s), 839 (vs), 762 (m), 754 (s), 725 (w), 707 (m), 694 (m), 647 (w), 615 (m), 554 (s), 545 (vs), 518 (w), 489 (m), 470 (w), 462 (w), 449 (w), 425 (w), 397 (w), 377 (s).

MS (FAB, 3-NBA): m/z (%) = 252 (64) $[M]^+$. HRMS (FAB, $C_{12}H_4N_2F_4$): calcd 252.0305, found 252.0305.

EA ($C_{12}H_4N_2F_4$) calc. C: 57.16, H: 1.60, N: 11.11; found C: 57.42, H: 1.55, N: 10.86. Mp 103 °C.

3,5-Bis(3,6-di-*tert*-butyl-9*H*-carbazol-9-yl)benzonitrile (60)

In a sealable vial, 3,5-difluorobenzonitrile (834 mg, 6.00 mmol, 1.00 equiv.), 3,6-di-*tert*-butyl-9*H*-carbazole (3.52 g, 12.6 mmol, 2.10 equiv.) and tripotassium phosphate (3.82 g, 18.0 mmol, 3.00 equiv.) were evacuated and backfilled with argon three times. Dry dimethyl sulfoxide (24 mL) was added, and the resulting mixture was heated at 130 °C for 72 h. After cooling to room temperature, the reaction mixture was poured into an excess of water (100 mL) and extracted with dichloromethane (3 × 100 mL). The combined organic layers were washed with brine (100 mL), dried over sodium sulfate, and reduced in a vacuum. The crude product was purified by flash column chromatography over silica gel (*c*Hex/CH_2Cl_2, 3:1 to 2:1) to yield 3.36 g of the title compound (5.11 mmol, 85%) as a colorless solid.

R_f (*c*Hex/CH_2Cl_2, 1:2) = 0.32.

^1H NMR (500 MHz, $CDCl_3$, ppm) δ = 8.14 (d, J = 1.8 Hz, 3H, H_{ar}), 8.08 (t, J = 2.1 Hz, 2H, H_{ar}), 7.92 (d, J = 1.9 Hz, 1H, H_{ar}), 7.51 (dd, J = 8.7, 1.9 Hz, 3H, H_{ar}), 7.48 – 7.42 (m, 5H, H_{ar}), 7.34 (d, J = 8.4 Hz, 1H, H_{ar}), 1.47 (s, 18H, CH_3), 1.45 (s, 18H, CH_3).

^{13}C NMR (126 MHz, $CDCl_3$, ppm) δ = 144.4 (C_q), 142.4 (C_q), 141.2 (C_q), 138.4 (C_q), 138.2 (C_q), 128.4 (+, CH), 127.5 (+, CH), 124.3 (+, CH), 124.2 (C_q), 123.7 (+, CH), 123.5 (C_q), 117.7 (C_q), 116.8 (+, CH), 116.3 (+, CH), 115.5 (C_q), 110.1 (+, CH), 109.0 (+, CH), 35.0 (C_q), 34.8 (C_q), 32.2 (+, CH_3), 32.1 (+, CH_3).

IR (ATR, cm^{-1}) $\tilde{\nu}$ = 3412 (w), 2953 (s), 2902 (w), 2864 (w), 1589 (m), 1487 (s), 1472 (vs), 1451 (m), 1392 (w), 1361 (m), 1344 (w), 1322 (m), 1295 (vs), 1259 (s), 1232 (s), 1201 (w), 1174 (w), 1149 (w), 1105 (w), 1034 (w), 891 (m), 880 (w), 805 (vs), 739 (w), 691 (w), 683 (w), 653 (w), 613 (vs), 594 (w), 472 (w), 456 (w), 450 (w), 422 (w), 408 (w).

MS (FAB, 3-NBA): m/z (%) = 659 (25) $[M+2H]^+$, 658 (68) $[M+H]^+$, 657 (100) $[M]^+$.

HRMS (FAB, $C_{47}H_{51}N_3$): calcd 657.4078, found 657.4075.

EA ($C_{47}H_{51}N_3$) calc. C: 85.80, H: 7.81, N: 6.39; found C: 85.41, H: 8.21, N: 5.88.

Mp 308 °C.

3',5'-Bis(3,6-di-*tert*-butyl-9*H*-carbazol-9-yl)-[1,1'-biphenyl]-4-carbonitrile (58a)

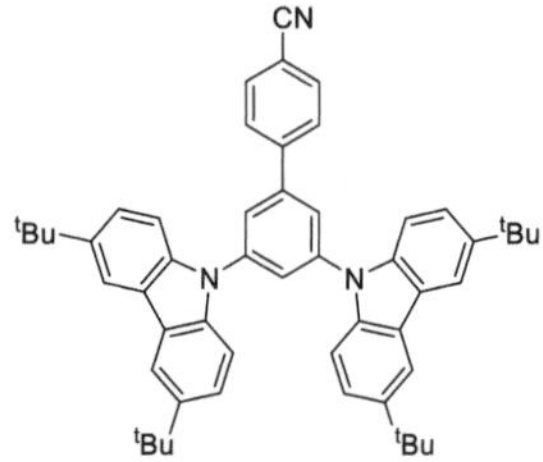

In a sealable vial, 3',5'-difluoro-[1,1'-biphenyl]-4-carbonitrile (**53a**) (861 mg, 4.00 mmol, 1.00 equiv.), 3,6-di-*tert*-butyl-9*H*-carbazole (2.35 g, 8.40 mmol, 2.10 equiv.) and tripotassium phosphate (2.55 g, 12.0 mmol, 3.00 equiv.) were evacuated and back-filled with argon three times. Dry dimethyl sulfoxide (30 mL) was added, and the resulting mixture was heated at 130 °C for 72 h. After cooling to room temperature, the reaction mixture was poured into an excess of water (100 mL) and extracted with dichloromethane (3 × 100 mL). The combined organic layers were washed with brine (100 mL), dried over sodium sulfate, and reduced in a vacuum. The crude product was purified by flash column chromatography over silica gel (*c*Hex/CH_2Cl_2, 2:1 to 0:1) to yield 2.94 g of the title compound (4.00 mmol, quant.) as a colorless solid.

R_f (*c*Hex/CH_2Cl_2, 1:2) = 0.18.

^{1}H NMR (500 MHz, $CDCl_3$, ppm) δ = 8.16 (t, J = 1.3 Hz, 4H, H_{ar}), 7.87 (q, J = 1.5 Hz, 3H, H_{ar}), 7.79 (d, J = 1.9 Hz, 4H, H_{ar}), 7.52 – 7.49 (m, 8H, H_{ar}), 1.47 (s, 36H, CH_3).

^{13}C NMR (126 MHz, $CDCl_3$, ppm) δ = 144.0 (C_q), 143.7 (C_q), 142.5 (C_q), 140.8 (C_q), 138.9 (C_q), 133.1 (+, CH), 128.0 (+, CH), 124.3 (C_q), 124.1 (+, CH), 123.9 (+, CH),

123.7 (+, CH), 123.5 (+, CH), 118.8 (C_q), 116.7 (+, CH), 112.2 (C_q), 110.1 (C_q), 109.2 (+, CH), 34.9 (C_q), 32.1 (+, CH_3).

IR (ATR, ṽ, cm^{-1}) = 3403 (w), 2952 (vs), 2901 (m), 2864 (w), 1594 (s), 1476 (vs), 1455 (s), 1394 (w), 1361 (s), 1343 (w), 1326 (m), 1296 (vs), 1262 (vs), 1242 (m), 1230 (m), 1201 (w), 1170 (w), 1103 (w), 1033 (w), 925 (w), 899 (w), 878 (s), 847 (w), 836 (m), 820 (vs), 812 (vs), 741 (w), 713 (w), 693 (w), 686 (w), 657 (w), 615 (vs), 571 (w), 535 (w), 516 (w), 470 (w), 448 (w), 422 (m).

MS (ESI): m/z (%) = 734 (32) $[M+H]^+$, 733 (55) $[M]^+$. HRMS (ESI, $C_{53}H_{55}N_3$): calc. 733.4396, found 733.4383.

EA ($C_{53}H_{55}N_3$) calc. C: 86.72, H: 7.55, N: 5.72; found C: 86.14, H: 5.56, N: 7.67.

Mp > 350 °C.

5-(3,5-Bis(3,6-di-*tert*-butyl-9*H*-carbazol-9-yl)phenyl)picolinonitrile (58b)

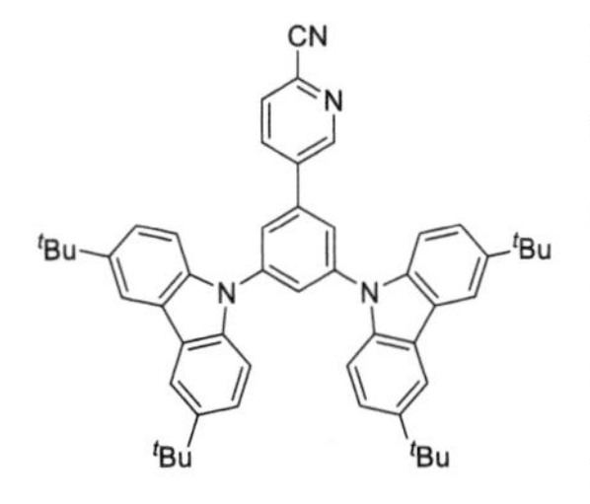

In a sealable vial, 5-(3,5-difluorophenyl)picolinonitrile (**53b**) (840 mg, 3.89 mmol, 1.00 equiv.), 3,6-di-*tert*-butyl-9*H*-carbazole (2.28 g, 8.16 mmol, 2.10 equiv.) and tripotassium phosphate (2.06 g, 9.71 mmol, 2.50 equiv.) were evacuated and backfilled with argon three times. Dry dimethyl sulfoxide (16 mL) was added, and the resulting mixture was heated at 130 °C for 16 h. After cooling to room temperature, the reaction mixture was poured into an excess of water (100 mL) and extracted with dichloromethane (3 × 50 mL). The combined organic layers were washed with brine (50 mL), dried over sodium sulfate, and reduced in a vacuum. The crude product was purified by flash column chromatography over silica gel (*n*-pentane/CH_2Cl_2, 3:1 to 1:2) to yield 1.32 g of the title compound (1.80 mmol, 92%) as a yellow solid.

R_f (*n*-pentane/CH_2Cl_2, 1:3) = 0.29.

^{1}H NMR (500 MHz, $CDCl_3$, ppm) δ = 9.09 (d, J = 2.3 Hz, 1H, H_{ar}), 8.18 (d, J = 1.5 Hz, 4H, H_{ar}), 8.12 (dd, J = 8.1, 2.3 Hz, 1H, H_{ar}), 7.96 (t, J = 1.8 Hz, 1H, H_{ar}), 7.89 (d, J = 1.9 Hz, 2H, H_{ar}), 7.83 (d, J = 8.1 Hz, 1H, H_{ar}), 7.52 (d, J = 1.4 Hz, 8H, H_{ar}), 1.48 (s, 36H, CH_3).

^{13}C NMR (126 MHz, $CDCl_3$, ppm) δ = 149.7 (+, CH), 143.9 (C_q), 141.2 (C_q), 139.2 (C_q), 138.8 (C_q), 138.4 (C_q), 135.3 (+, CH), 133.4 (C_q), 128.7 (+, CH), 124.9 (+, CH), 124.2 (+, CH), 124.0 (+, CH), 123.4 (+, CH), 117.2 (+, CH), 116.7 (+, CH), 109.1 (+, CH), 35.0 (C_q), 32.1 (+, CH_3, 12C).

IR (ATR, cm^{-1}) $\tilde{\nu}$ = 2953 (m), 2902 (w), 2864 (w), 1594 (s), 1575 (w), 1472 (vs), 1452 (vs), 1392 (w), 1363 (s), 1344 (w), 1322 (m), 1295 (vs), 1261 (s), 1232 (s), 1203 (w), 1171 (w), 1105 (w), 1034 (w), 1023 (w), 928 (w), 898 (w), 878 (s), 850 (w), 810 (vs), 762 (w), 741 (w), 718 (w), 694 (w), 686 (w), 657 (w), 613 (s), 571 (w), 469 (w), 421 (w).

MS (ESI): m/z (%) = 735 (60) $[M+H]^+$, 734 (100) $[M]^+$. HRMS (ESI, $C_{52}H_{54}N_4$): calc. 734.4348, found 734.4341.

EA ($C_{52}H_{54}N_4$) calc. C: 84.97, H: 7.41, N: 7.62; found C: 84.93, H: 7.36, N: 7.59.

Mp 393 °C.

4-(2,6-Bis(3,6-di-*tert*-butyl-9*H*-carbazol-9-yl)pyridin-4-yl)benzonitrile (58c)

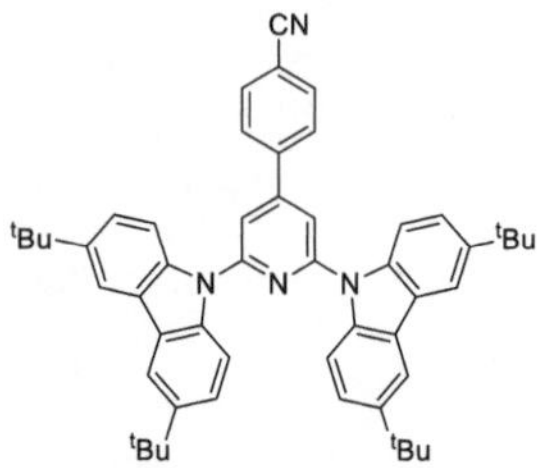

In a sealable vial, 4-(2,6-difluoropyridin-4-yl)benzonitrile (**53c**) (322 mg, 1.49 mmol, 1.00 equiv.), 3,6-di-*tert*-butyl-9*H*-carbazole (874 mg, 3.13 mmol, 2.10 equiv.) and tripotassium phosphate (948 mg, 4.47 mmol, 3.00 equiv.) were evacuated and backfilled with argon three times. Dry dimethyl sulfoxide (10 mL) was added, and the resulting mixture was heated at 130 °C for 72 h. After cooling to room temperature, the reaction mixture was poured into an excess of water (100 mL) and extracted with dichloromethane (3 × 100 mL). The combined organic layers were washed with brine (100 mL), dried over sodium sulfate, and reduced in a vacuum. The crude product was purified by flash column chromatography over silica gel (*c*Hex/CH_2Cl_2, 3:1 to 1:1) to yield 969 mg of the title compound (1.32 mmol, 89%) as a yellow solid.

R_f (*c*Hex/CH_2Cl_2, 1:2) = 0.32.

^{1}H NMR (500 MHz, $CDCl_3$, ppm) δ = 8.14 (d, *J* = 1.9 Hz, 4H, H_{ar}), 8.01 (d, *J* = 8.7 Hz, 4H, H_{ar}), 7.86 (s, 4H, H_{ar}), 7.74 (s, 2H, H_{ar}), 7.48 (dd, *J* = 8.7, 2.0 Hz, 4H, H_{ar}), 1.47 (s, 36H, CH_3).

^{13}C NMR (126 MHz, $CDCl_3$, ppm) δ = 152.9 (C_q), 150.9 (C_q), 144.7 (C_q), 142.6 (C_q), 138.0 (C_q), 133.3 (+, CH), 128.1 (+, CH), 125.0 (C_q), 124.3 (+, CH), 118.5 (+, CH), 116.4 (C_q), 113.4 (C_q), 111.9 (+, CH), 111.5 (+, CH), 34.9 (C_q), 32.1 (+, CH_3).

IR (ATR, ṽ, cm^{-1}) = 2955 (m), 2904 (w), 2867 (w), 1602 (s), 1581 (w), 1544 (m), 1511 (w), 1489 (m), 1473 (s), 1465 (vs), 1451 (vs), 1428 (vs), 1400 (s), 1363 (m), 1339 (w), 1324 (w), 1309 (s), 1298 (vs), 1262 (s), 1227 (m), 1194 (m), 1166 (w), 1139 (w), 1109 (w), 1098 (w), 1085 (w), 1047 (w), 1034 (w), 1018 (w), 901 (w), 881 (m), 841 (w), 820 (w), 810 (s), 803 (vs), 764 (w), 745 (w), 739 (w), 701 (w), 691 (w), 670 (w), 654 (w), 629 (w), 612 (m), 602 (w), 575 (w), 555 (w), 543 (w), 520 (w), 500 (w), 469 (w), 448 (w), 422 (w), 411 (w), 404 (w).

MS (FAB, 3-NBA): m/z (%) = 737 (49) $[M+2H]^+$, 736 (100) $[M+H]^+$, 735 (93) $[M]^+$.

HRMS (FAB, $C_{52}H_{54}N_4$): calcd 735.4421, found 735.4423.

EA ($C_{52}H_{54}N_4$) calc. C: 84.97, H: 7.40, N: 7.62; found C: 84.29, H: 7.41, N: 7.43.

Mp >400 °C.

4-((2r,3s,5s,6s)-2,3,5,6-Tetrakis(3,6-di-*tert*-butyl-9*H*-carbazol-9-yl)pyridin-4-yl)benzonitrile (44)

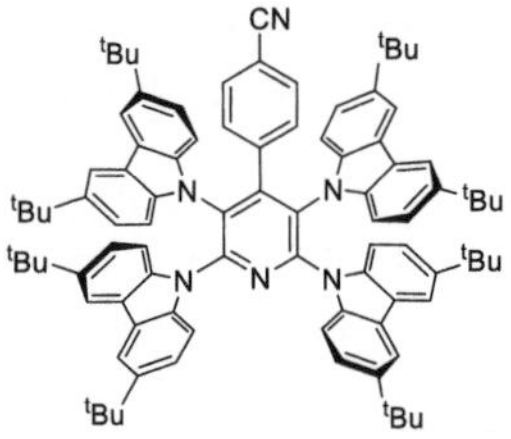

In a sealable vial, 4-(perfluoropyridin-4-yl)benzonitrile (**43**) (217 mg, 860 µmol, 1.00 equiv.), 3,6-di-*tert*-butyl-9*H*-carbazole (986 mg, 3.53 mmol, 4.10 equiv.) and tripotassium phosphate (913 g, 4.30 mmol, 5.00 equiv.) were evacuated and back-filled with argon three times. Dry dimethyl sulfoxide (24 mL) was added, and the resulting mixture was heated at 130 °C for 72 h. After cooling to room temperature, the reaction mixture was poured into an excess of water (150 mL) and extracted with dichloromethane (3 × 100 mL). The combined organic layers were washed with brine (100 mL), dried over sodium sulfate, and reduced in a vacuum. The crude product was purified by flash column chromatography over silica gel (*n*-pentane/EtOAc, 10:1) to yield 953 mg of the title compound (739 µmol, 86%) as a yellow solid.

R_f (*n*-pentane/EtOAc, 10:1) = 0.35.

^{1}H NMR (500 MHz, $CDCl_3$, ppm) δ = 7.56 (dd, J = 22.0, 1.9 Hz, 8H, H_{ar}), 7.10 (d, J = 8.6 Hz, 4H, H_{ar}), 6.99 (d, J = 8.4 Hz, 2H, H_{ar}), 6.92 (d, J = 8.4 Hz, 2H, H_{ar}), 6.81 (dt, J = 8.6, 2.0 Hz, 8H, H_{ar}), 6.59 (d, J = 8.5 Hz, 4H, H_{ar}), 1.32 (d, J = 1.2 Hz, 72H, CH_3).

^{13}C NMR (126 MHz, $CDCl_3$, ppm) δ = 151.0 (C_q), 148.8 (C_q), 143.7 (C_q), 143.3 (C_q), 138.2 (C_q), 137.7 (C_q), 137.2 (C_q), 131.9 (+, CH), 129.2 (+, CH), 126.4 (C_q), 124.3 (C_q), 123.9 (C_q), 122.9 (+, CH), 122.8 (+, CH), 118.2 (C_q), 115.7 (+, CH), 115.2 (+, CH), 112.0 (C_q), 110.5 (+, CH), 109.2 (+, CH), 34.6 (C_q), 32.0 (+, CH_3).

IR (ATR, cm^{-1}) $\tilde{\nu}$ = 2952 (s), 2902 (w), 2866 (w), 1487 (s), 1469 (vs), 1425 (vs), 1404 (m), 1394 (m), 1361 (s), 1323 (m), 1295 (vs), 1262 (vs), 1231 (s), 1201 (w), 1149 (w), 1034 (w), 874 (m), 841 (w), 805 (vs), 755 (w), 739 (w), 613 (s), 499 (w), 424 (w).

MS (ESI): m/z (%) = 1291 (51) $[M+2H]^+$, 1290 (100) $[M+H]^+$, 1289 (97) $[M]^+$.

HRMS (ESI, $C_{92}H_{100}N_6$): calc. 1288.8009, found 1288.7996.

EA ($C_{92}H_{100}N_6$) calc. C: 85.67, H: 7.81, N: 6.52; found C: 85.46, H: 7.71, N: 6.45.
Mp 396 °C.

(2r,3r,4r,5s,6r)-2,3,4,5,6-Pentakis(3,6-di-*tert*-butyl-9*H*-carbazol-9-yl)benzonitrile (9)

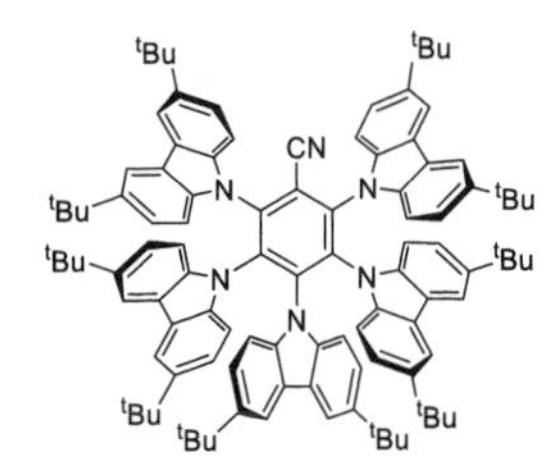

In a sealable vial, 2',3',4',5',6'-pentafluoro-4-carbonitrile (772 mg, 4.00 mmol, 1.00 equiv.), 3,6-di-tert-butyl-9*H*-carbazole (5.59 g, 20.0 mmol, 1.00 equiv.) and tripotassium phosphate (5.94 g, 28.0 mmol, 7.00 equiv.) were evacuated and backfilled with argon three times. Dry dimethyl sulfoxide (16 mL) was added, and the resulting mixture was heated at 110 °C for 16 h. After cooling to room temperature, the reaction mixture was poured into an excess of water (100 mL) and extracted with dichloromethane (3 × 50 mL). The combined organic layers were washed with brine (100 mL), dried over sodium sulfate, and reduced in a vacuum. The crude product was purified by flash column chromatography over silica gel (*c*Hex/CH_2Cl_2, 15:1 to 5:1) to yield 3.59 g of the title compound (2.41 mmol, 60%) as a yellow solid.
R_f (*c*Hex/CH_2Cl_2, 3.5:1) = 0.42.
^{1}H NMR (500 MHz, $CDCl_3$, ppm) δ = 7.58 (t, *J* = 1.3 Hz, 4H, H_{ar}), 7.20 (t, *J* = 2.0 Hz, 6H, H_{ar}), 7.04 (d, *J* = 8.6 Hz, 2H, H_{ar}), 6.96 (s, 8H, H_{ar}), 6.67 (d, *J* = 8.6 Hz, 4H, H_{ar}), 6.58 (dd, *J* = 8.7, 2.0 Hz, 2H, H_{ar}), 6.53 (dd, *J* = 8.6, 1.9 Hz, 4H, H_{ar}), 1.33 (s, 36H, CH_3), 1.23 (s, 36H, CH_3), 1.12 (s, 18H, CH_3).
^{13}C NMR (126 MHz, $CDCl_3$, ppm) δ = 143.6 (C_q), 143.3 (C_q), 143.0 (C_q), 142.8 (C_q), 140.2 (C_q), 138.7 (C_q), 137.8 (C_q), 137.5 (C_q), 136.7 (C_q), 135.9 (C_q), 124.7 (C_q), 124.5 (C_q), 124.5 (+, CH), 124.4 (C_q), 124.1 (C_q), 122.9 (+, CH), 122.4 (+, CH), 121.9 (+, CH), 117.2 (C_q), 116.5 (C_q), 115.8 (+, CH), 115.2 (+, CH), 114.9 (+, CH), 113.6

(C_q), 110.7 (+, CH), 110.1 (+, CH), 109.8 (+, CH), 34.6 (C_q, 4C), 34.4 (C_q, 4C), 34.3 (C_q, 2C), 32.0 (+, CH_3, 12C, ^{t}Bu), 31.9 (+, CH_3, 12C, ^{t}Bu), 31.8 (+, CH_3, 6C, ^{t}Bu).
IR (ATR, cm^{-1}) $\tilde{\nu}$ = 2956 (s), 2925 (m), 2904 (m), 2866 (w), 2850 (w), 1487 (s), 1470 (vs), 1441 (vs), 1392 (w), 1361 (s), 1324 (m), 1295 (vs), 1262 (vs), 1227 (s), 1201 (w), 1150 (w), 1106 (w), 1034 (w), 874 (m), 805 (vs), 764 (w), 737 (w), 611 (s), 557 (w), 499 (w), 469 (w), 421 (m).
MS (ESI): m/z (%) = 1491 (22) [M+2H]$^+$, 1490 (59) [M+H]$^+$, 1489 (100) [M]$^+$.
HRMS (ESI, $C_{107}H_{120}N_6$): calc. 1488.9574, found 1488.9585.
Mp > 400 °C.

(2's,3'r,4's,5's,6's)-2',3',4',5',6'-Pentakis(3,6-di-*tert*-butyl-9*H*-carbazol-9-yl)-[1,1'-biphenyl]-4-carbonitrile (37)

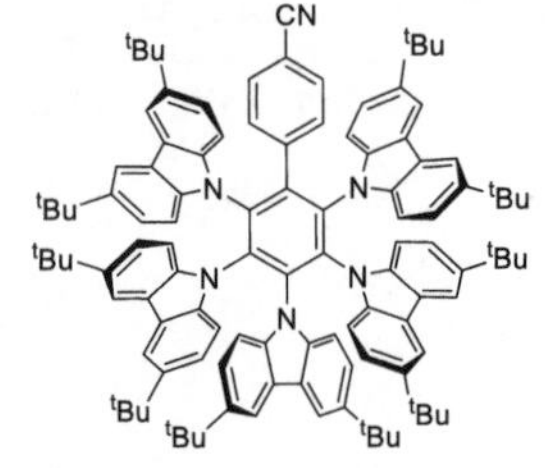

In a sealable vial, 2',3',4',5',6'-pentafluoro-[1,1'-biphenyl]-4-carbonitrile (**36**) (620 mg, 2.30 mmol, 1.00 equiv.), 3,6-di-*tert*-butyl-9*H*-carbazole (3.22 g, 11.5 mmol, 5.00 equiv.) and cesium carbonate (5.25 g, 16.1 mmol, 7.00 equiv.) were evacuated and back-filled with argon three times. Dry dimethylformamide (15 mL) was added, and the resulting mixture was heated at 150 °C for 72 h. After cooling to room temperature, the reaction mixture was poured into an excess of water (100 mL) and extracted with dichloromethane (3 × 50 mL). The combined organic layers were washed with brine (100 mL), dried over sodium sulfate, and reduced in a vacuum. The crude product was purified by flash column chromatography over silica gel (*c*Hex/EtOAc, 70:1 to 60:1) to yield 2.66 g of the title compound (1.70 mmol, 74%) as an off-white solid.
R_f (*c*Hex/EtOAc, 50:1) = 0.28.

^{1}H NMR (500 MHz, $CDCl_3$, ppm) δ = 7.49 (d, J = 1.9 Hz, 4H , H_{ar}), 7.22 (dd, J = 9.2, 2.0 Hz, 6H, H_{ar}), 7.11 (d, J = 8.7 Hz, 2H, H_{ar}), 6.87 – 6.83 (m, 10H, H_{ar}), 6.78 (d, J = 8.6 Hz, 4H, H_{ar}), 6.73 (d, J = 8.5 Hz, 2H, H_{ar}), 6.63 (dd, J = 8.6, 2.0 Hz, 2H, H_{ar}), 6.56 (dd, J = 8.7, 2.0 Hz, 4H, H_{ar}), 1.28 (s, 36H, CH_3), 1.21 (s, 36H, CH_3), 1.16 (s, 18H, CH_3).

^{13}C NMR (126 MHz, $CDCl_3$, ppm) δ = 142.8 (C_q), 142.7 (C_q), 142.4 (C_q), 142.0 (C_q), 139.4 (C_q), 138.5 (C_q), 138.0 (C_q), 137.9 (C_q), 137.5 (C_q), 137.3 (C_q), 135.7 (C_q), 131.2 (+, CH), 129.3 (+, CH), 124.3 (C_q), 124.0 (C_q), 123.6 (C_q), 122.5 (C_q), 122.2 (+, CH), 121.8 (+, CH), 118.5 (C_q), 115.6 (+, CH), 115.1 (C_q), 114.9 (+, CH), 111.0 (C_q), 110.6 (C_q), 110.3 (+, CH), 109.7 (+, CH), 34.6 (C_q, 2C), 34.4 (C_q, 8C), 32.0 (+, CH_3, 12C, ^{t}Bu), 31.9 (+, CH_3, 12C, ^{t}Bu), 31.9 (+, CH_3, 6C, ^{t}Bu).

IR (ATR, cm^{-1}) $\tilde{v}$ = 2952 (s), 2902 (w), 2864 (w), 1486 (s), 1472 (vs), 1446 (vs), 1404 (w), 1392 (w), 1361 (s), 1324 (m), 1295 (vs), 1262 (vs), 1231 (s), 1201 (w), 1150 (m), 1106 (w), 1034 (w), 874 (m), 841 (w), 803 (vs), 766 (w), 739 (m), 611 (s), 596 (w), 562 (s), 496 (w), 469 (w), 418 (m).

MS (ESI): m/z (%) = 1567 (25) $[M+2H]^+$, 1566 (62) $[M+H]^+$, 1565 (100) $[M]^+$.

HRMS (ESI, $C_{113}H_{124}N_6$): calc. 1564.9887, found 1564.9904.

Mp > 400 °C.

5.2.2.2. Tetrazole Precursors

9,9',9'',9''',9''''-((2s,3r,4s,5s,6s)-4'-(2*H*-Tetrazol-5-yl)-[1,1'-biphenyl]-2,3,4,5,6-pentayl)pentakis(3,6-di-*tert*-butyl-9*H*-carbazole) (38)

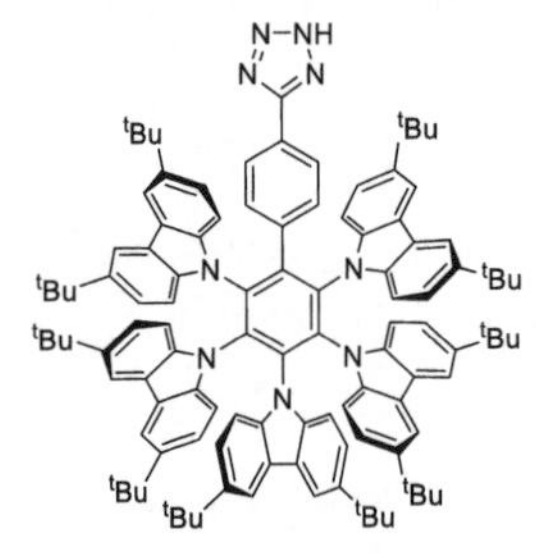

In a sealable vial, (2's,3'r,4's,5's,6's)-2',3',4',5',6'-pentakis(3,6-di-*tert*-butyl-9*H*-carbazol-9-yl)-[1,1'-biphenyl]-4-carbonitrile (**37**) (548 mg, 350 µmol, 1.00 equiv.), sodium azide (68.2 mg, 1.05 mmol, 3.00 equiv.) and ammonium chloride (56.0 mg, 1.05 mmol, 3.00 equiv.) were evacuated and backfilled with argon three times. Dry dimethylformamide (5 mL) was added, and the resulting mixture was heated at 130 °C for 16 h until completion. After cooling to room temperature, the reaction mixture was poured into an excess of 1 M hydrochloric acid (50 mL) and mixed thoroughly. The colorless solid was filtered off, washed several times with water, and thoroughly dried to yield 549 mg of the title compound (341 µmol, 98%) as an off-white solid.

R_f (CH_2Cl_2/CH_3OH, 200:1) = 0.44.

^{1}H NMR (500 MHz, $CDCl_3$, ppm) δ = 7.46 (d, J = 1.6 Hz, 4H, H_{ar}), 7.22 (dd, J = 8.1, 2.0 Hz, 6H, H_{ar}), 7.13 (d, J = 8.6 Hz, 2H, H_{ar}), 7.02 (s, 2H, H_{ar}), 6.94 (d, J = 8.1 Hz, 2H, H_{ar}), 6.85 (dd, J = 8.6, 1.8 Hz, 4H, H_{ar}), 6.82 (d, J = 8.6 Hz, 8H, H_{ar}), 6.64 (dd, J = 8.7, 2.0 Hz, 2H, H_{ar}), 6.56 (dd, J = 8.7, 2.0 Hz, 4H, H_{ar}), 1.24 (s, 36H, CH_3), 1.22 (s, 36H, CH_3), 1.15 (s, 18H, CH_3). Missing signal (1H) NH due to H/D exchange in $CDCl_3$.

^{13}C NMR (126 MHz, $CDCl_3$, ppm) δ = 142.6 (C_q), 142.5 (C_q), 142.3 (C_q), 138.4 (C_q), 138.1 (C_q), 137.5 (C_q), 137.3 (C_q), 135.9 (C_q), 129.6 (+, CH), 126.3 (+, CH), 124.3 (C_q), 124.0 (C_q), 123.6 (C_q), 122.5 (+, CH), 122.2 (+, CH), 121.7 (+, CH), 115.5 (+, CH), 115.0 (C_q), 114.8 (+, CH), 110.7 (+, CH), 110.3 (+, CH), 109.8 (+, CH), 34.5 (C_q,

2C), 34.4 (C_q, 8C), 32.0 (+, CH_3, 12C, ^{t}Bu), 31.9 (+, CH_3, 12C, ^{t}Bu), 31.8 (+, CH_3, 6C, ^{t}Bu).

IR (ATR, cm^{-1}) $\tilde{\nu}$ = 2952 (s), 2901 (w), 2864 (w), 1680 (w), 1656 (w), 1487 (s), 1472 (vs), 1445 (vs), 1411 (w), 1391 (w), 1361 (s), 1326 (m), 1295 (vs), 1262 (vs), 1232 (s), 1201 (w), 1152 (m), 1106 (w), 1095 (w), 1033 (m), 1021 (w), 873 (m), 849 (w), 840 (w), 805 (vs), 754 (w), 738 (m), 656 (w), 611 (s), 591 (w), 560 (s), 493 (w), 470 (w), 422 (m).

MS (ESI): m/z (%) = 1609 (7) $[M+H]^+$, 1608 (5) $[M]^+$. HRMS (ESI, $C_{113}H_{125}N_9$): calc. 1608.0058, found 1608.0061.

Mp 285 °C.

9,9',9'',9''',9''''-((1r,2r,3r,4s,5r)-6-(2*H*-Tetrazol-5-yl)benzene-1,2,3,4,5-pentayl)pentakis(3,6-di-*tert*-butyl-9*H*-carbazole) (33)

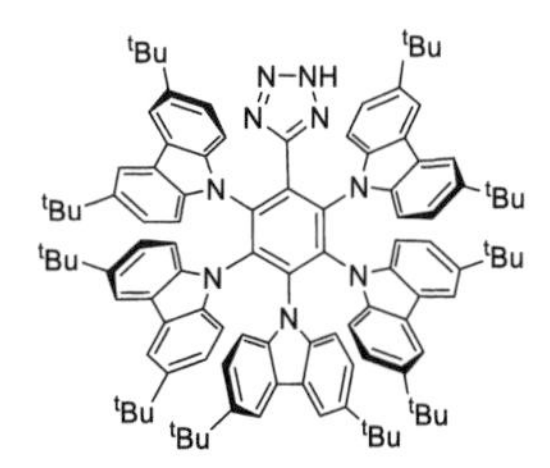

In a sealable vial, (2r,3r,4r,5s,6r)-2,3,4,5,6-pentakis(3,6-di-*tert*-butyl-9*H*-carbazol-9-yl)benzonitrile (**9**) (2.09 g, 1.40 mmol, 1.00 equiv.), sodium azide (273 mg, 4.20 mmol, 3.00 equiv.) and ammonium chloride (225 mg, 4.20 mmol, 3.00 equiv.) were evacuated and backfilled with argon three times. Dry dimethylformamide (10 mL) was added, and the resulting mixture was heated at 130 °C for 16 h until completion. After cooling to room temperature, the reaction mixture was poured into an excess of 1 M hydrochloric acid (50 mL) and mixed thoroughly. The brownish solid was filtered off, washed several times with water, and thoroughly dried to yield 1.70 g of the title compound (1.11 mmol, 79%) as a brownish solid.

R_f (CH_2Cl_2/*c*Hex, 4:1) = 0.29.

^{1}H NMR (500 MHz, $CDCl_3$, ppm) δ = 7.52 (d, *J* = 1.9 Hz, 4H, H_{ar}), 7.23 (dd, *J* = 9.0, 2.0 Hz, 6H, H_{ar}), 7.09 (d, *J* = 8.7 Hz, 2H, H_{ar}), 6.88 (dd, *J* = 8.7, 1.9 Hz, 4H, H_{ar}), 6.82 (dd, *J* = 8.6, 3.4 Hz, 8H, H_{ar}), 6.63 (dd, *J* = 8.6, 2.0 Hz, 2H, H_{ar}), 6.56 (dd, *J* = 8.6, 2.0 Hz, 4H, H_{ar}), 1.28 (s, 36H, CH_3), 1.22 (s, 36H, CH_3), 1.16 (s, 18H, CH_3). Missing signal (1H) NH due to H/D exchange in $CDCl_3$.

^{13}C NMR (126 MHz, $CDCl_3$, ppm) δ = 143.1 (C_q), 143.0 (C_q), 142.7 (C_q), 137.7 (C_q), 137.3 (C_q), 137.1 (C_q), 124.5 (C_q), 124.2 (C_q), 123.8 (C_q), 123.0 (+, CH), 122.3 (+, CH), 121.9 (+, CH), 115.8 (+, CH), 115.2 (C_q), 114.9 (+, CH), 110.6 (+, CH), 110.1 (+, CH), 109.0 (+, CH), 34.6 (C_q, 2C), 34.4 (C_q, 8C), 32.0 (+, CH_3, 12C, ^{t}Bu), 31.9 (+, CH_3, 12C, ^{t}Bu), 31.8 (+, CH_3, 6C, ^{t}Bu).

IR (ATR, cm^{-1}) $\tilde{\nu}$ = 2952 (s), 2902 (m), 2864 (w), 1487 (s), 1472 (vs), 1449 (vs), 1392 (w), 1361 (s), 1324 (m), 1295 (vs), 1262 (vs), 1230 (s), 1201 (w), 1152 (w), 1106 (w), 1034 (m), 1020 (w), 895 (w), 874 (m), 841 (w), 803 (vs), 761 (w), 739 (m), 653 (w), 609 (s), 594 (w), 560 (w), 551 (m), 490 (w), 466 (w), 421 (w).

MS (ESI): m/z (%) = 1533 (59) $[M+H]^+$, 1532 (100) $[M]^+$. HRMS (ESI, $C_{107}H_{121}N_9$): calc. 1531.9745, found 1531.9755.

Mp 208 °C.

9,9',9'',9'''-((2r,3s,5s,6s)-4-(4-(2*H*-Tetrazol-5-yl)phenyl)pyridine-2,3,5,6-tetrayl)tetrakis(3,6-di-*tert*-butyl-9*H*-carbazole) (45)

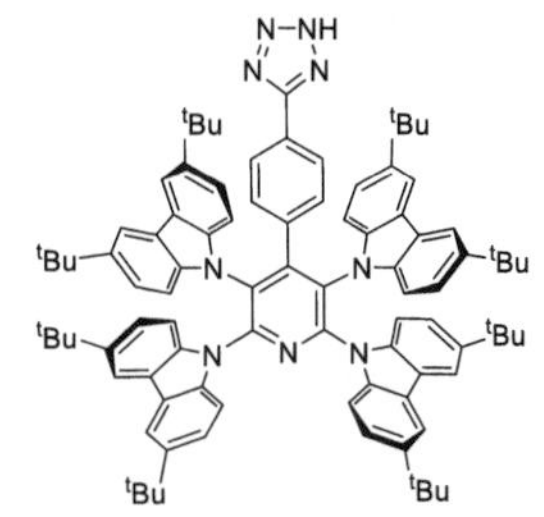

In a sealable vial, 4-((2r,3s,5s,6s)-2,3,5,6-tetrakis(3,6-di-tert-butyl-9*H*-carbazol-9-yl)pyridin-4-yl)benzonitrile (**44**) (774 mg, 600 μmol, 1.00 equiv.), sodium azide (117 mg, 1.80 mmol, 3.00 equiv.) and ammonium chloride (96.0 mg, 1.80 mmol, 3.00 equiv.) were evacuated and backfilled with argon three times. Dry dimethylformamide (10 mL) was added, and the resulting mixture was heated at 130 °C for 16 h until completion. After cooling to room temperature, the reaction mixture was poured into an excess of 1 M hydrochloric acid (50 mL) and mixed thoroughly. The yellow solid was filtered off, washed several times with water, extracted with dichloromethane (3 × 100 mL) and washed with brine (100 mL), reduced in a vacuum, and thoroughly dried to yield 748 mg of the title compound (562 μmol, 94%) as a yellow solid.

^{1}H NMR (500 MHz, $CDCl_3$, ppm) δ = 7.56 (d, *J* = 2.0 Hz, 4H, H_{ar}), 7.51 (d, *J* = 1.9 Hz, 4H, H_{ar}), 7.11 (s, 2H, H_{ar}), 7.05 (dd, *J* = 8.5, 2.2 Hz, 4H, H_{ar}), 7.01 (s, 2H, H_{ar}), 6.80 (td, *J* = 7.8, 6.9, 1.9 Hz, 8H, H_{ar}), 6.64 (d, *J* = 8.6 Hz, 4H, H_{ar}), 1.31 (s, 36H, CH_3), 1.27 (s, 36H, CH_3). Missing signal (1H) NH due to H/D exchange in $CDCl_3$.

^{13}C NMR (126 MHz, $CDCl_3$, ppm) δ = 152.0 (C_q), 148.7 (C_q), 143.6 (C_q), 143.1 (C_q), 137.7 (C_q), 137.4 (C_q), 129.3 (+, CH), 126.9 (+, CH), 126.9 (+, CH), 124.2 (+, CH), 124.0 (+, CH), 122.9 (+, CH), 122.8 (+, CH), 115.7 (+, CH), 115.2 (+, CH), 110.4 (+, CH), 109.3 (+, CH), 34.6 (C_q), 34.6 (C_q), 32.0 (+, CH_3), 31.9 (+, CH_3).

IR (ATR, cm^{-1}) $\tilde{\nu}$ = 2952 (s), 2902 (w), 2866 (w), 1489 (s), 1470 (vs), 1434 (vs), 1392 (w), 1363 (s), 1324 (m), 1295 (vs), 1265 (s), 1252 (m), 1232 (m), 1201 (w), 1149 (w), 1106 (w), 1034 (w), 874 (m), 843 (w), 809 (vs), 761 (w), 739 (m), 686 (w), 666 (w), 656 (w), 613 (m), 599 (w), 575 (w), 550 (w), 470 (w), 465 (w), 422 (w).

MS (ESI): m/z (%) = 1334 (49) $[M+2H]^+$, 1333 (100) $[M+H]^+$, 1332 (94) $[M]^+$.
HRMS (ESI, $C_{92}H_{101}N_9$): calc. 1331.8180, found 1332.8197.
EA ($C_{92}H_{101}N_9$) calc. C: 82.90, H: 7.64, N: 9.46; found C: 82.56, H: 7.50, N: 9.33.
Mp > 400 °C.

9,9'-(4'-(2*H*-Tetrazol-5-yl)-[1,1'-biphenyl]-3,5-diyl)bis(3,6-di-*tert*-butyl-9*H*-carbazole (61a)

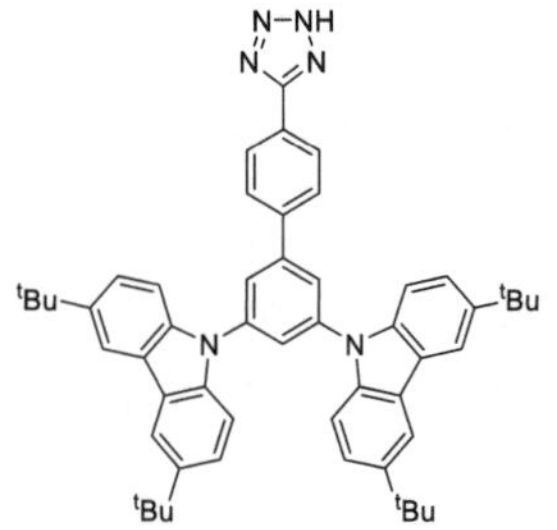

N–NH
N N
tBu
N
N
tBu
tBu
tBu

In a sealable vial, 3',5'-bis(3,6-di-*tert*-butyl-9*H*-carbazol-9-yl)-[1,1'-biphenyl]-4-carbonitrile (**58a**) (1.47 g, 2.00 mmol, 1.00 equiv.), sodium azide (390 mg, 6.00 mmol, 3.00 equiv.) and ammonium chloride (321 mg, 6.00 mmol, 3.00 equiv.) were evacuated and backfilled with argon three times. Dry dimethylformamide (20 mL) was added, and the resulting mixture was heated at 130 °C for 16 h until completion. After cooling to room temperature, the reaction mixture was poured into an excess of 1 M hydrochloric acid (50 mL) and mixed thoroughly. The colorless solid was filtered off, washed several times with water, extracted with dichloromethane (3 × 100 mL) and washed with brine (100 mL), reduced in a vacuum, and thoroughly dried to yield 1.60 g of the title compound (2.06 mmol, quant.) as a colorless solid.

^1H NMR (500 MHz, THF-d_8, ppm) δ = 8.25 – 8.21 (m, 6H, H_{ar}), 8.12 (d, *J* = 1.9 Hz, 2H, H_{ar}), 8.07 (d, *J* = 8.4 Hz, 2H, H_{ar}), 7.91 (s, 1H, H_{ar}), 7.57 (d, *J* = 8.6 Hz, 4H, H_{ar}), 7.51 (dd, *J* = 8.6, 1.9 Hz, 4H, H_{ar}), 1.46 (s, 36H, CH_3). Missing signal (1H) NH due to H/D exchange in THF-d_8.

^{13}C NMR (126 MHz, THF-d_8, ppm) δ = 144.1 (C_q), 141.6 (C_q), 140.2 (C_q), 128.9 (+, CH), 128.6 (+, CH), 124.9 (+, CH), 124.8 (+, CH), 124.7 (+, CH), 124.5 (+, CH), 117.4 (+, CH), 110.3 (+, CH), 35.6 (C_q), 32.6 (+, CH_3), 32.5 (+, CH_3).

IR (ATR, ṽ, cm^{-1}) = 2959 (s), 2902 (w), 2866 (w), 1619 (w), 1599 (m), 1587 (s), 1565 (w), 1502 (m), 1487 (s), 1472 (vs), 1452 (vs), 1425 (w), 1392 (m), 1363 (s), 1343 (w), 1323 (s), 1295 (vs), 1261 (s), 1254 (s), 1230 (vs), 1203 (w), 1171 (w), 1035 (w), 898 (w), 877 (s), 841 (s), 810 (vs), 754 (m), 739 (m), 720 (w), 696 (w), 683 (w), 654 (w), 613 (s), 562 (w), 470 (w), 421 (m).

MS (FAB, 3-NBA): m/z (%) = 779 (9) $[M+2H]^+$, 778 (18) $[M+H]^+$,777 (27) $[M]^+$.

HRMS (FAB, $C_{53}H_{56}N_6$): calcd 776.4561, found 776.4563.

Mp 347 °C.

9,9'-(5-(6-(2*H*-Tetrazol-5-yl)pyridin-3-yl)-1,3-phenylene)bis(3,6-di-*tert*-butyl-9*H*-carbazole) (61b)

In a sealable vial, 5-(3,5-bis(3,6-di-*tert*-butyl-9*H*-carbazol-9-yl)phenyl)picolinonitrile (**58b**) (440 mg, 600 µmol, 1.00 equiv.), sodium azide (117 mg, 1.80 mmol, 3.00 equiv.) and ammonium chloride (96.0 mg, 1.80 mmol, 3.00 equiv.) were evacuated and backfilled with argon three times. Dry dimethylformamide (5 mL) was added, and the resulting mixture was heated at 130 °C for 16 h until completion. After cooling to room temperature, the reaction mixture was poured into an excess of 1 M hydrochloric acid (50 mL) and mixed thoroughly. The off-white solid was filtered off, washed several times with water, and thoroughly dried to yield 465 mg of the title compound (598 µmol, quant.) as an off-white solid.

^{1}H NMR (500 MHz, $CDCl_3$, ppm) δ = 9.12 (s, 1H, H_{ar}), 8.49 (d, *J* = 8.2 Hz, 1H, H_{ar}), 8.23 (d, *J* = 8.2 Hz, 1H, H_{ar}), 8.10 (s, 4H, H_{ar}), 7.91 – 7.84 (m, 3H, H_{ar}), 7.50 – 7.40 (m, 9H, H_{ar}), 1.40 (s, 36H, CH_3).

^{13}C NMR (126 MHz, $CDCl_3$, ppm) δ = 148.2 (C_q), 142.5 (C_q), 141.1 (C_q), 139.4 (C_q), 138.7 (C_q), 138.0 (C_q), 136.8 (C_q), 124.6 (+, CH), 124.1 (+, CH), 123.9 (+, CH), 123.6 (+, CH), 123.2 (+, CH), 116.6 (+, CH), 109.1 (+, CH), 34.8 (C_q), 32.0 (+, CH_3), 29.7 (C_q).

IR (ATR, cm^{-1}) $\tilde{\nu}$ = 2958 (m), 1599 (w), 1588 (m), 1487 (s), 1472 (vs), 1452 (vs), 1392 (w), 1363 (s), 1322 (m), 1295 (vs), 1261 (s), 1231 (s), 1035 (w), 878 (s), 810 (vs), 741 (w), 720 (w), 696 (w), 684 (w), 654 (w), 613 (s), 469 (w), 419 (m), 398 (w).

MS (ESI): m/z (%) = 778 (64) $[M+H]^+$, 777 (100) $[M]^+$. HRMS (ESI, $C_{52}H_{55}N_7$): calcd 777.4519, found 777.4508.

Mp 320 °C.

9,9'-(4-(4-(2*H*-Tetrazol-5-yl)phenyl)pyridine-2,6-diyl)bis(3,6-di-*tert*-butyl-9*H*-carbazole) (61c)

In a sealable vial, 4-(2,6-bis(3,6-di-*tert*-butyl-9*H*-carbazol-9-yl)pyridin-4-yl)benzonitrile (**58c**) (809 mg, 1.10 mmol, 1.00 equiv.), sodium azide (215 mg, 3.30 mmol, 3.00 equiv.) and ammonium chloride (177 mg, 3.30 mmol, 3.00 equiv.) were evacuated and backfilled with argon three times. Dry dimethylformamide (20 mL) was added, and the resulting mixture was heated at 130 °C for 16 h until completion. After cooling to room temperature, the reaction mixture was poured into an excess of 1 M hydrochloric acid (50 mL) and mixed thoroughly. The yellow solid was filtered off, washed several times with water, extracted with dichloromethane (3 × 100 mL) and washed with brine (100 mL), reduced in a vacuum, and thoroughly dried to yield 854 mg of the title compound (1.10 mmol, quant.) as a yellow solid.

^{1}H NMR (500 MHz, $CDCl_3$, ppm) δ = 8.27 (d, J = 8.0 Hz, 2H, H_{ar}), 8.14 (d, J = 2.0 Hz, 4H, H_{ar}), 8.02 (d, J = 8.7 Hz, 4H, H_{ar}), 7.92 (d, J = 8.4 Hz, 2H, H_{ar}), 7.81 (s, 2H, H_{ar}), 7.47 (dd, J = 8.8, 2.0 Hz, 4H, H_{ar}), 1.46 (s, 36H, CH_3). Missing signal (1H) NH due to H/D exchange in $CDCl_3$.

^{13}C NMR (126 MHz, $CDCl_3$, ppm) δ = 152.8 (C_q), 144.5 (C_q), 138.0 (C_q), 128.3 (+, CH), 128.2 (+, CH), 124.9 (C_q), 124.3 (+, CH), 116.3 (+, CH), 111.9 (+, CH), 111.7 (+, CH), 34.9 (C_q), 32.1 (+, CH_3).

IR (ATR, cm^{-1}) $\tilde{\nu}$ = 2952 (m), 2902 (w), 2866 (w), 1599 (vs), 1575 (w), 1543 (s), 1489 (m), 1463 (vs), 1449 (vs), 1434 (vs), 1417 (vs), 1408 (vs), 1361 (s), 1310 (s), 1295 (vs), 1261 (vs), 1227 (s), 1193 (s), 1162 (w), 1106 (w), 1096 (w), 1085 (w), 1034 (m), 1021 (w), 994 (w), 921 (w), 901 (w), 875 (s), 840 (s), 810 (vs), 764 (m), 752 (m), 739 (m), 704 (w), 691 (m), 681 (m), 669 (w), 654 (m), 640 (w), 612 (vs), 599 (m), 574 (w), 551 (w), 524 (w), 509 (w), 501 (w), 469 (m), 453 (w), 448 (w), 421 (m), 394 (w), 377 (w).

MS (FAB, 3-NBA): m/z (%) = 780 (44) $[M+H]^+$, 779 (100) $[M]^+$. HRMS (FAB, $C_{52}H_{55}N_7$): calcd 778.4592, found 778.4590.

EA ($C_{52}H_{55}N_7$) calc. C: 80.27, H: 7.13, N: 12.60; found C: 78.88, H: 7.00, N: 12.22.

Mp 267 °C.

9,9'-(5-(2*H*-Tetrazol-5-yl)-1,3-phenylene)bis(3,6-di-*tert*-butyl-9*H*-carbazole) (62)

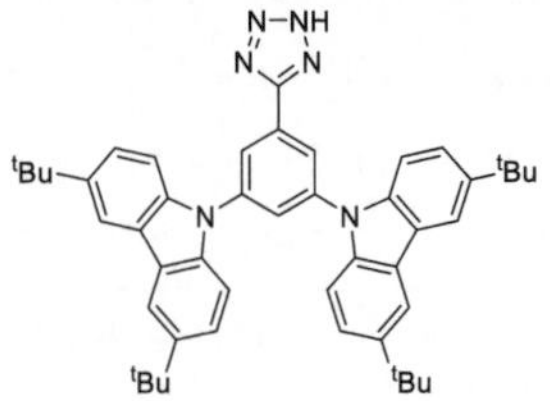

In a sealable vial, 3,5-bis(3,6-di-*tert*-butyl-9*H*-carbazol-9-yl)benzonitrile (**60**) (1.97 g, 3.00 mmol, 1.00 equiv.), sodium azide (585 mg, 9.00 mmol, 3.00 equiv.) and ammonium chloride (481 mg, 9.00 mmol, 3.00 equiv.) were evacuated and backfilled with argon three times. Dry dimethylformamide (20 mL) was added, and the resulting mixture was heated at 130 °C for 16 h until completion. After cooling to room temperature, the reaction mixture was poured into an excess of 1 M hydrochloric acid (50 mL) and mixed thoroughly. The colorless solid was filtered off, washed several times with water, extracted with dichloromethane (3 × 100 mL) and washed with brine (100 mL), reduced in a vacuum, and thoroughly dried to yield 2.19 g of the title compound (3.12 mmol, quant.) as a colorless solid.

^{1}H NMR (500 MHz, $CDCl_3$, ppm) δ = 8.44 (d, J = 2.0 Hz, 1H, H_{ar}), 8.16 (d, J = 1.1 Hz, 2H, H_{ar}), 8.08 (d, J = 1.3 Hz, 2H, H_{ar}), 7.98 – 7.92 (m, 1H, H_{ar}), 7.86 (s, 1H, NH), 7.54 – 7.45 (m, 7H, H_{ar}), 7.33 (d, J = 8.5 Hz, 2H, H_{ar}), 1.47 (s, 18H, CH_3), 1.45 (s, 18H, CH_3).

^{13}C NMR (126 MHz, $CDCl_3$, ppm) δ = 143.8 (C_q), 142.4 (C_q), 140.9 (C_q), 138.8 (C_q), 138.2 (C_q), 124.1 (+, CH), 124.0 (+, CH), 123.7 (+, CH), 123.5 (+, CH), 123.2 (+, CH), 116.6 (+, CH), 116.3 (+, CH), 110.1 (+, CH), 109.3 (+, CH), 34.9 (C_q), 34.8 (C_q), 32.2 (+, CH_3), 32.1 (+, CH_3).

IR (ATR, $\tilde{\nu}$, cm^{-1}) = 2952 (s), 2901 (w), 2864 (w), 1647 (m), 1594 (m), 1475 (vs), 1391 (w), 1361 (s), 1322 (m), 1295 (vs), 1262 (vs), 1231 (s), 1201 (w), 1173 (w), 1103 (w), 1034 (w), 922 (w), 899 (w), 877 (s), 839 (w), 812 (vs), 764 (w), 741 (w), 730 (w), 717 (w), 690 (m), 615 (vs), 467 (w), 422 (m), 382 (w).

MS (ESI): m/z (%) = 701 (2) $[M+H]^+$, 700 (4) $[M]^+$. HRMS (ESI, $C_{47}H_{52}N_6$): calcd 700.4253, found 700.4243.

Mp 232 °C.

5.2.2.3. Arylations of the Cz-Tetrazoles **38** and **33**

9,9',9'',9''',9''''-((2s,3r,4s,5s,6s)-4'-(2-(3,5-Di-*tert*-butylphenyl)-2*H*-tetrazol-5-yl)-[1,1'-biphenyl]-2,3,4,5,6-pentayl)pentakis(3,6-di-*tert*-butyl-9*H*-carbazole) (39a)

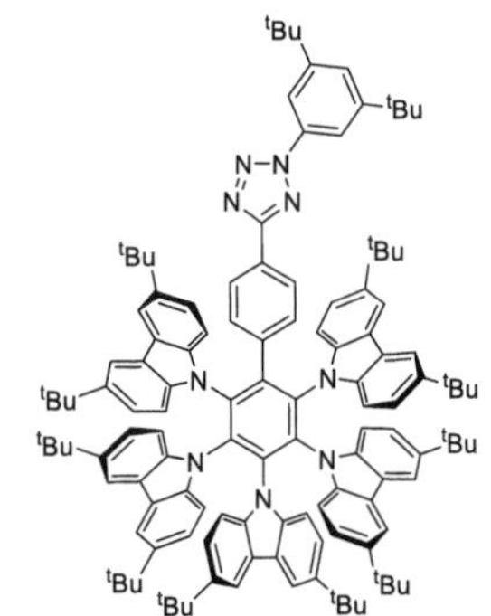

A sealable vial was charged with 9,9',9'',9''',9''''-((2s,3r,4s,5s,6s)-4'-(2*H*-tetrazol-5-yl)-[1,1'-biphenyl]-2,3,4,5,6-pentayl)pentakis(3,6-di-*tert*-butyl-9*H*-carbazole) (**38**) (161 mg, 100 µmol, 1.00 equiv.), (3,5-di-*tert*-butylphenyl)boronic acid (58.6 mg, 250 µmol, 2.50 equiv.), potassium carbonate (21.0 mg, 150 µmol, 1.50 equiv.), and Cu(TMEDA) (5.60 mg, 12.0 µmol, 0.12 equiv.). Dichloromethane (5 mL) was added, and the resulting mixture was stirred under an O_2 atmosphere at room temperature for 16 h. Subsequently, a 10% aqueous ammonia solution (100 mL) was added, and the mixture was extracted with dichloromethane (3 × 100 mL). The combined organic layers were washed with brine (100 mL), dried over sodium sulfate, and reduced in a vacuum. The crude product was purified by flash column chromatography over silica gel (*c*Hex/CH_2Cl_2, 6:1 to 3:1) to yield 133 mg of the title compound (74.0 µmol, 74%) as a light brown solid.

R_f (CH_2Cl_2/*c*Hex, 3:1) = 0.23.

^{1}H NMR (500 MHz, $CDCl_3$, ppm) δ = 7.76 (d, *J* = 1.8 Hz, 1H, H_{ar}), 7.46 (dt, *J* = 5.2, 1.3 Hz, 4H, H_{ar}), 7.45 (t, *J* = 1.8 Hz, 1H, H_{ar}), 7.40 (d, *J* = 8.6 Hz, 2H, H_{ar}), 7.21 (dd, *J* = 8.6, 2.0 Hz, 6H, H_{ar}), 7.16 (d, *J* = 8.6 Hz, 2H, H_{ar}), 6.99 (d, *J* = 8.5 Hz, 2H, H_{ar}), 6.90 – 6.80 (m, 13H, H_{ar}), 6.65 (dd, *J* = 8.7, 2.0 Hz, 2H, H_{ar}), 6.56 (dd, *J* = 8.7, 2.0 Hz, 4H, H_{ar}), 1.31 (s, 18H, CH_3), 1.25 (s, 36H, CH_3), 1.22 (s, 36H, CH_3), 1.15 (s, 18H, CH_3).

^{13}C NMR (126 MHz, $CDCl_3$) δ 164.4 (C_q), 152.8 (C_q), 143.5 (C_q), 142.5 (C_q), 142.3 (C_q), 142.1 (C_q), 138.2 (C_q), 138.1 (C_q), 138.0 (C_q), 137.6 (C_q), 137.1 (C_q), 136.9 (C_q), 136.6 (C_q), 136.1 (C_q), 129.2 (+, CH), 126.3 (+, CH), 126.1 (C_q), 124.3 (+, CH), 123.9 (+, CH), 123.6 (+, CH), 122.4 (+, CH), 122.2 (+, CH), 121.6 (C_q), 115.4 (+, CH),

115.0 (+, CH), 114.7 (+, CH), 114.4 (C_q), 110.8 (+, CH), 110.4 (+, CH), 109.9 (+, CH), 35.3 (C_q, 12C), 34.5 (+, CH_3, 3C, ^{t}Bu), 34.4 (+, CH_3, 3C, ^{t}Bu), 32.0 (+, CH_3, 12C, ^{t}Bu), 31.9 (+, CH_3, 12C, ^{t}Bu), 31.4 (+, CH_3, 6C, ^{t}Bu).
IR (ATR, cm^{-1}) $\tilde{v}$ = 2952 (vs), 2902 (m), 2866 (m), 1486 (s), 1470 (vs), 1449 (vs), 1412 (m), 1392 (w), 1361 (s), 1324 (m), 1295 (vs), 1262 (vs), 1231 (s), 1201 (m), 1150 (m), 1106 (w), 1034 (m), 1017 (w), 874 (s), 850 (w), 841 (w), 802 (vs), 778 (w), 761 (w), 751 (w), 738 (m), 707 (w), 688 (w), 653 (w), 609 (s), 592 (w), 558 (s), 490 (w), 463 (w), 421 (m), 375 (w).
MS (ESI): m/z (%) = 1798 (70) $[M+2H]^+$, 1797 (100) $[M+H]^+$, 1796 (70) $[M]^+$.
HRMS (ESI, $C_{127}H_{145}N_9$): calcd 1796.1623, found 1796.1653.
Mp 248 °C.

9,9',9'',9''',9''''-((2s,3r,4s,5s,6s)-4'-(2-(3,4,5-Trimethoxyphenyl)-2*H*-tetrazol-5-yl)-[1,1'-biphenyl]-2,3,4,5,6-pentayl)pentakis(3,6-di-*tert*-butyl-9*H*-carbazole) (39b)

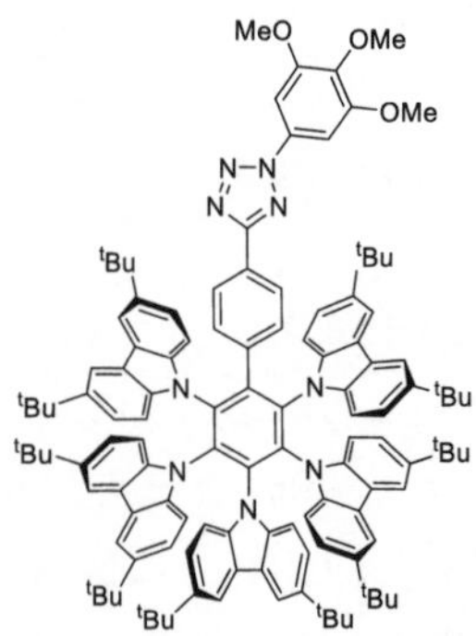

A sealable vial was charged with 9,9',9'',9''',9''''-((2s,3r,4s,5s,6s)-4'-(2*H*-tetrazol-5-yl)-[1,1'-biphenyl]-2,3,4,5,6-pentayl)pentakis(3,6-di-*tert*-butyl-9*H*-carbazole) (**38**) (161 mg, 100 µmol, 1.00 equiv.), (3,4,5-trimethoxyphenyl)boronic acid (53.0 mg, 250 µmol, 2.50 equiv.), potassium carbonate (21.0 mg, 150 µmol, 1.50 equiv.), and Cu(TMEDA) (5.60 mg, 12.0 µmol, 0.12 equiv.). Dichloromethane (5 mL) was added, and the resulting mixture was stirred under an O_2 atmosphere at room temperature for 16 h. Subsequently, a 10% aqueous ammonia solution (100 mL) was added, and the mixture was extracted with dichloromethane (3 × 100 mL). The combined organic layers were washed with brine (100 mL), dried over sodium sulfate, and reduced in a vacuum.

The crude product was purified by flash column chromatography over silica gel (cHex/CH_2Cl_2, 3:2 to 1:1) to yield 125 mg of the title compound (70.6 µmol, 71%) as a yellow-brownish solid.

R_f (cHex/CH_2Cl_2, 1:1) = 0.28.

^{1}H-NMR (500 MHz, $CDCl_3$, ppm) δ = 7.46 (t, J = 1.2 Hz, 4H, H_{ar}), 7.38 (d, J = 8.5 Hz, 2H, H_{ar}), 7.23 – 7.20 (m, 8H, H_{ar}), 7.15 (d, J = 8.6 Hz, 2H, H_{ar}), 6.99 (d, J = 8.6 Hz, 2H, H_{ar}), 6.85 (d, J = 2.1 Hz, 7H, H_{ar}), 6.83 (d, J = 8.7 Hz, 5H, H_{ar}), 6.64 (dd, J = 8.7, 2.0 Hz, 2H, H_{ar}), 6.56 (dd, J = 8.7, 2.0 Hz, 4H, H_{ar}) 3.87 (s, 6H, CH_3), 3.84 (s, 3H, CH_3), 1.25 (s, 36H, CH_3), 1.22 (s, 36H, CH_3), 1.15 (s, 18H, CH_3).

^{13}C-NMR (126 MHz, $CDCl_3$, ppm) δ = 164.5 (C_q), 153.8 (C_q), 142.5 (C_q), 142.3 (C_q), 142.2 (C_q), 138.2 (C_q), 138.1 (C_q), 137.6 (C_q), 137.1 (C_q), 136.0 (C_q), 129.3 (+, CH), 126.3 (+, CH), 125.8 (C_q), 124.3 (C_q), 123.9 (C_q), 123.6 (C_q), 122.4 (+,CH), 122.2 (+, CH), 121.7 (+, CH), 115.4 (+,CH), 115.0 (C_q), 114.8 (+, CH), 110.6 (+, CH), 110.4 (+, CH), 109.9 (+, CH), 97.3 (+, CH), 61.2 (+, CH_3, 1C, OMe), 56.5 (+, CH_3, 2C, OMe), 34.5 (C_q), 34.4 (C_q), 32.0 (+, CH_3, 12C, ^{t}Bu), 31.9 (+, CH_3, 12C, ^{t}Bu), 31.8 (+, CH_3, 6C, ^{t}Bu).

IR (ATR, cm^{-1}) $\tilde{\nu}$ = 2952 (s), 2902 (m), 2864 (m), 1608 (w), 1487 (s), 1470 (vs), 1449 (vs), 1424 (s), 1392 (m), 1361 (s), 1324 (m), 1293 (vs), 1262 (vs), 1230 (vs), 1203 (m), 1150 (m), 1128 (s), 1106 (s), 1034 (m), 1014 (m), 929 (w), 922 (w), 873 (s), 849 (w), 840 (w), 833 (w), 802 (vs), 775 (w), 762 (w), 749 (w), 738 (s), 717 (w), 688 (w), 679 (w), 653 (w), 609 (s), 592 (m), 560 (s), 524 (w), 490 (w), 479 (w), 466 (m), 452 (w), 441 (w), 422 (m), 395 (w), 385 (w).

MS (ESI): m/z (%) = 1776 (66) $[M+2H]^+$, 1775 (99) $[M+H]^+$, 1774 (73) $[M]^+$. HRMS (ESI, $C_{122}H_{135}N_9O_3$): calcd 1774.0688, found 1774.0712.

EA ($C_{122}H_{135}N_9O_3$) calcd: C: 82.53, H: 7.66, N: 7.10; found: C: 82.57, H: 7.87, N: 6.64.

Mp 332 °C.

9,9',9'',9''',9''''-((2s,3r,4s,5s,6s)-4'-(2-(4-(3,6-Di-*tert*-butyl-9*H*-carbazol-9-yl)phenyl)-2*H*-tetrazol-5-yl)-[1,1'-biphenyl]-2,3,4,5,6-pentayl)pentakis(3,6-di-*tert*-butyl-9*H*-carbazole) (39c)

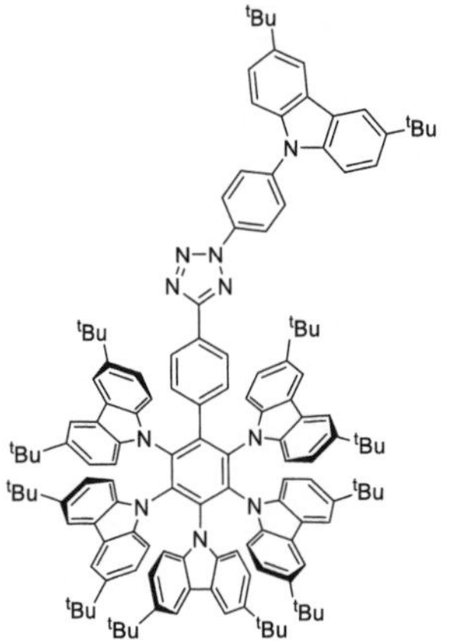

A sealable vial was charged with 9,9',9'',9''',9''''-((2s,3r,4s,5s,6s)-4'-(2*H*-tetrazol-5-yl)-[1,1'-biphenyl]-2,3,4,5,6-pentayl)pentakis(3,6-di-*tert*-butyl-9*H*-carbazole) (**38**) (161 mg, 100 µmol, 1.00 equiv.), (4-(3,6-di-*tert*-butyl-9*H*-carbazol-9-yl)phenyl)boronic acid (84.9 mg, 250 µmol, 2.50 equiv.), potassium carbonate (21.0 mg, 150 µmol, 1.50 equiv.), and Cu(TMEDA) (5.60 mg, 12.0 µmol, 0.12 equiv.). Dichloromethane (5 mL) was added, and the resulting mixture was stirred under an O_2 atmosphere at room temperature for 16 h. Subsequently, a 10% aqueous ammonia solution (100 mL) was added, and the mixture was extracted with dichloromethane (3 × 100 mL). The combined organic layers were washed with brine (100 mL), dried over sodium sulfate, and reduced in a vacuum. The crude product was purified by flash column chromatography over silica gel (*c*Hex/CH_2Cl_2, 3:1) to yield 83.9 mg of the title compound (42.7 mmol, 43%) as a red solid.

R_f (*c*Hex/CH_2Cl_2, 1:1) = 0.37.

^{1}H NMR (500 MHz, $CDCl_3$, ppm) δ = 8.17 (d, *J* = 8.9 Hz, 2H, H_{ar}), 8.12 (d, *J* = 1.9 Hz, 2H, H_{ar}), 7.67 (d, *J* = 8.9 Hz, 2H, H_{ar}), 7.47 (t, *J* = 1.3 Hz, 4H, H_{ar}), 7.45 (dd, *J* = 8.7, 1.9 Hz, 2H, H_{ar}), 7.39 (d, *J* = 8.6 Hz, 2H, H_{ar}), 7.33 (d, *J* = 8.6 Hz, 2H, H_{ar}), 7.22 (dd, *J* = 7.8, 1.9 Hz, 6H, H_{ar}), 7.15 (d, *J* = 8.6 Hz, 2H, H_{ar}), 7.01 – 6.97 (m, 2H, H_{ar}), 6.87 (t, *J* = 1.6 Hz, 6H, H_{ar}), 6.86 (d, *J* = 8.7 Hz, 6H, H_{ar}), 6.65 (dd, *J* = 8.7, 2.0 Hz, 2H, H_{ar}), 6.57 (dd, *J* = 8.6, 2.0 Hz, 4H, H_{ar}), 1.45 (s, 18H, CH_3), 1.26 (s, 36H, CH_3), 1.23 (s, 36H, CH_3), 1.16 (s, 18H, CH_3).

^{13}C NMR (126 MHz, $CDCl_3$, ppm) δ = 143.6 (C_q), 142.5 (C_q), 142.3 (C_q), 138.5 (C_q), 138.2 (C_q), 137.6 (C_q), 136.1 (C_q), 129.3 (+, CH), 127.5 (+, CH), 126.2 (+, CH), 124.3 (C_q), 124.0 (C_q), 123.8 (+, CH), 123.6 (C_q), 122.4 (+, CH), 122.2 (+, CH), 121.7 (+, CH), 121.2 (+, CH), 116.5 (+, CH), 115.4 (+, CH), 115.0 (C_q), 114.8 (+, CH), 110.8 (+, CH), 110.5 (+, CH), 110.4 (+, CH), 109.9 (+, CH), 109.2 (+, CH), 34.9 (C_q), 34.5 (C_q), 34.4 (C_q), 32.2 (C_q), 32.1 (+, CH_3, ^{t}Bu), 32.0 (+, CH_3, ^{t}Bu), 31.9 (+, CH_3, 12C, ^{t}Bu), 31.8 (+, CH_3, 12C, ^{t}Bu), 31.7 (C_q, 6C, ^{t}Bu).

IR (ATR, cm^{-1}) $\tilde{\nu}$ = 2955 (s), 2902 (w), 2866 (w), 1514 (w), 1487 (s), 1472 (vs), 1452 (s), 1414 (w), 1392 (w), 1363 (s), 1324 (m), 1295 (vs), 1262 (vs), 1232 (s), 1203 (w), 1152 (w), 1106 (w), 1034 (w), 1011 (w), 874 (m), 843 (w), 805 (vs), 749 (w), 739 (w), 653 (w), 611 (s), 594 (w), 560 (w), 490 (w), 470 (w), 422 (w).

MS (ESI): m/z (%) = 1963 (76) $[M+2H]^+$, 1962 (100) $[M+H]^+$, 1961 (64) $[M]^+$.

HRMS (ESI, $C_{139}H_{152}N_{10}$): calcd 1961.2201, found 1961.2250.

Mp 323 °C.

9,9',9'',9''',9''''-((2s,3r,4s,5s,6s)-4'-(2-(4-(Trifluoromethyl)phenyl)-2*H*-tetrazol-5-yl)-[1,1'-biphenyl]-2,3,4,5,6-pentayl)pentakis(3,6-di-*tert*-butyl-9*H*-carbazole) (39d)

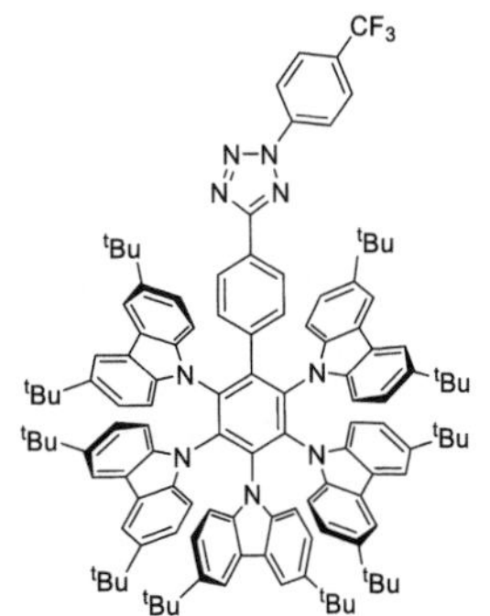

A sealable vial was charged with 9,9',9'',9''',9''''-((2s,3r,4s,5s,6s)-4'-(2*H*-tetrazol-5-yl)-[1,1'-biphenyl]-2,3,4,5,6-pentayl)pentakis(3,6-di-*tert*-butyl-9*H*-carbazole) (**38**) (161 mg, 100 µmol, 1.00 equiv.), (4-(trifluoromethyl)phenyl)boronic acid (47.5 mg, 250 µmol, 2.50 equiv.), potassium carbonate (21.0 mg, 150 µmol, 1.50 equiv.), and Cu(TMEDA) (5.60 mg, 12.0 µmol, 0.12 equiv.). Dichloromethane (5 mL) was added, and the resulting mixture was stirred under an O_2 atmosphere at room temperature for 16 h. Subsequently, a

10% aqueous ammonia solution (100 mL) was added, and the mixture was extracted with dichloromethane (3 × 100 mL). The combined organic layers were washed with brine (100 mL), dried over sodium sulfate, and reduced in a vacuum. The crude product was purified by flash column chromatography over silica gel (*c*Hex/CH_2Cl_2, 4:1) to yield 65.3 mg of the title compound (37.2 µmol, 37%) as a pale-yellow solid.

R_f (*c*Hex/CH_2Cl_2, 2:1) = 0.28.

^{1}H NMR (500 MHz, $CDCl_3$, ppm) δ = 8.11 (d, *J* = 8.3 Hz, 2H, H_{ar}), 7.73 (d, *J* = 8.6 Hz, 2H, H_{ar}), 7.46 – 7.45 (m, 4H, H_{ar}), 7.36 (d, *J* = 8.5 Hz, 2H, H_{ar}), 7.22 (dd, *J* = 8.4, 2.0 Hz, 6H, H_{ar}), 7.15 (d, *J* = 8.6 Hz, 2H, H_{ar}), 6.99 (d, *J* = 8.5 Hz, 2H, H_{ar}), 6.87 – 6.83 (m, 12H, H_{ar}), 6.64 (dd, *J* = 8.7, 2.0 Hz, 2H, H_{ar}), 6.56 (dd, *J* = 8.7, 2.0 Hz, 4H, H_{ar}), 1.24 (s, 36H, CH_3), 1.22 (s, 36H, CH_3), 1.16 (s, 18H, CH_3).

^{13}C NMR (126 MHz, $CDCl_3$, ppm) δ = 142.9 (C_q), 142.5 (C_q), 142.4 (C_q), 142.2 (C_q), 138.1 (C_q), 138.1 (C_q), 137.6 (C_q), 137.2 (C_q), 136.0 (C_q), 129.3 (+, CH), 127.1 (+, CH), 126.3 (+, CH), 125.4 (C_q), 124.3 (C_q), 124.0 (C_q), 123.6 (C_q), 122.4 (+, CH), 122.2 (+, CH), 121.7 (+, CH), 119.9 (+, CH), 115.4 (+, CH), 115.0 (C_q), 114.8 (+, CH), 110.8 (+, CH), 110.4 (+, CH), 109.9 (+, CH), 34.5 (C_q), 34.4 (C_q), 32.0 (+, CH_3, 24C, ^{t}Bu), 31.9 (+, CH_3, 6C, ^{t}Bu).

^{19}F NMR (376 MHz, $CDCl_3$, ppm) δ = -59.5 – -65.0 (m, CF_3).

IR (ATR, cm^{-1}) ṽ = 2953 (s), 2904 (w), 2866 (w), 1487 (s), 1472 (vs), 1449 (vs), 1414 (w), 1392 (w), 1363 (s), 1323 (vs), 1295 (vs), 1264 (vs), 1231 (s), 1214 (w), 1203 (w), 1170 (s), 1150 (m), 1132 (s), 1106 (m), 1058 (m), 1034 (m), 1011 (m), 989 (w), 875 (m), 847 (m), 803 (vs), 749 (w), 738 (m), 688 (w), 609 (vs), 592 (m), 560 (s), 484 (m), 469 (m), 422 (m), 395 (w), 387 (w).

MS (ESI): m/z (%) = 1754 (66) $[M+2H]^+$, 1753 (100) $[M+H]^+$, 1752 (74) $[M]^+$.

HRMS (ESI, $C_{120}H_{128}N_9F_3$): calcd 1752.0245, found 1752.0280.

EA ($C_{122}H_{135}N_9O_3$) calcd: C: 82.20, H: 7.36, N: 7.19; found: C: 82.43, H: 7.54, N: 6.72.

Mp 252 °C.

9,9',9'',9''',9''''-((2s,3r,4s,5s,6s)-4'-(2-(1*H*-Indol-5-yl)-2*H*-tetrazol-5-yl)-[1,1'-biphenyl]-2,3,4,5,6-pentayl)pentakis(3,6-di-*tert*-butyl-9*H*-carbazole) (39e)

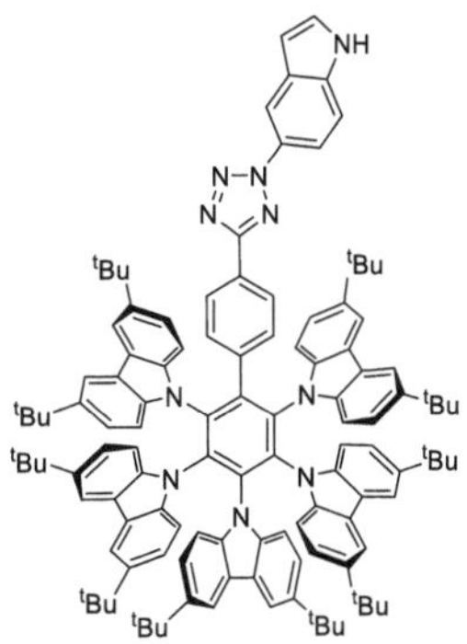

A sealable vial was charged with 9,9',9'',9''',9''''-((2s,3r,4s,5s,6s)-4'-(2*H*-tetrazol-5-yl)-[1,1'-biphenyl]-2,3,4,5,6-pentayl)pentakis(3,6-di-*tert*-butyl-9*H*-carbazole) (**38**) (161 mg, 100 µmol, 1.00 equiv.), (1*H*-indol-5-yl)boronic acid (40.3 mg, 250 µmol, 2.50 equiv.), potassium carbonate (21.0 mg, 0.150 mmol, 1.50 equiv.), and Cu(TMEDA) (5.60 mg, 12.0 µmol, 0.12 equiv.). Dichloromethane (5 mL) was added, and the resulting mixture was stirred under an O_2 atmosphere at room temperature for 16 h. Subsequently, a 10% aqueous ammonia solution (100 mL) was added, and the mixture was extracted with dichloromethane (3 × 100 mL). The combined organic layers were washed with brine (100 mL), dried over sodium sulfate, and reduced in a vacuum. The crude product was purified by flash column chromatography over silica gel (*c*Hex/CH_2Cl_2, 2:1 to 1:1) to yield 93.6 mg of the title compound (54.2 mmol, 54%) as a yellow solid.

R_f (*c*Hex/CH_2Cl_2, 1:1) = 0.36.

^1H NMR (500 MHz, $CDCl_3$, ppm) δ = 7.99 (m, 1H, H_{ar}), 7.63 (m, 1H, H_{ar}), 7.51 – 7.42 (m, 4H, H_{ar}), 7.27 (s, 2H, H_{ar}), 7.21 (dd, *J* = 8.3, 2.0 Hz, 6H, H_{ar}), 7.16 (d, *J* = 8.7 Hz, 2H, H_{ar}), 7.06 (m, 1H, H_{ar}), 6.99 – 6.94 (m, 2H, H_{ar}), 6.94 – 6.80 (m, 13H, H_{ar}), 6.65 (dd, *J* = 8.7, 2.0 Hz, 2H, H_{ar}), 6.57 (dd, *J* = 8.7, 2.0 Hz, 4H, H_{ar}), 6.45 (d, *J* = 18.3 Hz, 1H, H_{ar}), 1.24 (s, 36H, CH_3), 1.22 (s, 36H, CH_3), 1.15 (s, 18H, CH_3). Missing signal (1H) NH due to H/D exchange in $CDCl_3$.

^{13}C NMR (126 MHz, $CDCl_3$, ppm) δ = 143.6 (C_q), 142.5 (C_q), 142.4 (C_q), 142.2 (C_q), 138.2 (C_q), 137.6 (C_q), 137.2 (C_q), 136.7 (C_q), 136.1 (C_q), 129.1 (+, CH), 127.6 (C_q), 126.3 (+, CH), 126.0 (+, CH), 124.3 (C_q), 124.0 (C_q), 123.6 (C_q), 122.5 (+, CH), 122.2 (+, CH), 121.7 (+, CH), 115.5 (+, CH), 115.0 (+, CH), 114.8 (+, CH),114.2 (+, CH),

111.5 (+, CH), 110.8 (+, CH), 110.4 (+, CH), 109.9 (+, CH), 103.6 (+, CH), 34.5 (C_q), 34.4 (C_q), 32.0 (+, CH_3, 24C, ${}^{t}Bu$), 31.9 (+, CH_3, 6C, ${}^{t}Bu$).

IR (ATR, cm^{-1}) $\tilde{\nu}$ = 2952 (s), 2901 (w), 2864 (w), 1487 (s), 1470 (vs), 1449 (vs), 1414 (m), 1392 (w), 1361 (s), 1324 (m), 1295 (vs), 1262 (vs), 1231 (s), 1201 (m), 1150 (m), 1106 (w), 1094 (w), 1034 (w), 1016 (w), 894 (w), 874 (m), 847 (w), 841 (w), 802 (vs), 759 (w), 738 (m), 722 (m), 690 (w), 679 (w), 654 (w), 609 (s), 594 (w), 558 (s), 521 (w), 490 (w), 466 (m), 446 (w), 422 (m), 385 (w).

MS (ESI): m/z (%) = 1725 (68) $[M+2H]^+$, 1724 (100) $[M+H]^+$, 1723 (73) $[M]^+$.

HRMS (ESI, $C_{121}H_{130}N_{10}$): calcd 1723.0480, found 1723.0514.

5.2.2.4. Alkylation of the Cz-Tetrazoles **38** and **33**

9,9',9'',9''',9''''-((2s,3r,4s,5s,6s)-4'-(2-Propyl-2*H*-tetrazol-5-yl)-[1,1'-biphenyl]-2,3,4,5,6-pentayl)pentakis(3,6-di-*tert*-butyl-9*H*-carbazole) (39f)

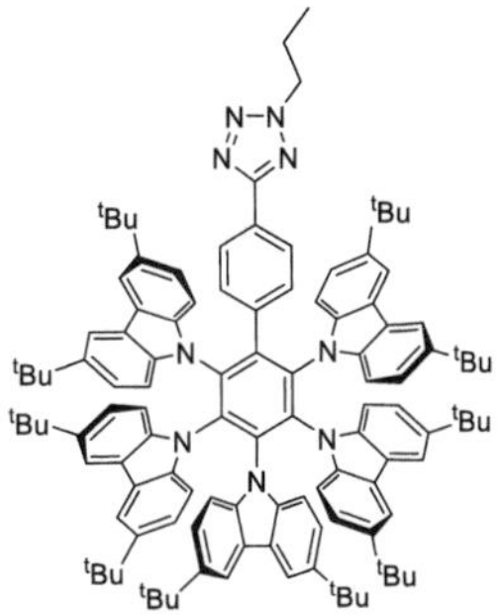

A sealable vial was charged with 9,9',9'',9''',9''''-((2s,3r,4s,5s,6s)-4'-(2*H*-tetrazol-5-yl)-[1,1'-biphenyl]-2,3,4,5,6-pentayl)pentakis(3,6-di-*tert*-butyl-9*H*-carbazole) (**38**) (210 mg, 130 µmol, 1.00 equiv.), 1-bromopropane (32.1 mg, 261 µmol, 2.00 equiv.) and triethylamine (17.0 mg, 163 µmol, 1.25 equiv.). Acetonitrile (5 mL) was added, and the resulting mixture was stirred at room temperature for 24 h. Subsequently, the mixture was added to an excess of water (200 mL) and extracted with dichloromethane (3 × 100 mL). The combined organic layers were washed with brine (100 mL), dried over sodium sulfate, and reduced in a vacuum. The crude product was purified by flash column chromatography over silica gel (*c*Hex/CH_2Cl_2, 1:1) to yield 203 mg of the title compound (124 µmol, 95%) as an off-white solid.

R_f (*c*Hex/CH_2Cl_2, 1:1) = 0.27.

^{1}H NMR (500 MHz, $CDCl_3$, ppm) δ = 7.45 (t, *J* = 1.3 Hz, 4H, H_{ar}), 7.24 – 7.19 (m, 8H, H_{ar}), 7.13 (d, *J* = 8.7 Hz, 2H, H_{ar}), 6.89 (d, *J* = 8.5 Hz, 2H, H_{ar}), 6.86 – 6.83 (m, 12H, H_{ar}), 6.64 (dd, *J* = 8.7, 2.0 Hz, 2H, H_{ar}), 6.56 (dd, *J* = 8.7, 2.0 Hz, 4H, H_{ar}), 4.37 (t, *J* = 7.0 Hz, 2H, CH_2), 1.87 (h, *J* = 7.3 Hz, 2H, CH_2), 1.24 (s, 36H, CH_3), 1.22 (s, 36H, CH_3), 1.16 (s, 18H, CH_3), 0.83 (t, *J* = 7.4 Hz, 3H, CH_3).

^{13}C NMR (126 MHz, $CDCl_3$, ppm) δ = 143.3 (C_q), 142.5 (C_q), 142.3 (C_q), 142.2 (C_q), 138.2 (C_q), 138.1 (C_q), 137.9 (C_q), 137.6 (C_q), 137.1 (C_q), 136.4 (C_q), 136.1 (C_q), 129.1 (+, CH), 126.2 (C_q), 125.9 (+, CH), 124.2 (C_q), 124.0 (C_q), 123.6 (C_q), 122.4 (+, CH), 122.1 (+, CH), 121.7 (+, CH), 115.4 (+, CH), 115.0 (C_q), 114.8 (+, CH), 110.7

(+, CH), 110.4 (C_q), 109.9 (+, CH), 54.6 (-, CH_2, 1C), 34.5 (C_q), 34.4 (C_q), 32.0 (+, CH_3, 24C, ^{t}Bu), 31.9 (+, CH_3, 6C, ^{t}Bu), 22.9 (-, CH_2, 1C), 11.0 (+, CH_3, 1C).

IR (ATR, cm^{-1}) $\tilde{\nu}$ = 2952 (vs), 2927 (s), 2904 (m), 2864 (m), 2853 (m), 1487 (s), 1472 (vs), 1448 (vs), 1415 (m), 1392 (w), 1361 (s), 1324 (m), 1295 (vs), 1262 (vs), 1231 (s), 1201 (m), 1152 (m), 1106 (w), 1034 (m), 1021 (w), 897 (w), 874 (s), 850 (w), 841 (w), 802 (vs), 775 (w), 761 (w), 752 (m), 738 (m), 690 (w), 679 (w), 653 (w), 609 (vs), 592 (m), 560 (s), 523 (w), 489 (w), 465 (m), 421 (m), 398 (w), 390 (w).

MS (ESI): m/z (%) = 1652 (63) $[M+2H]^+$, 1651 (100) $[M+H]^+$, 1650 (78) $[M]^+$.

HRMS (ESI, $C_{116}H_{131}N_9$): calcd 1650.0527, found 1650.0531.

Mp 310 °C.

9,9',9'',9''',9''''-((2s,3r,4s,5s,6s)-4'-(2-Allyl-2*H*-tetrazol-5-yl)-[1,1'-biphenyl]-2,3,4,5,6-pentayl)pentakis(3,6-di-*tert*-butyl-9*H*-carbazole) (39g)

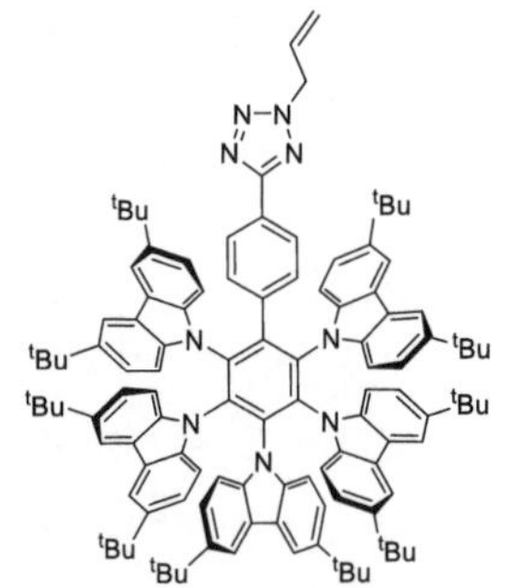

A sealable vial was charged with 9,9',9'',9''',9''''-((2s,3r,4s,5s,6s)-4'-(2*H*-tetrazol-5-yl)-[1,1'-biphenyl]-2,3,4,5,6-pentayl)pentakis(3,6-di-*tert*-butyl-9*H*-carbazole) (**38**) (161 mg, 100 µmol, 1.00 equiv.), 3-bromoprop-1-ene (24.2 mg, 200 µmol, 2.00 equiv.) and triethylamine (13.0 mg, 125 µmol, 1.25 equiv.). Acetonitrile (5 mL) was added, and the resulting mixture was stirred at room temperature for 24 h. Subsequently, the mixture was added to an excess of water (200 mL) and extracted with dichloromethane (3 × 100 mL). The combined organic layers were washed with brine (100 mL), dried over sodium sulfate, and reduced in a vacuum. The crude product was purified by flash column chromatography over silica gel

(cHex/CH_2Cl_2, 2:1) to yield 134 mg of the title compound (81.4 µmol, 81%) as an off-white solid.

R_f (cHex/CH_2Cl_2, 1:1) = 0.33.

^{1}H NMR (500 MHz, $CDCl_3$, ppm) δ = 7.45 (t, J = 1.3 Hz, 4H, H_{ar}), 7.25 – 7.19 (m, 8H, H_{ar}), 7.13 (d, J = 8.7 Hz, 2H, H_{ar}), 6.92 – 6.89 (m, 2H) , H_{ar}, 6.86 – 6.82 (m, 12H, H_{ar}), 6.64 (dd, J = 8.7, 2.0 Hz, 2H, H_{ar}), 6.56 (dd, J = 8.7, 2.0 Hz, 4H, H_{ar}), 5.90 (ddt, J = 16.6, 10.3, 6.2 Hz, 1H, CH), 5.32 – 5.17 (m, 2H, CH_2), 5.01 (dt, J = 6.2, 1.4 Hz, 2H, CH_2), 1.24 (s, 36H, CH_3), 1.22 (s, 36H, CH_3), 1.15 (s, 18H, CH_3).

^{13}C NMR (126 MHz, $CDCl_3$, ppm) δ = 142.3 (C_q), 142.1 (C_q), 142.0 (C_q), 138.0 (C_q), 137.9 (C_q), 137.5 (C_q), 137.0 (C_q), 136.8 (C_q), 135.9 (C_q), 129.7 (C_q), 129.0 (+, CH), 125.9 (+, CH), 124.1 (C_q), 123.8 (C_q), 123.5 (C_q), 122.3 (+, CH), 122.0 (+, CH), 121.5 (+, CH), 120.7 (-, CH_2, 1C), 115.3 (+, CH), 114.8 (C_q), 114.6 (+, CH), 110.6 (+, CH), 110.2 (C_q), 109.7 (+, CH), 55.1 (-, CH_2, 1C), 34.4 (C_q), 34.2 (C_q), 31.8 (+, CH_3, 24C, ^{t}Bu), 31.7 (+, CH_3, 6C, ^{t}Bu).

IR (ATR, cm^{-1}) $\tilde{v}$ = 2951 (s), 2902 (w), 2864 (w), 1487 (s), 1470 (vs), 1448 (vs), 1414 (m), 1392 (w), 1361 (s), 1324 (m), 1295 (vs), 1262 (vs), 1231 (s), 1201 (w), 1150 (m), 1106 (w), 1035 (w), 1021 (w), 932 (w), 919 (w), 874 (m), 850 (w), 802 (vs), 775 (w), 761 (w), 752 (w), 738 (s), 718 (w), 690 (w), 653 (w), 609 (s), 592 (w), 560 (s), 493 (w), 469 (w), 465 (w), 422 (w).

MS (ESI): m/z (%) = 1650 (65) $[M+2H]^+$, 1649 (100) $[M+H]^+$, 1648 (76) $[M]^+$.

HRMS (ESI, $C_{116}H_{129}N_9$): calcd 1648.0371, found 1648.0388.

9,9',9'',9''',9''''-(4'-(2-(Prop-2-yn-1-yl)-2*H*-tetrazol-5-yl)-[1,1'-biphenyl]-2,3,4,5,6-pentayl)pentakis(3,6-di-*tert*-butyl-9*H*-carbazole) (39h)

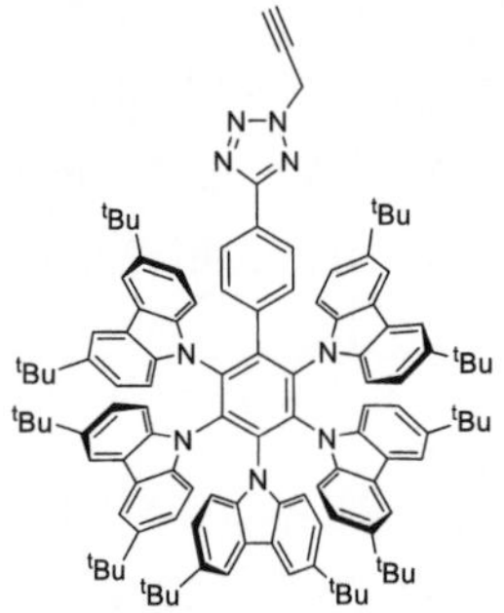

A sealable vial was charged with 9,9',9'',9''',9''''-((2s,3r,4s,5s,6s)-4'-(2*H*-tetrazol-5-yl)-[1,1'-biphenyl]-2,3,4,5,6-pentayl)pentakis(3,6-di-*tert*-butyl-9*H*-carbazole) (**38**) (100 mg, 62.0 µmol, 1.00 equiv.), 3-bromoprop-1-yne (7.40 mg, 62.0 µmol, 1.00 equiv.) and triethylamine (7.80 mg, 78.0 µmol, 1.25 equiv.). Acetonitrile (3 mL) was added, and the resulting mixture was stirred at 60 °C for 24 h. Subsequently, the mixture was added to an excess of water (200 mL) and extracted with dichloromethane (3 × 100 mL). The combined organic layers were washed with brine (100 mL), dried over sodium sulfate, and reduced in a vacuum. The crude product was purified by flash column chromatography over silica gel (*c*Hex/CH_2Cl_2, 3:1) to yield 53.5 mg of the title compound (32.8 µmol, 53%) as an off-white solid.

R_f (*c*Hex/CH_2Cl_2, 1:1) = 0.22.

^{1}H NMR (500 MHz, $CDCl_3$, ppm) δ = 7.45 (t, *J* = 1.4 Hz, 4H, H_{ar}), 7.24 (d, *J* = 1.9 Hz, 1H, H_{ar}), 7.22 – 7.18 (m, 7H, H_{ar}), 7.14 (dd, *J* = 8.7, 3.3 Hz, 2H, H_{ar}), 6.93 (t, *J* = 8.8 Hz, 2H, H_{ar}), 6.83 (ddt, *J* = 8.1, 4.8, 2.8 Hz, 12H, H_{ar}), 6.64 (dd, *J* = 8.6, 2.0 Hz, 2H, H_{ar}), 6.56 (dt, *J* = 8.5, 2.3 Hz, 4H, H_{ar}), 5.20 (d, *J* = 2.6 Hz, 2H, CH_2), 2.44 (t, *J* = 2.6 Hz, 1H, CH), 1.24 (s, 36H, CH_3), 1.22 (s, 36H, CH_3), 1.15 (s, 18H, CH_3).

^{13}C NMR (126 MHz, $CDCl_3$, ppm) δ = 143.1 (C_q), 142.5 (C_q), 142.3 (C_q), 142.2 (C_q), 138.2 (C_q), 138.1 (C_q), 137.6 (C_q), 137.2 (C_q), 136.1 (C_q), 129.2 (+, CH), 126.1 (+, CH), 126.0 (C_q), 124.2 (C_q), 124.0 (C_q), 123.6 (C_q), 122.4 (+, CH), 122.1 (+, CH), 121.7 (+, CH), 115.5 (+, CH), 115.0 (+, CH), 114.8 (+, CH), 110.8 (+, CH), 110.4 (+, CH), 109.8 (+, CH), 42.5 (-, CH_2, 1C), 34.5 (C_q), 34.4 (C_q), 32.0 (+, CH_3, 12C, ^{t}Bu), 31.9 (+, CH_3, 12C, ^{t}Bu), 31.8 (+, CH_3, 6C, ^{t}Bu).

IR (ATR, cm^{-1}) $\tilde{\nu}$ = 2952 (s), 2902 (m), 2864 (w), 1487 (s), 1470 (vs), 1449 (vs), 1415 (m), 1392 (w), 1361 (s), 1324 (m), 1295 (vs), 1262 (vs), 1231 (s), 1201 (m),

1150 (m), 1123 (w), 1106 (m), 1079 (w), 1034 (m), 1020 (w), 931 (w), 922 (w), 895 (w), 874 (s), 849 (w), 841 (w), 802 (vs), 776 (m), 761 (w), 752 (m), 738 (s), 718 (w), 688 (w), 679 (w), 653 (w), 609 (vs), 592 (m), 560 (vs), 523 (w), 489 (m), 466 (m), 422 (s), 405 (w), 395 (w), 387 (w), 380 (w).

MS (ESI): m/z (%) = 1648 (64) $[M+2H]^+$, 1647 (100) $[M+H]^+$, 1646 (75) $[M]^+$.

HRMS (ESI, $C_{116}H_{127}N_9$): calcd 1646.0214, found 1646.0245.

Mp 284 °C.

9,9',9'',9''',9''''-((1r,2r,3r,4s,5r)-6-(2-Propyl-2*H*-tetrazol-5-yl)benzene-1,2,3,4,5-pentayl)pentakis(3,6-di-*tert*-butyl-9*H*-carbazole) (40a)

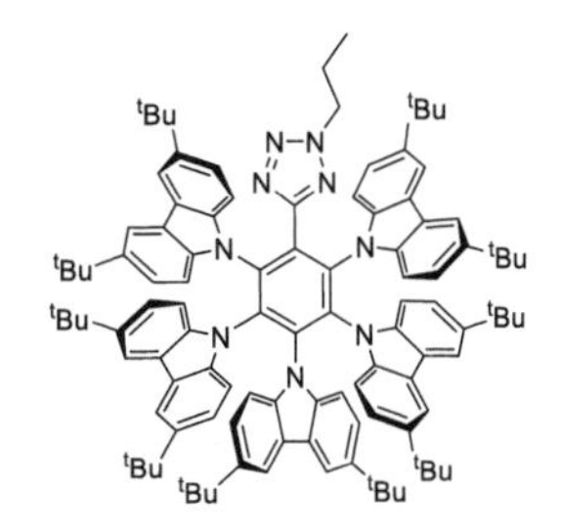

A sealable vial was charged with 9,9',9'',9''',9''''-((1r,2r,3r,4s,5r)-6-(2*H*-tetrazol-5-yl)benzene-1,2,3,4,5-pentayl)pentakis(3,6-di-*tert*-butyl-9*H*-carbazole) (**33**) (153 mg, 100 µmol, 1.00 equiv.), 1-bromopropane (24.5 mg, 200 µmol, 2.00 equiv.) and triethylamine (13.0 mg, 125 µmol, 1.25 equiv.). Acetonitrile (5 mL) was added, and the resulting mixture was stirred at room temperature for 24 h. Subsequently, the mixture was added to an excess of water (200 mL) and extracted with dichloromethane (3 × 100 mL). The combined organic layers were washed with brine (100 mL), dried over sodium sulfate, and reduced in a vacuum. The crude product was purified by flash column chromatography over silica gel (*c*Hex/CH_2Cl_2, 3:1 to 1:1) to yield 45.7 mg of the title compound (29.0 µmol, 29%) as an off-white solid.

R_f (*c*Hex/CH_2Cl_2, 1:1) = 0.31.

^1H NMR (500 MHz, $CDCl_3$, ppm) δ = 7.48 (d, J = 1.8 Hz, 4H, H_{ar}), 7.22 (dd, J = 12.5, 2.0 Hz, 6H, H_{ar}), 7.08 (d, J = 8.6 Hz, 2H, H_{ar}), 6.90 – 6.81 (m, 12H, H_{ar}), 6.62 (dd, J = 8.6, 2.0 Hz, 2H, H_{ar}), 6.55 (dd, J = 8.7, 1.9 Hz, 4H, H_{ar}), 3.73 (t, J = 6.8 Hz, 2H, CH_2),

1.26 (s, 36H, CH_3), 1.21 (s, 36H, CH_3), 1.16 (s, 18H, CH_3), 0.99 (q, J = 7.2 Hz, 2H, CH_2), 0.10 (t, J = 7.4 Hz, 3H, CH_3).

^{13}C NMR (126 MHz, $CDCl_3$, ppm) δ = 142.7 (C_q), 142.4 (C_q), 139.7 (C_q), 138.0 (C_q), 137.9 (C_q), 137.5 (C_q), 137.4 (C_q), 136.6 (C_q), 130.8 (C_q), 124.4 (C_q), 124.1 (C_q), 123.7 (C_q), 122.6 (+, CH), 122.3 (+, CH), 121.8 (+, CH), 115.3 (C_q), 115.1 (+, CH), 114.8 (+, CH), 110.6 (+, CH), 110.3 (+, CH), 109.6 (+, CH), 54.0 (-, CH_2, 1C), 34.5 (C_q), 34.4 (C_q), 32.0 (+, CH_3, 12C, ^{t}Bu), 31.9 (+, CH_3, 12C, ^{t}Bu), 31.8 (+, CH_3, 6C, ^{t}Bu), 22.4 (-, CH_2, 1C), 10.3 (+, CH_3, 1C).

IR (ATR, cm^{-1}) $\tilde{\nu}$ = 2952 (s), 1487 (s), 1473 (vs), 1449 (vs), 1361 (s), 1324 (m), 1296 (vs), 1264 (vs), 1231 (s), 874 (m), 807 (vs), 800 (vs), 741 (m), 612 (s), 561 (m), 551 (m), 470 (w), 421 (m).

MS (ESI): m/z (%) = 1576 (55) $[M+2H]^+$, 1575 (86) $[M+H]^+$, 1574 (65) $[M]^+$. HRMS (ESI, $C_{110}H_{127}N_9$): calcd 1574.0214, found 1574.0217.

Mp 253 °C.

9,9',9'',9''',9''''-(6-(2-(Prop-2-yn-1-yl)-2*H*-tetrazol-5-yl)benzene-1,2,3,4,5-pentayl)pentakis(3,6-di-*tert*-butyl-9*H*-carbazole) (40b)

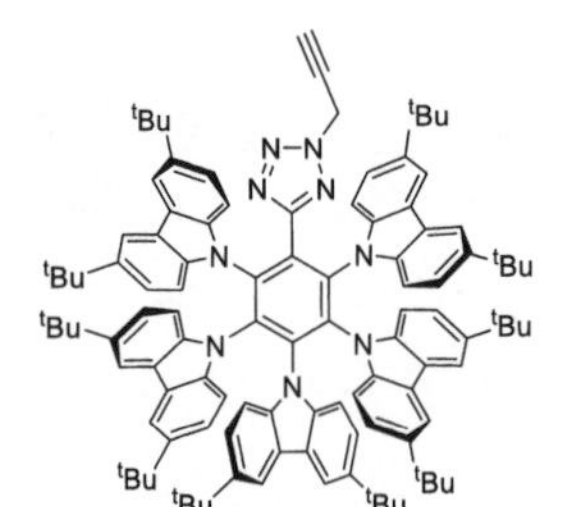

A sealable vial was charged with 9,9',9'',9''',9''''-((1r,2r,3r,4s,5r)-6-(2*H*-tetrazol-5-yl)benzene-1,2,3,4,5-pentayl)pentakis(3,6-di-*tert*-butyl-9*H*-carbazole) (**33**) (100 mg, 65.2 µmol, 1.00 equiv.), 3-bromoprop-1-yne (7.76 mg, 65.2 µmol, 1.00 equiv.) and triethylamine (8.25 mg, 81.5 µmol, 1.25 equiv.). Acetonitrile (3 mL) was added, and the resulting mixture was stirred at 60 °C for 24 h. Subsequently, the mixture was added to an excess of water (200 mL) and extracted with dichloromethane (3 × 100 mL). The combined organic layers were washed with brine (100 mL), dried over sodium sulfate, and reduced in a vacuum.

The crude product was purified by flash column chromatography over silica gel (*c*Hex/CH_2Cl_2, 3:1) to yield 58.4 mg of the title compound (37.1 µmol, 57%) as an off-white solid.

R_f (*c*Hex/CH_2Cl_2, 2:1) = 0.21.

^{1}H NMR (500 MHz, $CDCl_3$, ppm) δ = 7.49 (dd, *J* = 1.9, 0.7 Hz, 4H, H_{ar}), 7.22 (dd, *J* = 10.1, 2.0 Hz, 6H, H_{ar}), 7.07 (d, *J* = 8.5 Hz, 2H, H_{ar}), 6.88 (d, *J* = 8.5 Hz, 4H, H_{ar}), 6.85 (dd, *J* = 8.6, 1.9 Hz, 4H, H_{ar}), 6.82 (d, *J* = 8.6 Hz, 4H, H_{ar}), 6.62 (dd, *J* = 8.7, 2.0 Hz, 2H, H_{ar}), 6.56 (dd, *J* = 8.7, 2.0 Hz, 4H, H_{ar}), 4.51 (d, *J* = 2.6 Hz, 2H, CH_2), 2.02 (t, *J* = 2.6 Hz, 1H, CH), 1.27 (s, 36H, CH_3), 1.22 (s, 36H, CH_3), 1.17 (s, 18H, CH_3).

^{13}C NMR (126 MHz, $CDCl_3$, ppm) δ = 142.7 (C_q), 142.4 (C_q), 142.3 (C_q), 140.0 (C_q), 138.0 (C_q), 137.8 (C_q), 137.5 (C_q), 137.4 (C_q), 136.6 (C_q), 130.2 (C_q), 124.4 (C_q), 124.1 (C_q), 123.8 (C_q), 122.6 (+, CH), 122.3 (+, CH), 121.9 (+, CH), 115.3 (C_q), 115.1 (+, CH), 114.8 (+, CH), 110.6 (+, CH), 110.2 (+, CH), 109.6 (+, CH), 75.3 (C_q), 73.1 (C_q), 42.0 (-, CH_2, 1C), 34.5 (C_q), 34.4 (C_q), 32.0 (+, CH_3, 12C, ^{t}Bu), 31.9 (+, CH_3, 12C, ^{t}Bu), 31.8 (+, CH_3, 6C, ^{t}Bu).

IR (ATR, cm^{-1}) $\tilde{\nu}$ = 2952 (s), 2902 (m), 2864 (m), 1487 (s), 1472 (vs), 1449 (vs), 1392 (w), 1361 (s), 1324 (m), 1295 (vs), 1264 (vs), 1231 (s), 1201 (m), 1152 (w), 1106 (w), 1071 (w), 1034 (m), 955 (w), 932 (w), 922 (w), 897 (w), 874 (m), 841 (w), 806 (vs), 762 (w), 741 (m), 688 (w), 676 (w), 654 (w), 611 (vs), 592 (w), 561 (m), 551 (m), 523 (w), 493 (w), 469 (m), 446 (w), 421 (m), 405 (w), 398 (w).

MS (ESI): m/z (%) = 1571 (61) $[M+H]^+$, 1570 (100) $[M]^+$. HRMS (ESI, $C_{110}H_{123}N_9$): calcd 1569.9901, found 1569.9917.

Mp 278 °C.

5.2.2.5. Oxadiazoles

2-Mesityl-5-((2's,3'r,4's,5's,6's)-2',3',4',5',6'-pentakis(3,6-di-*tert*-butyl-9*H*-carbazol-9-yl)-[1,1'-biphenyl]-4-yl)-1,3,4-oxadiazole (41a)

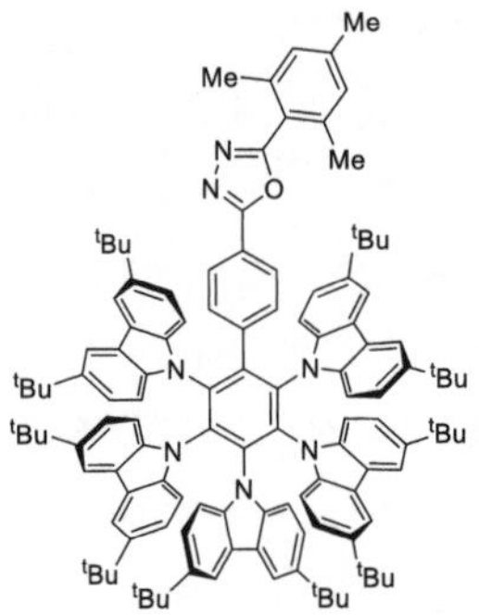

A sealable vial was charged with 9,9',9'',9''',9''''-((2s,3r,4s,5s,6s)-4'-(2*H*-tetrazol-5-yl)-[1,1'-biphenyl]-2,3,4,5,6-pentayl)pentakis(3,6-di-*tert*-butyl-9*H*-carbazole) (**38**) (150 mg, 93.0 μmol, 1.00 equiv.) and 2,4,6-trimethylbenzoyl chloride (68.1 mg, 373 μmol, 4.00 equiv.). Chloroform (10 mL) was added, and the resulting mixture was heated at 100 °C for 16 h. After cooling to room temperature, the reaction mixture was poured into an excess of saturated aqueous sodium hydrogen carbonate solution (50 mL) and stirred for 15 min. Subsequently, the reaction mixture was extracted with dichloromethane (3 × 50 mL). The combined organic layers were washed with brine (50 mL), dried over sodium sulfate, and reduced in a vacuum. The crude product was purified by flash column chromatography over silica gel (*c*Hex/CH_2Cl_2, 1:1 to CH_2Cl_2/CH_3OH, 50:1) to yield 152 mg of the title compound (88.0 mmol, 94%) as an off-white solid.

R_f (*c*Hex/CH_2Cl_2, 1:3) = 0.32.

^{1}H NMR (500 MHz, $CDCl_3$, ppm) δ = 7.47 (t, *J* = 1.3 Hz, 4H, H_{ar}), 7.22 (dd, *J* = 7.4, 1.9 Hz, 6H, H_{ar}), 7.16 (d, *J* = 8.6 Hz, 2H, H_{ar}), 7.12 (d, *J* = 8.7 Hz, 2H, H_{ar}), 6.89 (d, *J* = 8.6 Hz, 2H, H_{ar}), 6.89 – 6.82 (m, 14H, H_{ar}), 6.64 (dd, *J* = 8.7, 2.0 Hz, 2H, H_{ar}), 6.57 (dd, *J* = 8.6, 2.0 Hz, 4H, H_{ar}), 2.27 (s, 3H, CH_3), 2.08 (s, 6H, CH_3), 1.25 (s, 36H, CH_3), 1.22 (s, 36H, CH_3), 1.17 (s, 18H, CH_3).

^{13}C NMR (126 MHz, $CDCl_3$, ppm) δ = 164.4 (C_q), 163.6 (C_q), 142.9 (C_q), 142.5 (C_q), 142.5 (C_q), 142.3 (C_q), 141.0 (C_q), 138.7 (C_q), 138.2 (C_q), 138.1 (C_q), 138.0 (C_q), 137.9 (C_q), 137.6 (C_q), 137.3 (C_q), 136.0 (C_q), 129.3 (+, CH), 128.8 (+, CH), 126.0

(+, CH), 124.3 (C_q), 124.0 (C_q), 123.6 (C_q), 122.6 (+, CH), 122.5 (+, CH), 122.1 (+, CH), 121.8 (+, CH), 121.1 (C_q), 115.5 (+, CH), 115.0 (C_q), 114.8 (+, CH), 110.7 (+, CH), 110.3 (+, CH), 109.9 (+, CH), 34.5 (C_q, 2C), 34.4 (C_q, 8C), 32.0 (+, CH_3, 24C, ^{t}Bu), 31.9 (+, CH_3, 6C, ^{t}Bu), 21.4 (+, CH_3, 1C, Me), 20.4 (+, CH_3, 2C, Me).

IR (ATR, cm^{-1}) $\tilde{\nu}$ =2952 (s), 2932 (m), 2902 (m), 2864 (w), 1487 (s), 1470 (vs), 1449 (vs), 1408 (w), 1392 (w), 1361 (s), 1324 (m), 1295 (vs), 1262 (vs), 1230 (s), 1201 (w), 1150 (m), 1106 (w), 1043 (w), 1034 (m), 1018 (w), 874 (m), 847 (w), 802 (vs), 756 (w), 738 (m), 609 (s), 591 (w), 558 (s), 490 (w), 467 (w), 422 (w), 418 (w).

MS (ESI): m/z (%) = 1728 (69) $[M+2H]^+$, 1727 (100) $[M+H]^+$, 1726 (73) $[M]^+$.

HRMS (ESI, $C_{123}H_{135}N_7O$): calcd 1726.0728, found 1726.0729.

Mp 387 °C.

4-(5-((2's,3'r,4's,5's,6's)-2',3',4',5',6'-Pentakis(3,6-di-*tert*-butyl-9*H*-carbazol-9-yl)-[1,1'-biphenyl]-4-yl)-1,3,4-oxadiazol-2-yl)benzonitrile (41b)

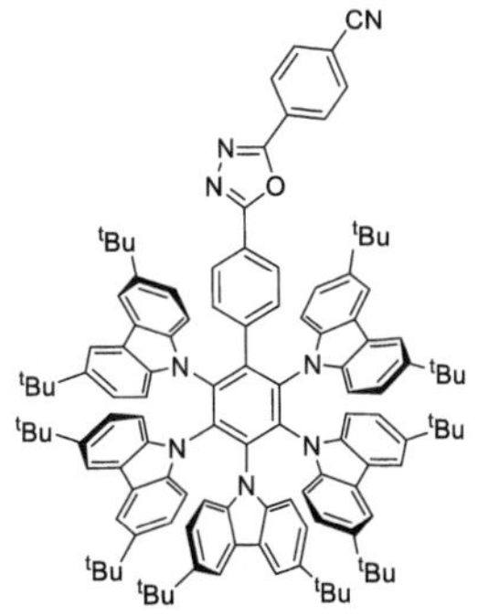

A sealable vial was charged with 9,9',9'',9''',9''''-((2s,3r,4s,5s,6s)-4'-(2*H*-tetrazol-5-yl)-[1,1'-biphenyl]-2,3,4,5,6-pentayl)pentakis(3,6-di-*tert*-butyl-9*H*-carbazole) (**38**) (150 mg, 93.0 µmol, 1.00 equiv.) and 4-cyanobenzoyl chloride (61.7 mg, 373 µmol, 4.00 equiv.). Chloroform (10 mL) was added, and the resulting mixture was heated at 100 °C for 16 h. After cooling to room temperature, the reaction mixture was poured into an excess of saturated aqueous sodium hydrogen carbonate solution (50 mL) and stirred for 15 min. Subsequently, the reaction mixture was extracted with dichloromethane (3 × 50 mL). The combined organic layers were washed with brine (50 mL), dried over sodium sulfate, and reduced in a vacuum.

The crude product was purified by flash column chromatography over silica gel (*c*Hex/CH_2Cl_2, 1:2) to yield 113 mg of the title compound (65.9 µmol, 71%) as an off-white solid.

R_f (*c*Hex/CH_2Cl_2, 1:5) = 0.23.

^{1}H NMR (500 MHz, $CDCl_3$, ppm) δ = 8.01 (d, *J* = 8.6 Hz, 2H, H_{ar}), 7.70 (d, *J* = 8.7 Hz, 2H, H_{ar}), 7.47 (t, *J* = 0.9 Hz, 4H, H_{ar}), 7.26 (d, *J* = 8.7 Hz, 2H, H_{ar}), 7.22 (dd, *J* = 8.7, 2.0 Hz, 6H, H_{ar}), 7.14 (d, *J* = 8.7 Hz, 2H, H_{ar}), 7.01 (d, *J* = 8.6 Hz, 2H, H_{ar}), 6.87 – 6.82 (m, 12H, H_{ar}), 6.64 (dd, *J* = 8.7, 2.0 Hz, 2H, H_{ar}), 6.56 (dd, *J* = 8.7, 2.0 Hz, 4H, H_{ar}), 1.25 (s, 36H, CH_3), 1.22 (s, 36H, CH_3), 1.16 (s, 18H, CH_3).

^{13}C NMR (126 MHz, $CDCl_3$, ppm) δ = 164.9 (C_q), 162.7 (C_q), 142.7 (C_q), 142.6 (C_q), 142.6 (C_q), 142.3 (C_q), 138.9 (C_q), 138.4 (C_q), 138.1 (C_q), 137.5 (C_q), 137.3 (C_q), 135.9 (C_q), 132.9 (+, CH), 129.5 (+, CH), 127.7 (C_q), 127.3 (+, CH), 126.3 (+, CH), 124.3 (C_q), 124.0 (+, CH), 123.6 (+, CH), 122.5 (+, CH), 122.2 (C_q), 122.0 (C_q), 121.7 (+, CH), 118.0 (C_q), 115.5 (+, CH), 115.1 (C_q), 115.0 (C_q), 114.8 (+, CH), 110.7 (+, CH), 110.3 (+, CH), 109.8 (+, CH), 34.5 (C_q), 34.4 (C_q), 32.0 (+, CH_3, 24C, ^{t}Bu), 31.9 (+, CH_3, 6C, ^{t}Bu).

IR (ATR, cm^{-1}) $\tilde{\nu}$ = 2952 (s), 2902 (m), 2864 (w), 1487 (s), 1472 (vs), 1448 (vs), 1414 (w), 1392 (w), 1361 (s), 1324 (m), 1295 (vs), 1262 (vs), 1231 (s), 1201 (w), 1150 (m), 1105 (w), 1065 (w), 1045 (w), 1034 (w), 1017 (w), 874 (m), 849 (m), 803 (vs), 771 (w), 762 (w), 738 (m), 708 (w), 688 (w), 679 (w), 654 (w), 609 (s), 592 (w), 555 (s), 521 (w), 490 (w), 466 (w), 419 (w).

MS (ESI): m/z (%) = 1711 (68) $[M+2H]^+$, 1710 (100) $[M+H]^+$, 1709 (75) $[M]^+$.

HRMS (ESI, $C_{121}H_{128}N_8O$): calcd 1709.0211, found 1709.0215.

Mp 330 °C.

2-((2's,3'r,4's,5's,6's)-2',3',4',5',6'-Pentakis(3,6-di-*tert*-butyl-9*H*-carbazol-9-yl)-[1,1'-bi-phenyl]-4-yl)-5-(perfluorophenyl)-1,3,4-oxadiazole (41c)

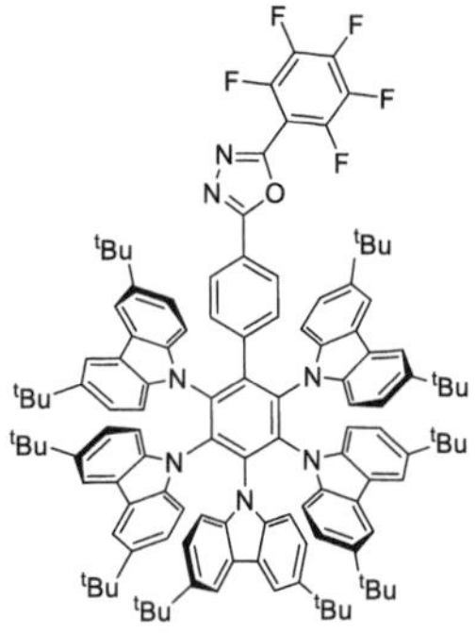

A sealable vial was charged with 9,9',9'',9''',9''''-((2s,3r,4s,5s,6s)-4'-(2*H*-tetrazol-5-yl)-[1,1'-bi-phenyl]-2,3,4,5,6-pentayl)pentakis(3,6-di-*tert*-butyl-9*H*-carbazole) (**38**) (150 mg, 93.0 µmol, 1.00 equiv.) and 2,3,4,5,6-pentafluorobenzoyl chloride (86.0 mg, 373 µmol, 4.00 equiv.). Chloroform (10 mL) was added, and the resulting mixture was heated at 100 °C for 16 h. After cooling to room temperature, the reaction mixture was poured into an excess of saturated aqueous sodium hydrogen carbonate solution (50 mL) and stirred for 15 min. Subsequently, the reaction mixture was extracted with dichloromethane (3 × 50 mL). The combined organic layers were washed with brine (50 mL), dried over sodium sulfate, and reduced in a vacuum. The crude product was purified by flash column chromatography over silica gel (*c*Hex/CH_2Cl_2, 1:2) to yield 135 mg of the title compound (75.7 µmol, 81%) as an off-white solid.

R_f (*c*Hex/CH_2Cl_2, 1:2) = 0.29.

^{1}H NMR (500 MHz, $CDCl_3$, ppm) δ = 7.47 (d, *J* = 1.7 Hz, 4H, H_{ar}), 7.25 – 7.20 (m, 8H, H_{ar}), 7.13 (d, *J* = 8.6 Hz, 2H, H_{ar}), 6.97 (d, *J* = 8.6 Hz, 2H, H_{ar}), 6.89 – 6.81 (m, 12H, H_{ar}), 6.64 (dd, *J* = 8.7, 2.0 Hz, 2H, H_{ar}), 6.57 (dd, *J* = 8.6, 2.0 Hz, 4H, H_{ar}), 1.25 (s, 36H, CH_3), 1.22 (s, 36H, CH_3), 1.16 (s, 18H, CH_3).

^{13}C NMR (126 MHz, $CDCl_3$, ppm) δ = 165.4 (C_q), 142.6 (C_q), 142.6 (C_q), 142.5 (C_q), 142.3 (C_q), 139.1 (C_q), 138.4 (C_q), 138.1 (C_q), 138.0 (C_q), 137.6 (C_q), 137.3 (C_q), 135.9 (C_q), 129.5 (+, CH), 126.4 (+, CH), 124.3 (C_q), 124.0 (C_q), 123.6 (C_q), 122.5 (+, CH), 122.2 (+, CH), 121.7 (+, CH), 121.6 (+, CH), 115.5 (+, CH), 115.0 (+, CH), 114.8 (+, CH), 110.7 (+, CH), 110.3 (+, CH), 109.8 (+, CH), 34.5 (C_q), 34.4 (C_q), 31.9 (+, CH_3, 24C, ^{t}Bu), 31.8 (+, CH_3, 6C, ^{t}Bu).

^{19}F NMR (376 MHz, $CDCl_3$, ppm) δ = -135.6 (qd), -147.4 (tt), -159.5 – -159.8 (m).

IR (ATR, cm^{-1}) $\tilde{\nu}$ =2952 (s), 2902 (w), 2864 (w), 1519 (m), 1487 (vs), 1470 (vs), 1448 (vs), 1408 (m), 1392 (w), 1361 (s), 1324 (m), 1295 (vs), 1262 (vs), 1230 (s), 1201 (w), 1150 (m), 1103 (m), 1034 (w), 1010 (w), 994 (s), 874 (m), 847 (w), 833 (m), 802 (vs), 771 (w), 762 (w), 738 (m), 711 (w), 688 (w), 677 (w), 654 (w), 609 (s), 592 (w), 558 (s), 523 (w), 490 (w), 462 (w), 419 (m), 398 (w).

MS (ESI): m/z (%) = 1775 (65) $[M+H]^+$, 1774 (100) $[M]^{+\cdot}$. HRMS (ESI, $C_{120}H_{124}N_7OF_5$): calcd 1773.9788, found 1773.9807.

Mp 390°C.

2-((2's,3'r,4's,5's,6's)-2',3',4',5',6'-Pentakis(3,6-di-*tert*-butyl-9*H*-carbazol-9-yl)-[1,1'-biphenyl]-4-yl)-5-(perfluoroheptyl)-1,3,4-(41d)

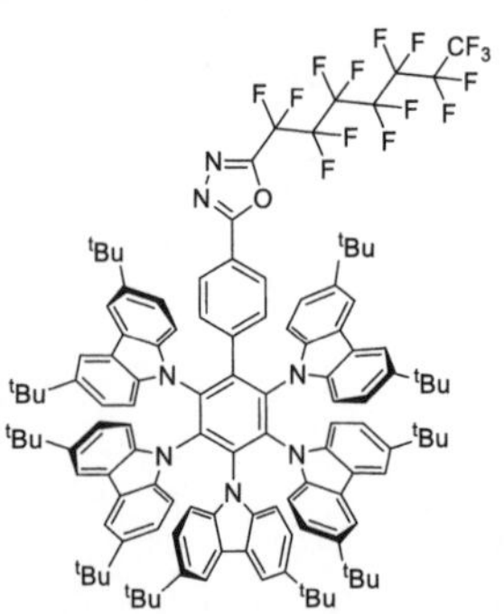

A sealable vial was charged with 9,9',9'',9''',9''''-((2s,3r,4s,5s,6s)-4'-(2*H*-tetrazol-5-yl)-[1,1'-biphenyl]-2,3,4,5,6-pentayl)pentakis(3,6-di-*tert*-butyl-9*H*-carbazole) (**38**) (150 mg, 93.0 µmol, 1.00 equiv.) and pentadecafluorooctanoyl chloride (80.6 mg, 186 µmol, 2.00 equiv.). Chloroform (10 mL) was added, and the resulting mixture was heated at 100 °C for 16 h. After cooling to room temperature, the reaction mixture was poured into an excess of saturated aqueous sodium hydrogen carbonate solution (50 mL) and stirred for 15 min. Subsequently, the reaction mixture was extracted with dichloromethane (3 × 50 mL). The combined organic layers were washed with brine (50 mL), dried over sodium sulfate, and reduced in a vacuum. The crude product was purified by flash column chromatography over silica gel (*c*Hex/CH_2Cl_2, 1:2) to yield 151 mg of the title compound (76.4 µmol, 82%) as a yellow solid.

R_f (*c*Hex/CH_2Cl_2, 1:1) = 0.25.

^{1}H NMR (400 MHz, $CDCl_3$, ppm) δ = 7.47 (d, *J* = 1.8 Hz, 4H, H_{ar}), 7.23 (dt, *J* = 6.5, 1.7 Hz, 6H, H_{ar}), 7.19 (dd, *J* = 8.5, 1.6 Hz, 2H, H_{ar}), 7.13 (d, *J* = 8.6 Hz, 2H, H_{ar}), 6.94 (dd, *J* = 8.5, 1.6 Hz, 2H, H_{ar}), 6.90 – 6.81 (m, 12H, H_{ar}), 6.65 (dt, *J* = 8.6, 1.7 Hz, 2H, H_{ar}), 6.58 (dt, *J* = 8.6, 1.7 Hz, 4H, H_{ar}), 1.25 (s, 36H, CH_3), 1.22 (s, 36H, CH_3), 1.17 (s, 18H, CH_3).

^{13}C NMR (101 MHz, $CDCl_3$, ppm) δ = 166.7 (C_q), 142.7 (C_q), 142.4 (C_q), 139.9 (C_q), 138.1 (C_q), 138.0 (C_q), 137.6 (C_q), 137.4 (C_q), 135.9 (C_q), 129.6 (+, CH), 126.7 (+, CH), 124.3 (C_q), 124.1 (C_q), 123.6 (C_q), 122.6 (+, CH), 122.2 (+, CH), 121.8 (+, CH), 120.7 (C_q), 115.5 (+, CH), 115.1 (+, CH), 114.9 (+, CH), 110.7 (+, CH), 110.3 (+, CH), 109.8 (+, CH), 34.5 (C_q), 34.4 (C_q), 32.0 (+, CH_3, 12C, ^{t}Bu), 31.9 (+, CH_3, 12C, ^{t}Bu), 31.8 (+, CH_3, 6C, ^{t}Bu).

^{19}F NMR (376 MHz, $CDCl_3$, ppm) δ = -80.73 (t), -112.82 (t), -121.54, -121.80 – -122.38 (m), -122.68, -126.10 (td).

IR (ATR, cm^{-1}) $\tilde{\nu}$ = 2953 (s), 2904 (w), 2866 (w), 1487 (s), 1472 (vs), 1449 (s), 1411 (w), 1392 (w), 1363 (s), 1324 (m), 1295 (vs), 1264 (s), 1232 (vs), 1203 (vs), 1149 (vs), 1106 (m), 1094 (w), 1069 (w), 1034 (m), 1017 (w), 874 (m), 847 (w), 803 (vs), 772 (w), 762 (w), 738 (m), 707 (w), 700 (w), 690 (w), 677 (w), 646 (w), 609 (s), 591 (w), 558 (s), 530 (w), 490 (w), 469 (w), 421 (m).

MS (ESI): m/z (%) = 1977 (68) $[M+2H]^+$, 1976 (100) $[M+H]^+$, 1975 (75) $[M]^+$.

HRMS (ESI, $C_{121}H_{124}N_7OF_{15}$): calcd 1975.9628, found 1975.9655.

Mp 350 °C.

2-Mesityl-5-((2r,3r,4r,5s,6r)-2,3,4,5,6-pentakis(3,6-di-*tert*-butyl-9*H*-carbazol-9-yl)phenyl)-1,3,4-oxadiazole (42a)

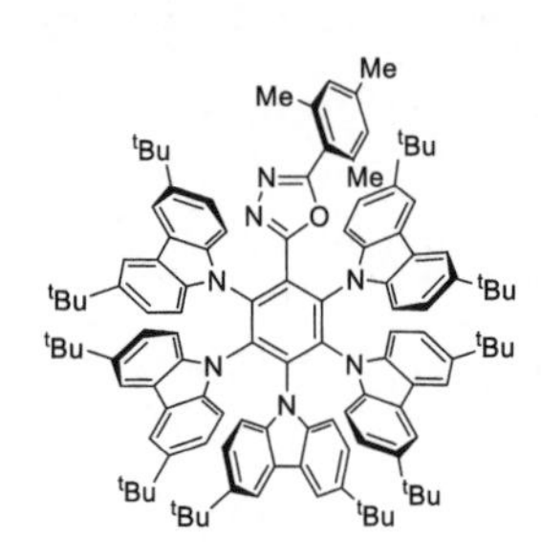

A sealable vial was charged with 9,9',9'',9''',9''''-((1r,2r,3r,4s,5r)-6-(2*H*-tetrazol-5-yl)benzene-1,2,3,4,5-pentayl)pentakis(3,6-di-tert-butyl-9*H*-carbazole) (**33**) (150 mg, 98.0 µmol, 1.00 equiv.) and 2,4,6-trimethylbenzoyl chloride (71.5 mg, 391 µmol, 4.00 equiv.). Chloroform (10 mL) was added, and the resulting mixture was heated at 100 °C for 16 h. After cooling to room temperature, the reaction mixture was poured into an excess of saturated aqueous sodium hydrogen carbonate solution (50 mL) and stirred for 15 min. Subsequently, the reaction mixture was extracted with dichloromethane (3 × 50 mL). The combined organic layers were washed with brine (50 mL), dried over sodium sulfate, and reduced in a vacuum. The crude product was purified by flash column chromatography over silica gel (*c*Hex/CH_2Cl_2, 2:1 to 1:1) to yield 157 mg of the title compound (95.1 mmol, 97%) as a brown solid.

R_f (*c*Hex/CH_2Cl_2, 1:1) = 0.55.

^{1}H NMR (500 MHz, $CDCl_3$, ppm) δ = 7.47 (d, *J* = 1.9 Hz, 4H, H_{ar}), 7.21 (dd, *J* = 14.0, 2.0 Hz, 6H, H_{ar}), 7.14 (d, *J* = 8.7 Hz, 2H, H_{ar}), 6.94 (d, *J* = 8.6 Hz, 4H, H_{ar}), 6.85 (d, *J* = 7.9 Hz, 4H, H_{ar}), 6.75 (d, *J* = 8.6 Hz, 4H, H_{ar}), 6.63 (dd, *J* = 8.7, 2.0 Hz, 2H, H_{ar}), 6.55 – 6.50 (m, 6H, H_{ar}), 2.09 (s, 3H, CH_3), 1.28 (s, 36H, CH_3), 1.22 (s, 36H, CH_3), 1.13 (s, 18H, CH_3), 0.99 (s, 6H, CH_3).

^{13}C NMR (126 MHz, $CDCl_3$, ppm) δ = 164.1 (C_q), 160.0 (C_q), 143.0 (C_q), 142.8 (C_q), 142.5 (C_q), 141.6 (C_q), 140.5 (C_q), 139.1 (C_q), 137.9 (C_q), 137.6 (C_q), 137.1 (C_q), 136.5 (C_q), 128.0 (+, CH), 127.9 (+, CH), 124.6 (C_q), 124.1 (C_q), 124.0 (C_q), 122.9 (+, CH), 122.4 (+, CH), 121.7 (+, CH), 120.5 (C_q), 115.5 (+, CH), 115.1 (+, CH), 114.8 (+, CH), 110.8 (+, CH), 110.1 (+, CH), 109.4 (+, CH), 34.5 (C_q), 34.4 (C_q), 32.0 (+, CH_3, 12C, ^{t}Bu), 31.9 (+, CH_3, 12C , ^{t}Bu), 31.8 (+, CH_3, 6C, ^{t}Bu), 21.2 (+, CH_3, 1C, Me), 19.2 (+, CH_3, 2C, Me).

IR (ATR, cm^{-1}) $\tilde{\nu}$ = 2952 (s), 2929 (m), 2902 (m), 2864 (w), 1487 (s), 1472 (vs), 1448 (vs), 1392 (w), 1380 (w), 1361 (s), 1324 (m), 1295 (vs), 1264 (vs), 1231 (s), 1201 (w), 1166 (w), 1152 (w), 1136 (w), 1106 (w), 1064 (w), 1034 (m), 1006 (w), 932 (w), 897 (w), 874 (m), 850 (w), 841 (w), 805 (vs), 761 (w), 739 (m), 673 (w), 653 (w), 611 (s), 592 (w), 560 (m), 551 (m), 523 (w), 492 (w), 469 (w), 421 (m).

MS (ESI): m/z (%) = 1652 (62) $[M+2H]^+$, 1651 (90) $[M+H]^+$, 1650 (64) $[M]^+$. HRMS (ESI, $C_{117}H_{131}N_7O$): calcd 1650.0415, found 1650.0428.

Mp 356 °C.

2-((2r,3r,4r,5s,6r)-2,3,4,5,6-Pentakis(3,6-di-*tert*-butyl-9*H*-carbazol-9-yl)phenyl)-5-(perfluorophenyl)-1,3,4-oxadiazole (42b)

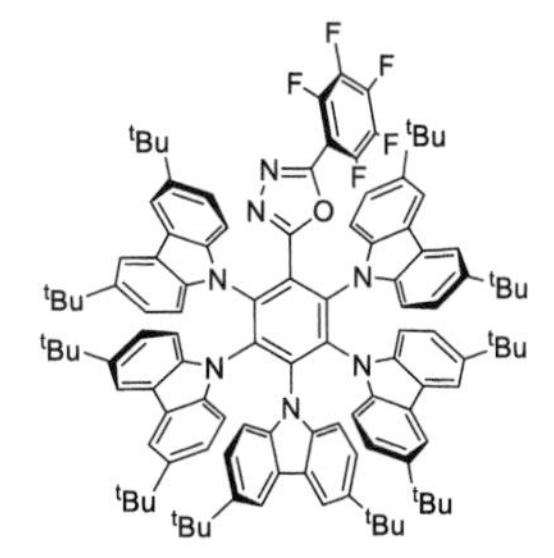

A sealable vial was charged with 9,9',9'',9''',9''''-((1r,2r,3r,4s,5r)-6-(2*H*-tetrazol-5-yl)benzene-1,2,3,4,5-pentayl)pentakis(3,6-di-tert-butyl-9*H*-carbazole) (**33**) (150 mg, 98.0 µmol, 1.00 equiv.) and 2,3,4,5,6-pentafluorobenzoyl chloride (90.2 mg, 391 µmol, 4.00 equiv.). Chloroform (10 mL) was added, and the resulting mixture was heated at 100 °C for 16 h. After cooling to room temperature, the reaction mixture was poured into an excess of saturated aqueous sodium hydrogen carbonate solution (50 mL) and stirred for 15 min. Subsequently, the reaction mixture was extracted with dichloromethane (3 × 50 mL). The combined organic layers were washed with brine (50 mL), dried over sodium sulfate, and reduced in a vacuum. The crude product was purified by flash column chromatography over silica gel (*c*Hex/CH_2Cl_2, 2:1 to 1:1) to yield 154 mg of the title compound (90.6 µmol, 93%) as a brown solid.

R_f (*c*Hex/CH_2Cl_2, 1:1) = 0.41.

^1H NMR (500 MHz, $CDCl_3$, ppm) δ = 7.50 (d, *J* = 1.7 Hz, 4H, H_{ar}), 7.25 (t, *J* = 2.3 Hz, 6H, H_{ar}), 7.01 (d, *J* = 8.7 Hz, 2H, H_{ar}), 6.97 – 6.89 (m, 12H, H_{ar}), 6.61 (ddd, *J* = 8.6, 6.6, 2.0 Hz, 6H, H_{ar}), 1.26 (s, 36H, CH_3), 1.22 (s, 36H, CH_3), 1.19 (s, 18H, CH_3).

^{13}C NMR (126 MHz, $CDCl_3$, ppm) δ = 143.2 (C_q), 142.9 (C_q), 142.8 (C_q), 141.1 (C_q), 138.3 (C_q), 138.1 (C_q), 137.5 (C_q), 137.3 (C_q), 137.2 (C_q), 125.9 (C_q), 124.4 (C_q), 124.3 (C_q), 124.1 (C_q), 123.1 (C_q), 122.3 (+, CH), 122.2 (C_q), 115.6 (+, CH), 115.2 (+, CH), 115.1 (+, CH), 110.4 (+, CH), 110.2 (+, CH), 109.2 (+, CH), 34.5 (C_q), 34.4 (C_q), 31.9 (+, CH_3, 24C, ^{t}Bu), 31.8 (+, CH_3, 6C, ^{t}Bu).

^{19}F NMR (376 MHz, $CDCl_3$, ppm) δ = -135.5 (qd), -148.0 (tt), -160.4 – -160.6 (m).

IR (ATR, cm^{-1}) $\tilde{v}$ = 2953 (s), 2904 (w), 2864 (w), 1519 (m), 1487 (s), 1470 (vs), 1448 (vs), 1392 (w), 1363 (s), 1324 (m), 1295 (vs), 1262 (vs), 1228 (s), 1201 (m), 1184 (w), 1152 (m), 1132 (w), 1103 (w), 1034 (w), 994 (s), 932 (w), 922 (w), 897 (w), 874 (m), 833 (m), 802 (vs), 773 (w), 761 (w), 739 (m), 671 (w), 654 (w), 609 (vs), 594 (w), 558 (m), 552 (m), 523 (w), 490 (w), 480 (w), 466 (m), 449 (w), 421 (m), 394 (w), 377 (w).

MS (ESI): m/z (%) = 1699 (67) $[M+H]^+$, 1698 (100) $[M]^+$. HRMS (ESI, $C_{114}H_{120}N_7OF_5$): calcd 1697.9475, found 1697.9488.

Mp 367 °C.

2-((2r,3r,4r,5s,6r)-2,3,4,5,6-Pentakis(3,6-di-*tert*-butyl-9*H*-carbazol-9-yl)phenyl)-5-(perfluoroheptyl)-1,3,4-oxadiazole (42c)

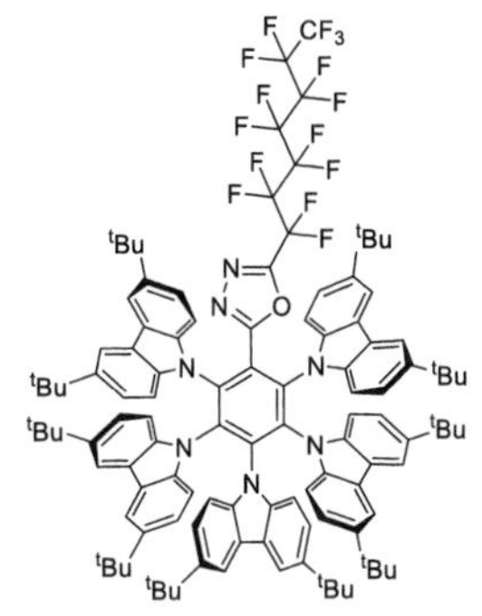

A sealable vial was charged with 9,9',9'',9''',9''''-((1r,2r,3r,4s,5r)-6-(2*H*-tetrazol-5-yl)benzene-1,2,3,4,5-pentayl)pentakis(3,6-di-tert-butyl-9*H*-carbazole) (**33**) (150 mg, 98.0 µmol, 1.00 equiv.) and 2,2,3,3,4,4,5,5,6,6,7,7,8,8,8-pentadecafluorooctanoyl chloride (169 mg, 391 µmol, 4.00 equiv.). Chloroform (10 mL) was added, and the resulting mixture was heated at 100 °C for 16 h. After cooling to room temperature, the reaction mixture was poured into an excess of saturated aqueous sodium hydrogen carbonate solution (50 mL) and stirred for 15 min. Subsequently, the reaction mixture was extracted with dichloromethane (3 × 50 mL). The combined organic layers were washed with brine (50 mL), dried over sodium sulfate, and reduced in a vacuum. The crude product was purified by flash column chromatography over silica gel (*c*Hex/CH_2Cl_2, 3:1) to yield 127 mg of the title compound (66.5 µmol, 68%) as a brown solid.

R_f (*c*Hex/CH_2Cl_2, 3:2) = 0.33.

^{1}H NMR (500 MHz, $CDCl_3$, ppm) δ = 7.57 (d, *J* = 2.0 Hz, 4H, H_{ar}), 7.26 (d, *J* = 3.0 Hz, 6H, H_{ar}), 7.00 (d, *J* = 8.7 Hz, 2H, H_{ar}), 6.96 – 6.86 (m, 12H, H_{ar}), 6.62 (td, *J* = 8.9, 2.0 Hz, 6H, H_{ar}), 1.27 (s, 36H, CH_3), 1.22 (s, 36H, CH_3), 1.20 (s, 18H, CH_3).

^{13}C NMR (126 MHz, $CDCl_3$, ppm) δ = 161.5 (C_q), 143.3 (C_q), 143.0 (C_q), 142.9 (C_q), 141.4 (C_q), 138.1 (C_q), 138.0 (C_q), 137.4 (C_q), 137.3 (C_q), 137.2 (C_q), 124.9 (C_q), 124.5 (C_q), 124.4 (C_q), 124.2 (C_q), 123.0 (+, CH), 122.3 (C_q), 122.2 (+, CH), 115.9 (+, CH), 115.2 (+, CH), 115.1 (+, CH), 110.4 (+, CH), 110.1 (+, CH), 108.9 (+, CH), 34.5 (C_q), 34.4 (C_q), 31.9 (+, CH_3, 12C, ^{t}Bu), 31.8 (+, CH_3, 6C, ^{t}Bu), 31.7 (+, CH_3, 12C, ^{t}Bu), 29.9 (C_q), 29.5 (C_q), 22.9 (C_q), 14.3 (C_q).

^{19}F NMR (376 MHz, $CDCl_3$, ppm) δ = -80.8 (t), -113.8, -122.1 (d), -122.8 (d), -125.9 – -126.4 (m).

IR (ATR, cm^{-1}) $\tilde{\nu}$ = 2955 (s), 2904 (w), 2866 (w), 1487 (s), 1470 (vs), 1448 (vs), 1392 (w), 1363 (s), 1324 (m), 1296 (vs), 1264 (s), 1231 (vs), 1203 (vs), 1149 (vs), 1106 (m), 1034 (m), 1000 (w), 875 (m), 841 (w), 803 (vs), 761 (w), 738 (m), 722 (w), 700 (w), 671 (w), 654 (w), 645 (w), 609 (vs), 592 (w), 561 (m), 551 (m), 524 (w), 489 (w), 467 (w), 421 (w).

MS (ESI): m/z (%) = 1901 (65) $[M+H]^+$, 1900 (75) $[M]^+$. HRMS (ESI, $C_{115}H_{120}N_7OF_{15}$): calcd 1899.9315, found 1899.9332.

Mp 172 °C.

2-(4-Methoxyphenyl)-5-((2r,3r,4r,5s,6r)-2,3,4,5,6-pentakis(3,6-di-*tert*-butyl-9*H*-carbazol-9-yl)phenyl)-1,3,4-oxadiazole (42d)

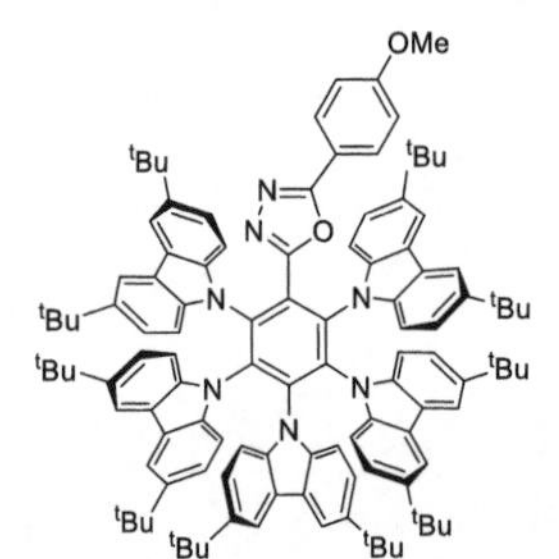

A sealable vial was charged with 9,9',9'',9''',9''''-((1r,2r,3r,4s,5r)-6-(2*H*-tetrazol-5-yl)benzene-1,2,3,4,5-pentayl)pentakis(3,6-di-*tert*-butyl-9*H*-carbazole) (**33**) (200 mg, 130 µmol, 1.00 equiv.) and 4-methoxybenzoyl chloride (89.0 mg, 522 µmol, 4.00 equiv.). Chloroform (10 mL) was added, and the resulting mixture was heated at 100 °C for 16 h. After cooling to room temperature, the reaction mixture was poured into an excess of saturated aqueous sodium hydrogen carbonate solution (50 mL) and stirred for 15 min. Subsequently, the reaction mixture was extracted with dichloromethane (3 × 50 mL). The combined organic layers were washed with brine (50 mL), dried over sodium sulfate, and reduced in a vacuum. The crude product was purified by flash column chromatography over silica gel (*n*-pentane/CH_2Cl_2, 1:1 to 1:3) to yield 183 mg of the title compound (111 µmol, 85%) as a yellow solid.

R_f (*n*-pentane/CH_2Cl_2, 3:2) = 0.21.

^1H NMR (500 MHz, CD_2Cl_2, ppm) δ = 7.65 (d, J = 1.9 Hz, 4H, H_{ar}), 7.37 (d, J = 1.9 Hz, 4H, H_{ar}), 7.33 (d, J = 1.9 Hz, 2H, H_{ar}), 7.07 (dd, J = 8.7, 2.3 Hz, 8H, H_{ar}), 7.04 – 6.98 (m, 6H, H_{ar}), 6.78 – 6.74 (m, 2H, H_{ar}), 6.71 (dd, J = 8.7, 1.9 Hz, 4H, H_{ar}), 6.67 – 6.60 (m, 4H, H_{ar}), 3.74 (s, 3H, CH_3), 1.26 (s, 36H, CH_3), 1.22 (s, 36H, CH_3), 1.22 (s, 18H, CH_3).

^{13}C NMR (126 MHz, CD_2Cl_2, ppm) δ = 164.3 (C_q), 162.6 (C_q), 158.0 (C_q), 143.8 (C_q), 143.6 (C_q), 140.3 (C_q), 139.1 (C_q), 138.6 (C_q), 138.0 (C_q), 137.9 (C_q), 137.4 (C_q), 128.7 (+, CH), 127.2 (+, CH), 124.6 (+, CH), 124.5 (+, CH), 124.4 (+, CH), 123.5 (+, CH), 122.7 (+, CH), 122.5 (+, CH), 116.6 (+, CH), 116.0 (+, CH), 115.9 (+, CH), 115.7 (+, CH), 114.1 (+, CH), 110.8 (+, CH), 110.7 (+, CH), 109.7 (+, CH), 55.8 (+, CH_3, OCH_3), 34.9 (C_q), 34.8 (C_q), 34.7 (C_q), 32.1 (+, CH_3), 32.0 (+, CH_3).

IR (ATR, cm^{-1}) $\tilde{\nu}$ = IR (ATR, $\tilde{\nu}$) = 2952 (s), 2902 (w), 2866 (w), 1612 (w), 1487 (s), 1472 (vs), 1449 (vs), 1392 (w), 1363 (m), 1324 (w), 1295 (vs), 1261 (vs), 1230 (s), 1201 (w), 1173 (m), 1152 (w), 1106 (w), 1034 (w), 874 (m), 837 (w), 802 (vs), 739 (m), 611 (s), 552 (w), 469 (w), 421 (w).

MS (ESI): m/z (%) = 1639 (62) $[M+H]^+$, 1638 (21) $[M]^+$. HRMS (ESI, $C_{115}H_{127}N_7O_2$): calcd 1638.0051, found 1637.9949.

Mp 375 °C.

2-(2,6-Dimethoxyphenyl)-5-((2r,3r,4r,5s,6r)-2,3,4,5,6-pentakis(3,6-di-*tert*-butyl-9*H*-carbazol-9-yl)phenyl)-1,3,4-oxadiazole (42e)

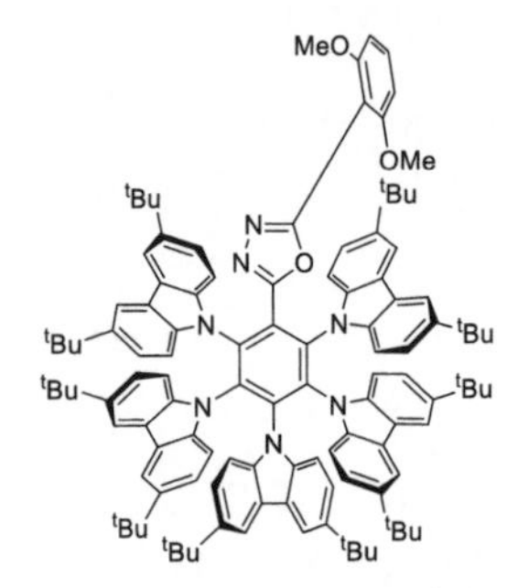

A sealable vial was charged with 9,9',9'',9''',9''''-((1r,2r,3r,4s,5r)-6-(2*H*-tetrazol-5-yl)benzene-1,2,3,4,5-pentayl)pentakis(3,6-di-*tert*-butyl-9*H*-carbazole) (**33**) (200 mg, 130 μmol, 1.00 equiv.) and 2,6-dimethoxybenzoyl chloride (105 mg, 522 μmol, 4.00 equiv.). Chloroform (10 mL) was added, and the resulting mixture was heated at 100 °C for 16 h. After cooling to room temperature, the reaction mixture was poured into an excess of saturated aqueous sodium hydrogen carbonate solution (50 mL) and stirred for 15 min. Subsequently, the reaction mixture was extracted with dichloromethane (3 × 50 mL). The combined organic layers were washed with brine (50 mL), dried over sodium sulfate, and reduced in a vacuum. The crude product was purified by flash column chromatography over silica gel (*n*-pentane/CH_2Cl_2, 3:1 to 3:2) to yield 166 mg of the title compound (99.2 μmol, 76%) as a yellow solid.

R_f (*n*-pentane/CH_2Cl_2, 1:3) = 0.32.

^{1}H NMR (500 MHz, $CDCl_3$, ppm) δ = 7.46 (d, *J* = 1.9 Hz, 4H, H_{ar}), 7.19 (dd, *J* = 3.9, 1.9 Hz, 6H, H_{ar}), 7.13 (d, *J* = 8.6 Hz, 2H, H_{ar}), 7.10 (d, *J* = 8.4 Hz, 1H, H_{ar}), 6.98 (d, *J* = 8.6 Hz, 4H, H_{ar}), 6.85 (dd, *J* = 8.6, 1.9 Hz, 4H, H_{ar}), 6.69 (d, *J* = 8.6 Hz, 4H, H_{ar}), 6.62 (dd, *J* = 8.7, 1.9 Hz, 2H, H_{ar}), 6.53 (dd, *J* = 8.6, 1.9 Hz, 4H, H_{ar}), 6.20 (d, *J* = 8.4 Hz, 2H, H_{ar}), 3.00 (s, 6H, OCH_3), 1.28 (s, 36H, CH_3), 1.22 (s, 36H, CH_3), 1.12 (s, 18H, CH_3).

^{13}C NMR (126 MHz, $CDCl_3$, ppm) δ = 160.4 (C_q), 159.9 (C_q), 159.7 (C_q), 142.8 (C_q), 142.6 (C_q), 142.3 (C_q), 141.8 (C_q), 138.3 (C_q), 138.1 (C_q), 138.0 (C_q), 137.1 (C_q), 136.4 (C_q), 132.6 (C_q), 128.7 (C_q), 124.5 (C_q), 124.2 (C_q), 123.9 (C_q), 122.7 (+, CH), 122.3 (+, CH), 121.7 (+, CH), 115.3 (+, CH), 115.0 (+, CH), 114.7 (+, CH), 110.9 (+,

CH), 110.1 (+, CH), 109.7, 104.6 (+, CH), 103.7 (+, CH), 102.5 (+, CH), 55.7 (+, OCH_3), 34.5 (C_q), 34.4 (C_q), 34.3 (C_q), 32.1 (+, CH_3), 32.0 (+, CH_3), 31.8 (+, CH_3).
IR (ATR, cm^{-1}) $\tilde{\nu}$ = 2952 (s), 2902 (w), 2864 (w), 1594 (w), 1472 (vs), 1449 (vs), 1392 (w), 1361 (s), 1324 (m), 1295 (vs), 1262 (vs), 1254 (vs), 1231 (s), 1201 (w), 1152 (w), 1116 (m), 1108 (m), 1034 (w), 1009 (w), 874 (m), 802 (vs), 739 (m), 609 (s), 594 (w), 561 (w), 550 (w), 467 (w), 419 (w).
MS (ESI): m/z (%) = 1670 (9) $[M+2H]^+$, 1668 (2) $[M]^+$. HRMS (ESI, $C_{112}H_{129}N_7O$): calcd 1670.0313 $[M+2H]^+$, found 1670.0129 $[M+2H]^+$.
Mp 392 °C.

2-Cyclohexyl-5-(2,3,4,5,6-pentakis(3,6-di-*tert*-butyl-9*H*-carbazol-9-yl)phenyl)-1,3,4-oxadiazole (42f)

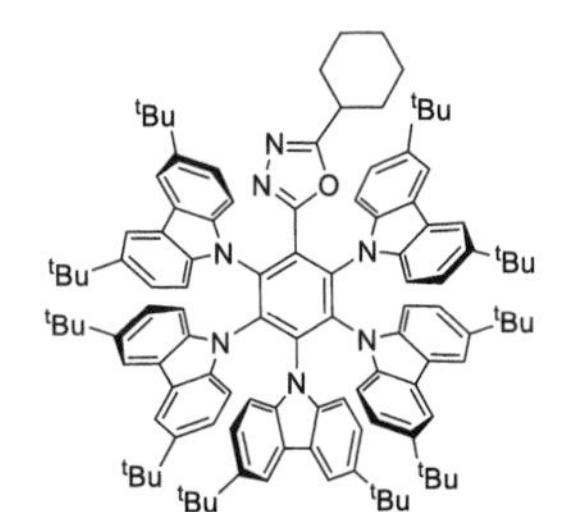

A sealable vial was charged with 9,9',9'',9''',9''''-((1r,2r,3r,4s,5r)-6-(2*H*-tetrazol-5-yl)benzene-1,2,3,4,5-pentayl)pentakis(3,6-di-*tert*-butyl-9*H*-carbazole) (**33**) (200 mg, 130 µmol, 1.00 equiv.) cyclohexanecarbonyl chloride (76.5 mg, 522 µmol, 4.00 equiv.). Chloroform (10 mL) was added, and the resulting mixture was heated at 100 °C for 16 h. After cooling to room temperature, the reaction mixture was poured into an excess of saturated aqueous sodium hydrogen carbonate solution (50 mL) and stirred for 15 min. Subsequently, the reaction mixture was extracted with dichloromethane (3 × 50 mL). The combined organic layers were washed with brine (50 mL), dried over sodium sulfate, and reduced in a vacuum. The crude product was purified by flash column chromatography over silica gel (*n*-pentane/CH_2Cl_2, 3:1 to 0:1) to yield 133 mg of the title compound (82.3 µmol, 63%) as a yellow solid.
R_f (*n*-pentane/CH_2Cl_2, 5:1) = 0.36.

^{1}H NMR (500 MHz, CD_2Cl_2, ppm) δ = 7.70 (s, 4H, H_{ar}), 7.34 (dd, J = 10.5, 2.0 Hz, 6H, H_{ar}), 7.06 – 6.98 (m, 14H, H_{ar}), 6.68 (dd, J = 8.6, 2.0 Hz, 4H, H_{ar}), 6.65 (dd, J = 8.6, 2.0 Hz, 2H, H_{ar}), 1.98 (tt, J = 10.9, 3.7 Hz, 1H, CH_2), 1.30 (s, 36H, CH_3), 1.21 (d, J = 1.9 Hz, 54H, CH_3), 1.11 – 0.80 (m, 8H, CH_2), 0.56 – 0.46 (m, 2H, CH_2).

^{13}C NMR (126 MHz, CD_2Cl_2, ppm) δ = 169.9 (C_q), 158.3 (C_q), 143.7 (C_q), 143.6 (C_q), 143.5 (C_q), 140.3 (C_q), 138.8 (C_q), 138.3 (C_q), 138.0 (C_q), 137.7 (C_q), 137.5 (C_q), 127.8 (C_q), 124.5 (+, CH), 124.4 (+, CH), 124.3 (+, CH), 123.4 (+, CH), 122.7 (+, CH), 122.6 (+, CH), 116.5 (+, CH), 115.9 (+, CH), 115.8 (+, CH), 110.8 (+, CH), 110.7 (+, CH), 109.7 (+, CH), 34.9 (C_q), 34.8 (C_q), 34.5 (C_q), 32.1 (+, CH_3), 32.0 (+, CH_3), 29.5 (-, CH_2), 25.9 (-, CH_2), 25.3 (-, CH_2).

IR (ATR, cm^{-1}) $\tilde{\nu}$ = 2952 (s), 2902 (w), 2864 (w), 1487 (s), 1472 (vs), 1449 (vs), 1392 (w), 1361 (s), 1324 (m), 1295 (vs), 1264 (vs), 1230 (s), 1201 (w), 1152 (w), 1106 (w), 1034 (w), 874 (m), 803 (vs), 739 (w), 609 (s), 594 (w), 551 (w), 463 (w), 421 (w).

MS (ESI): m/z (%) = 1616 (70) $[M+2H]^+$, 1615 (55) $[M+H]^+$, 1614 (4) $[M]^+$. HRMS (ESI, $C_{114}H_{131}N_7O$): calcd 1614.0415, found 1614.0309.

Mp 351 °C.

2-(*Tert*-butyl)-5-((2r,3r,4r,5s,6r)-2,3,4,5,6-pentakis(3,6-di-*tert*-butyl-9*H*-carbazol-9-yl)phenyl)-1,3,4-oxadiazole (42g)

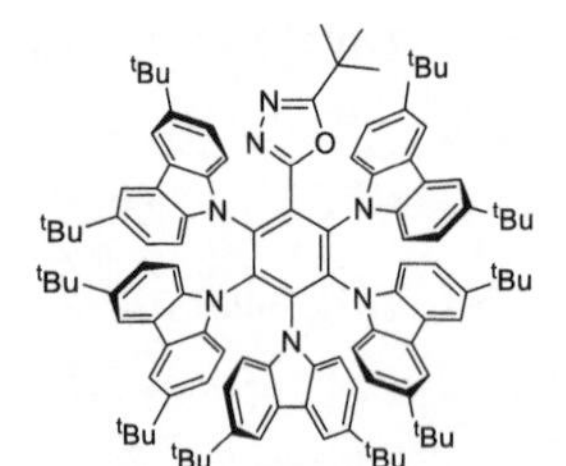

A sealable vial was charged with 9,9',9'',9''',9''''-((1r,2r,3r,4s,5r)-6-(2*H*-tetrazol-5-yl)benzene-1,2,3,4,5-pentayl)pentakis(3,6-di-*tert*-butyl-9*H*-carbazole) (**33**) (200 mg, 130 µmol, 1.00 equiv.) pivaloyl chloride (62.9 mg, 522 µmol, 4.00 equiv.). Chloroform (10 mL) was added, and the resulting mixture was heated at 100 °C for 16 h. After cooling to room temperature, the reaction mixture was poured into an excess of saturated aqueous sodium hydrogen carbonate solution (50 mL) and stirred for 15 min. Subsequently, the reaction mixture was extracted with dichloromethane (3 × 50 mL). The combined organic layers were washed with brine (50 mL), dried over sodium sulfate, and reduced in a vacuum. The crude product was purified by flash column chromatography over silica gel (*n*-pentane/CH_2Cl_2, 4:1 to 2:1) to yield 210 mg of the title compound (132 mmol, quant.) as a yellow solid.

R_f (*n*-pentane/CH_2Cl_2, 3:2) = 0.27.

^1H NMR (500 MHz, CD_2Cl_2, ppm) δ = 7.69 (t, *J* = 1.3 Hz, 4H, H_{ar}), 7.34 (dd, *J* = 5.8, 2.0 Hz, 6H, H_{ar}), 7.06 – 6.97 (m, 14H, H_{ar}), 6.67 (ddd, *J* = 11.6, 8.6, 2.0 Hz, 6H, H_{ar}), 1.29 (s, 36H, CH_3), 1.21 (s, 54H, CH_3), 0.49 (s, 9H, CH_3).

^{13}C NMR (126 MHz, CD_2Cl_2, ppm) δ = 173.3 (C_q), 158.4 (C_q), 143.7 (C_q), 143.6 (C_q), 143.5 (C_q), 140.3 (C_q), 138.7 (C_q), 138.2 (C_q), 137.9 (C_q), 137.7 (C_q), 137.5 (C_q), 127.9 (C_q), 124.5 (C_q), 124.2 (C_q), 123.3 (C_q), 122.7 (+, CH), 122.6 (+, CH), 116.5 (+, CH), 115.9 (+, CH), 115.8 (+, CH), 110.7 (+, CH), 109.7 (+, CH), 34.9 (C_q), 34.8 (C_q), 32.1 (+, CH_3), 32.0 (+, CH_3), 31.9 (C_q), 27.5 (+, CH_3).

IR (ATR, cm^{-1}) $\tilde{v}$ = 2953 (s), 2902 (w), 2866 (w), 1487 (s), 1472 (vs), 1449 (vs), 1392 (w), 1361 (s), 1324 (m), 1295 (vs), 1264 (vs), 1230 (s), 1201 (w), 1159 (w), 1152 (w), 1128 (w), 1106 (w), 1034 (w), 874 (m), 803 (vs), 739 (w), 611 (s), 552 (m), 467 (w), 419 (m).

MS (ESI): m/z (%) = 1590 (100) $[M+2H]^+$, 1589 (87) $[M+H]^+$, 1588 (13) $[M]^+$.

HRMS (ESI, $C_{112}H_{129}N_7O$): calcd 1588.0259, found 1588.0150.

Mp 334 °C.

2-Butyl-5-((2r,3r,4r,5s,6r)-2,3,4,5,6-pentakis(3,6-di-*tert*-butyl-9*H*-carbazol-9-yl)phenyl)-1,3,4-oxadiazole (42h)

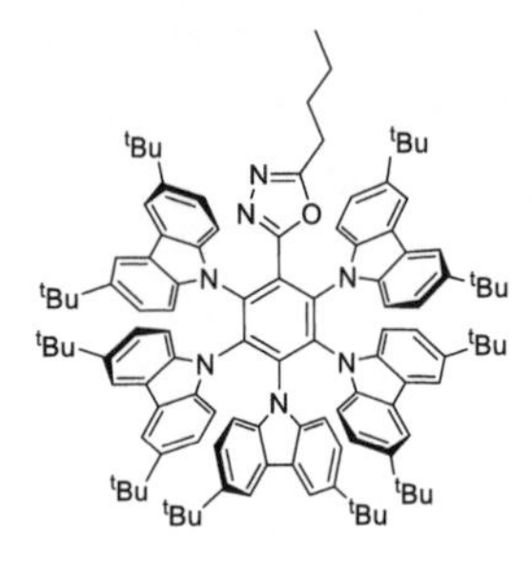

A sealable vial was charged with 9,9',9'',9''',9''''-((1r,2r,3r,4s,5r)-6-(2*H*-tetrazol-5-yl)benzene-1,2,3,4,5-pentayl)pentakis(3,6-di-*tert*-butyl-9*H*-carbazole) (**33**) (200 mg, 130 µmol, 1.00 equiv.) pentanoyl chloride (62.9 mg, 522 µmol, 4.00 equiv.). Chloroform (10 mL) was added, and the resulting mixture was heated at 100 °C for 16 h. After cooling to room temperature, the reaction mixture was poured into an excess of saturated aqueous sodium hydrogen carbonate solution (50 mL) and stirred for 15 min. Subsequently, the reaction mixture was extracted with dichloromethane (3 × 50 mL). The combined organic layers were washed with brine (50 mL), dried over sodium sulfate, and reduced in a vacuum. The crude product was purified by flash column chromatography over silica gel (*n*-pentane/CH_2Cl_2, 4:1 to 2:3) to yield 115 mg of the title compound (76.6 µmol, 56%) as a yellow solid.

R_f (*n*-pentane/CH_2Cl_2, 1:1) = 0.42.

^{1}H NMR (500 MHz, CD_2Cl_2, ppm) δ = 7.61 (s, 4H, H_{ar}), 7.26 (dd, *J* = 6.1, 1.8 Hz, 6H, H_{ar}), 6.96 – 6.90 (m, 14H, H_{ar}), 6.58 (td, *J* = 8.5, 2.0 Hz, 6H, H_{ar}), 1.82 (t, *J* = 7.0 Hz, 2H, CH_2), 1.22 (s, 36H, CH_3), 1.14 (s, 36H, CH_3), 1.12 (s, 18H, CH_3), 0.55 (p, *J* = 3.2 Hz, 4H, CH_2), 0.50 – 0.46 (m, 3H, CH_3).

^{13}C NMR (126 MHz, CD_2Cl_2, ppm) δ = 166.5 (C_q), 158.3 (C_q), 143.2 (C_q), 143.1 (C_q), 143.0 (C_q), 139.9 (C_q), 138.2 (C_q), 137.7 (C_q), 137.3 (C_q), 137.0 (C_q), 136.9 (C_q), 127.1 (C_q), 124.0 (C_q), 123.9 (C_q), 123.7 (C_q), 122.8 (+, CH), 122.1 (+, CH), 122.0 (+, CH), 115.9 (+, CH), 115.4 (+, CH), 115.3 (+, CH), 110.2 (+, CH), 110.1 (+, CH), 109.1 (+, CH), 34.4 (C_q), 34.2 (C_q), 31.5 (+, CH_3), 31.4 (+, CH_3), 27.6 (-, CH_2), 23.9 (-, CH_2), 21.4 (-, CH_2), 13.2 (+, CH_3).

IR (ATR, cm^{-1}) $\tilde{\nu}$ = 2952 (s), 2902 (w), 2864 (w), 1487 (s), 1472 (vs), 1449 (vs), 1392 (w), 1361 (s), 1324 (m), 1295 (vs), 1262 (vs), 1230 (s), 1201 (w), 1179 (w),

1152 (w), 1106 (w), 1034 (w), 874 (m), 803 (vs), 761 (w), 739 (m), 653 (w), 609 (s), 592 (w), 558 (w), 551 (w), 490 (w), 467 (w), 421 (w).

MS (ESI): m/z (%) = 1590 (17) $[M+2H]^+$, 1589 (13) $[M+H]^+$. HRMS (ESI, $C_{112}H_{129}N_7O$): calcd 1589.0337 $[M+H]^+$, found 1589.0232 $[M+H]^+$.

Mp 379 °C.

5.2.2.6. Oxadiazole Dimers

1,3-Bis(5-(2',3',4',5',6'-pentakis(3,6-di-*tert*-butyl-9*H*-carbazol-9-yl)-[1,1'-biphenyl]-4-yl)-1,3,4-oxadiazol-2-yl)benzene (49a)

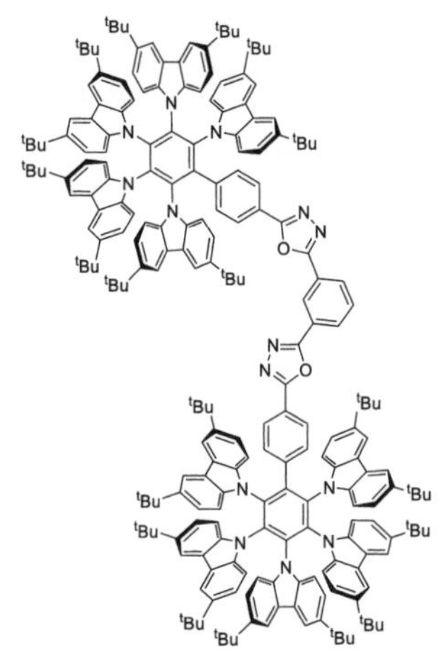

A sealable vial was charged with 9,9',9'',9''',9''''-((2s,3r,4s,5s,6s)-4'-(2*H*-tetrazol-5-yl)-[1,1'-biphenyl]-2,3,4,5,6-pentayl)pentakis(3,6-di-*tert*-butyl-9*H*-carbazole) (**38**) (390 mg, 242 μmol, 2.00 equiv.) and isophthaloyl dichloride (25.0 mg, 123 μmol, 1.00 equiv.). Chloroform (5 mL) was added, and the resulting mixture was heated at 100 °C for 16 h. After cooling to room temperature, the reaction mixture was poured into an excess of saturated aqueous sodium hydrogen carbonate solution (50 mL) and stirred for 15 min. Subsequently, the reaction mixture was extracted with dichloromethane (3 × 50 mL). The combined organic layers were washed with brine (50 mL), dried over sodium sulfate, and reduced in a vacuum. The crude product was purified by flash column chromatography over silica gel (*c*Hex/CH_2Cl_2, 2:1 to 1:1) to yield 48.6 mg of the title compound (14.8 μmol, 12%) as an off-white solid.

R_f (*c*Hex/CH_2Cl_2, 1:3) = 0.32.

^{1}H NMR (500 MHz, $CDCl_3$, ppm) δ = 8.35 (d, *J* = 1.8 Hz, 1H, H_{ar}), 8.00 (dd, *J* = 7.9, 1.8 Hz, 2H, H_{ar}), 7.66 (dd, *J* = 8.2, 1.3 Hz, 2H, H_{ar}), 7.61 (t, *J* = 7.6 Hz, 1H, H_{ar}), 7.46 (t, *J* = 1.3 Hz, 8H, H_{ar}), 7.24 (s, 2H, H_{ar}), 7.22 (dd, *J* = 8.5, 2.0 Hz, 12H, H_{ar}), 7.14 (d, *J* = 8.6 Hz, 4H, H_{ar}), 6.96 (d, *J* = 8.5 Hz, 4H, H_{ar}), 6.87 – 6.81 (m, 24H, H_{ar}), 6.65 (dd, *J* = 8.7, 2.0 Hz, 4H, H_{ar}), 6.57 (dd, *J* = 8.6, 2.0 Hz, 8H, H_{ar}), 1.25 (s, 72H, CH_3), 1.22 (s, 72H, CH_3), 1.16 (s, 36H, CH_3).

^{13}C NMR (126 MHz, $CDCl_3$, ppm) δ = 163.2, 142.6, 142.5, 142.3, 138.1, 138.1, 137.6, 137.3, 135.9, 132.9, 132.3, 129.3, 124.3, 124.0, 123.6, 122.5, 122.2, 121.7, 115.5, 114.8, 110.7, 110.3, 109.8, 34.5, 34.4, 32.0, 31.9, 31.8.
IR (ATR, cm^{-1}) ṽ = 2952 (s), 2902 (m), 2864 (w), 1487 (s), 1472 (vs), 1449 (vs), 1408 (w), 1392 (w), 1361 (s), 1324 (m), 1295 (vs), 1262 (vs), 1231 (s), 1201 (w), 1150 (m), 1106 (w), 1034 (w), 874 (m), 846 (w), 803 (vs), 771 (w), 761 (w), 737 (s), 714 (w), 687 (w), 609 (s), 592 (w), 558 (s), 490 (w), 466 (w), 421 (m).
MS (ESI): m/z (%) = 3293 (86) $[M+3H]^+$, 3292 (100) $[M+2H]^+$, 3291 (77) $[M+H]^+$, 3290 (31) $[M]^+$. HRMS (ESI, $C_{234}H_{252}N_{14}O_2$): calcd 3290.0048, found 3290.0220.

2,6-Bis(5-(2',3',4',5',6'-pentakis(3,6-di-*tert*-butyl-9*H*-carbazol-9-yl)-[1,1'-biphenyl]-4-yl)-1,3,4-oxadiazol-2-yl)pyridine (49b)

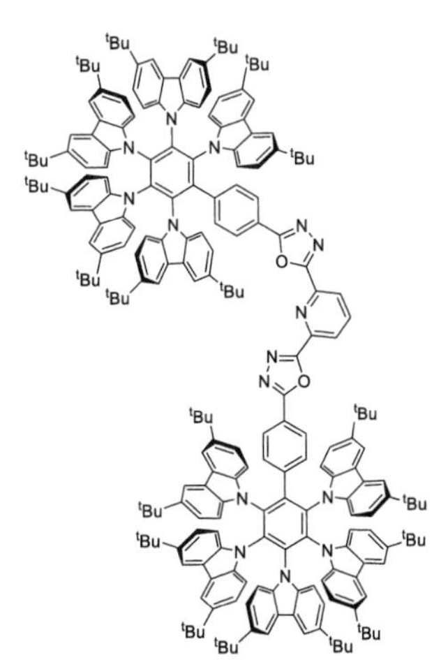

A sealable vial was charged with 9,9',9'',9''',9''''-((2s,3r,4s,5s,6s)-4'-(2*H*-tetrazol-5-yl)-[1,1'-biphenyl]-2,3,4,5,6-pentayl)pentakis(3,6-di-*tert*-butyl-9*H*-carbazole) (**38**) (222 mg, 138 µmol, 1.79 equiv.) and pyridine-2,6-dicarbonyl dichloride (15.7 mg, 77.0 µmol, 0.60 equiv.). Chloroform (5 mL) was added, and the resulting mixture was heated at 100 °C for 16 h. After cooling to room temperature, the reaction mixture was poured into an excess of saturated aqueous sodium hydrogen carbonate solution (50 mL) and stirred for 15 min. Subsequently, the reaction mixture was extracted with dichloromethane (3 × 50 mL). The combined organic layers were washed with brine (50 mL), dried over sodium sulfate, and reduced in a vacuum. The crude product was purified by flash column chromatography over silica gel (*c*Hex/CH_2Cl_2, 1:1 to 1:3) to yield 75.1 mg of the title compound (22.8 µmol, 30%) as a yellow solid.

R_f (CH_2Cl_2) = 0.33.

^{1}H NMR (500 MHz, $CDCl_3$, ppm) δ = 8.14 (d, *J* = 7.9 Hz, 2H, H_{ar}), 7.85 (t, *J* = 7.9 Hz, 1H, H_{ar}), 7.46 (t, *J* = 1.3 Hz, 8H, H_{ar}), 7.33 (d, *J* = 8.6 Hz, 4H, H_{ar}), 7.23 (dd, *J* = 7.6, 2.0 Hz, 12H, H_{ar}), 7.16 (d, *J* = 8.6 Hz, 4H, H_{ar}), 7.00 (d, *J* = 8.6 Hz, 4H, H_{ar}), 6.86 – 6.81 (m, 24H, H_{ar}), 6.67 (dd, *J* = 8.7, 2.0 Hz, 4H, H_{ar}), 6.58 (dd, *J* = 8.6, 2.0 Hz, 8H, H_{ar}), 1.29 (s, 72H, CH_3), 1.24 (s, 72H, CH_3), 1.17 (s, 36H, CH_3).

^{13}C NMR (126 MHz, $CDCl_3$, ppm) δ = 162.8, 144.2, 142.6, 142.4, 142.3, 138.1, 138.0, 137.5, 137.2, 135.9, 129.4, 124.3, 124.0, 123.6, 122.4, 122.2, 121.7, 115.5, 115.1, 114.8, 110.7, 110.3, 109.8, 34.5, 34.4, 32.1, 32.0, 31.9.

IR (ATR, cm^{-1}) $\tilde{\nu}$ = 2952 (s), 2931 (m), 2902 (m), 2864 (w), 1487 (s), 1470 (vs), 1448 (vs), 1409 (w), 1392 (w), 1361 (s), 1324 (m), 1295 (vs), 1262 (vs), 1230 (s), 1201 (w), 1150 (m), 1106 (w), 1092 (w), 1074 (w), 1034 (w), 1021 (w), 874 (m), 847 (w), 802 (vs), 782 (w), 771 (w), 761 (w), 738 (m), 714 (w), 688 (w), 653 (w), 609 (s), 592 (w), 558 (s), 490 (w), 469 (w), 419 (m).

MS (ESI): m/z (%) = 3293 (100) $[M+2H]^+$, 3292 (75) $[M+H]^+$, 3291 (28) $[M]^+$.

HRMS (ESI, $C_{233}H_{251}N_{15}O_2$): calcd 3291.0000, found 3291.0165.

Mp 280 °C.

1,2-Bis(5-(2',3',4',5',6'-pentakis(3,6-di-*tert*-butyl-9*H*-carbazol-9-yl)-[1,1'-biphenyl]-4-yl)-1,3,4-oxadiazol-2-yl)ethane (49c)

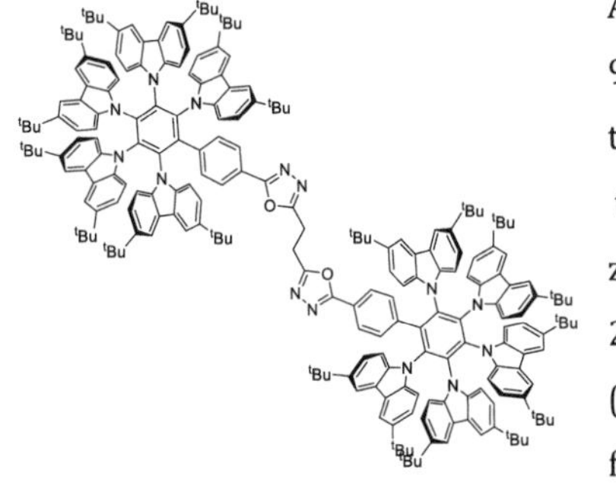

A sealable vial was charged with 9,9',9'',9''',9''''-((2s,3r,4s,5s,6s)-4'-(2*H*-tetrazol-5-yl)-[1,1'-biphenyl]-2,3,4,5,6-pentayl)pentakis(3,6-di-*tert*-butyl-9*H*-carbazole) (**38**) (305 mg, 190 µmol, 2.10 equiv.) and succinyl dichloride (14.0 mg, 90.0 µmol, 1.00 equiv.). Chloroform (5 mL) was added, and the resulting mixture was heated at 100 °C for 16 h. After cooling to room temperature, the reaction mixture was poured into an excess of saturated aqueous sodium hydrogen carbonate solution (50 mL) and stirred for 15 min. Subsequently, the reaction mixture was extracted with dichloromethane (3 × 50 mL). The combined organic layers were washed with brine (50 mL), dried over sodium sulfate, and reduced in a vacuum. The crude product was purified by flash column chromatography over silica gel (CH_2Cl_2/CH_3OH, 1:0 to 98:2) to yield 132 mg of the title compound (40.7 µmol, 45%) as a yellow solid.

R_f (*n*-pentane/CH_2Cl_2, 3:1) = 0.32.

^{1}H NMR (500 MHz, $CDCl_3$, ppm) δ = 7.44 (d, *J* = 1.7 Hz, 8H, H_{ar}), 7.21 (dd, *J* = 9.0, 2.0 Hz, 12H, H_{ar}), 7.12 (d, *J* = 8.7 Hz, 4H, H_{ar}), 7.07 (d, *J* = 8.6 Hz, 4H, H_{ar}), 6.89 (d, *J* = 8.5 Hz, 4H, H_{ar}), 6.85 – 6.79 (m, 24H, H_{ar}), 6.63 (dd, *J* = 8.7, 2.0 Hz, 4H, H_{ar}), 6.55 (dd, *J* = 8.6, 2.0 Hz, 8H, H_{ar}), 3.06 (s, 4H, CH_2), 1.23 (s, 72H, CH_3), 1.21 (s, 72H, CH_3), 1.15 (s, 36H, CH_3).

^{13}C NMR (126 MHz, $CDCl_3$, ppm) δ = 164.7 (C_q), 164.1 (C_q), 142.8 (C_q), 142.6 (C_q), 142.5 (C_q), 142.3 (C_q), 138.3 (C_q), 138.2 (C_q), 138.1 (C_q), 138.0 (C_q), 137.6 (C_q), 137.2 (C_q), 135.9 (C_q), 129.3 (+, CH), 126.0 (+, CH), 124.3 (C_q), 124.0 (C_q), 123.6 (C_q), 122.5 (+, CH), 122.3 (+, CH), 122.2 (+, CH), 121.7 (+, CH), 115.5 (+, CH), 115.0

(+, CH), 114.8 (+, CH), 110.7 (+, CH), 110.3 (+, CH), 109.8 (+, CH), 34.5 (C_q), 34.4 (C_q), 31.94 (+, CH_3), 31.9 (+, CH_3), 22.2 (-, CH_2).

IR (ATR, cm^{-1}) $\tilde{\nu}$ = 2952 (s), 2902 (w), 2866 (w), 1487 (s), 1470 (vs), 1449 (vs), 1409 (w), 1392 (w), 1361 (s), 1324 (m), 1295 (vs), 1262 (vs), 1230 (s), 1201 (m), 1150 (m), 1106 (w), 1034 (m), 1016 (w), 874 (m), 846 (w), 802 (vs), 771 (w), 762 (w), 738 (m), 708 (w), 690 (w), 679 (w), 654 (w), 609 (vs), 592 (w), 558 (s), 521 (w), 489 (w), 463 (m), 442 (w), 422 (m), 405 (w), 397 (w), 384 (w).

MS (ESI): m/z (%) = 3242 (2) $[M]^+$.

1-(5-(4-((2r,3s,6s)-2,3,5,6-Tetrakis(3,6-di-*tert*-butyl-9*H*-carbazol-9-yl)pyridin-4-yl)phenyl)-1,3,4-oxadiazol-2-yl)-3-(5-(4-((2r,5s,6s)-2,3,5,6-tetrakis(3,6-di-*tert*-butyl-9*H*-carbazol-9-yl)pyridin-4-yl)phenyl)-1,3,4-oxadiazol-2-yl)benzene (50a)

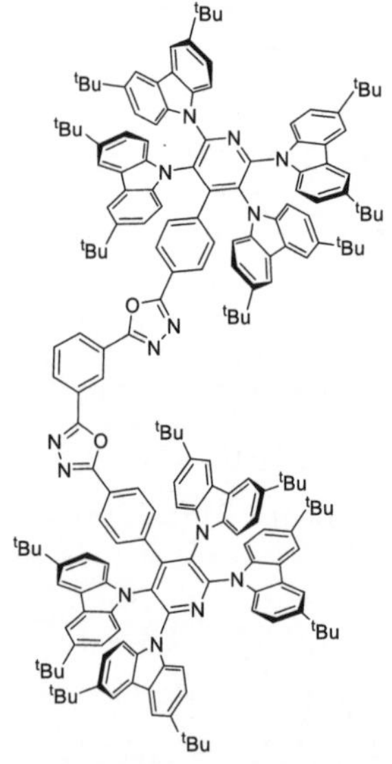

A sealable vial was charged with 9,9',9'',9'''-((2r,3s,5s,6s)-4-(4-(2*H*-tetrazol-5-yl)phenyl)pyridine-2,3,5,6-tetrayl)tetra-kis(3,6-di-*tert*-butyl-9*H*-carbazole) (**45**) (281 mg, 211 µmol, 2.10 equiv.) and pyridine-2,6-dicarbonyl dichloride (20.0 mg, 100 µmol, 1.00 equiv.). Chloroform (5 mL) was added, and the resulting mixture was heated at 100 °C for 16 h. After cooling to room temperature, the reaction mixture was poured into an excess of saturated aqueous sodium hydrogen carbonate solution (50 mL) and stirred for 15 min. Subsequently, the reaction mixture was extracted with dichloromethane (3 × 50 mL). The combined organic layers were washed with brine (50 mL), dried over sodium sulfate, and reduced in a vacuum. The crude product was purified by flash column chromatography over silica gel (toluene/EtOAc, 1:0 to 96:4) to yield 187 mg of the title compound (68.3 µmol, 69%) as a yellow solid.

R_f (toluene/EtOAc, 98:2) = 0.24.

^{1}H NMR (500 MHz, $CDCl_3$, ppm) δ = 8.43 (t, J = 1.8 Hz, 1H), 8.01 (dd, J = 7.9, 1.7 Hz, 2H), 7.58 (d, J = 2.0 Hz, 8H), 7.51 (d, J = 2.0 Hz, 8H), 7.41 (d, J = 8.2 Hz, 4H), 7.15 (d, J = 7.3 Hz, 4H), 7.08 (t, J = 8.7 Hz, 8H), 6.81 (ddd, J = 9.2, 7.6, 1.9 Hz, 17H), 6.66 (d, J = 8.6 Hz, 8H), 1.32 (s, 72H), 1.28 (s, 72H).

^{13}C NMR (126 MHz, $CDCl_3$, ppm) δ = 164.22 (C_q), 163.25 (C_q), 151.82 (C_q), 148.61 (C_q), 143.42 (C_q), 142.86 (C_q), 137.61 (C_q), 137.24 (C_q), 137.07 (C_q), 126.73 (+, CH), 126.66 (+, CH), 124.69 (+, CH), 124.11 (+, CH), 123.81 (+, CH), 122.96 (+, CH), 122.74 (+, CH), 122.66 (+, CH), 115.48 (+, CH), 115.08 (+, CH), 110.32 (+, CH), 109.17 (+, CH), 34.47 (C_q), 34.43 (C_q), 31.86 (+, CH_3), 31.83 (+, CH_3).

IR (ATR, cm^{-1}) $\tilde{v}$ = 2952 (s), 2901 (m), 2866 (w), 1489 (s), 1469 (vs), 1425 (vs), 1407 (vs), 1361 (s), 1322 (s), 1295 (vs), 1262 (vs), 1252 (s), 1231 (s), 1201 (m), 1147 (w), 1034 (m), 874 (m), 805 (vs), 739 (m), 727 (vs), 694 (m), 684 (m), 612 (s), 465 (m), 419 (m).

MS (ESI): m/z (%) = 2739 (7) $[M+H]^+$, 2738 (6) $[M]^+$. HRMS (ESI, $C_{192}H_{204}N_{14}O_2$): calcd 2737.6292, found 2737.6401.

2-(5-(4-((2r,3s,6s)-2,3,5,6-Tetrakis(3,6-di-*tert*-butyl-9*H*-carbazol-9-yl)pyridin-4-yl)phenyl)-1,3,4-oxadiazol-2-yl)-6-(5-(4-((2r,5s,6s)-2,3,5,6-tetrakis(3,6-di-*tert*-butyl-9*H*-carbazol-9-yl)pyridin-4-yl)phenyl)-1,3,4-oxadiazol-2-yl)pyridine (50b)

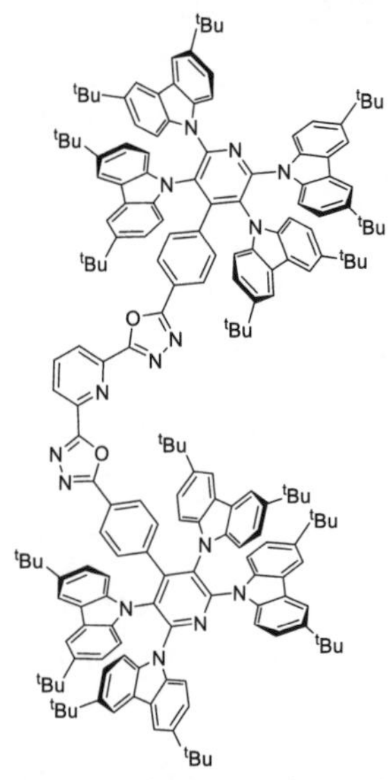

A sealable vial was charged with 9,9',9'',9'''-((2r,3s,5s,6s)-4-(4-(2*H*-tetrazol-5-yl)phenyl)pyridine-2,3,5,6-tetrayl)tetra-kis(3,6-di-*tert*-butyl-9*H*-carbazole) (**45**) (280 mg, 210 µmol, 2.10 equiv.) and pyridine-2,6-dicarbonyl dichloride (20.0 mg, 100 µmol, 1.00 equiv.). Chloroform (5 mL) was added, and the resulting mixture was heated at 100 °C for 16 h. After cooling to room temperature, the reaction mixture was poured into an excess of saturated aqueous sodium hydrogen carbonate solution (50 mL) and stirred for 15 min. Subsequently, the reaction mixture was extracted with dichloromethane (3 × 50 mL). The combined organic layers were washed with brine (50 mL), dried over sodium sulfate, and reduced in a vacuum. The crude product was purified by flash column chromatography over silica gel (*n*-pentane/CH_2Cl_2, 1:2 to 0:1) to yield 220 mg of the title compound (80.3 µmol, 82%) as a yellow solid.

R_f (CH_2Cl_2) = 0.28.

^{1}H NMR (500 MHz, $CDCl_3$, ppm) δ = 8.17 (d, *J* = 7.9 Hz, 2H, H_{ar}), 7.89 (t, *J* = 8.0 Hz, 1H, H_{ar}), 7.59 (d, *J* = 1.9 Hz, 8H, H_{ar}), 7.53 – 7.46 (m, 12H, H_{ar}), 7.08 (dd, *J* = 8.5, 3.1 Hz, 12H, H_{ar}), 6.83 (dd, *J* = 8.3, 1.8 Hz, 8H, H_{ar}), 6.79 (dd, *J* = 8.2, 1.8 Hz, 8H, H_{ar}), 6.64 (d, *J* = 8.5 Hz, 8H, H_{ar}), 1.33 (s, 72H, CH_3), 1.29 (s, 72H, CH_3).

^{13}C NMR (126 MHz, $CDCl_3$, ppm) δ = 165.3 (C_q), 162.9 (C_q), 148.7 (C_q), 144.1 (C_q), 143.6 (C_q), 143.0 (C_q), 137.7 (+, CH), 137.3 (+, CH), 129.1 (+, CH), 127.2 (+, CH), 126.8 (+, CH), 124.3 (+, CH), 123.9 (+, CH), 122.8 (+, CH), 122.8 (+, CH), 115.6 (+, CH), 115.2 (+, CH), 110.4 (+, CH), 109.3 (+, CH), 34.6 (C_q), 34.5 (C_q), 32.0 (+, CH_3).

IR (ATR, cm^{-1}) $\tilde{\nu}$ = 2952 (s), 2902 (m), 2864 (w), 1489 (s), 1469 (vs), 1425 (vs), 1408 (vs), 1361 (vs), 1322 (s), 1295 (vs), 1262 (vs), 1231 (vs), 1201 (m), 1166 (w), 1147 (m), 1105 (w), 1089 (w), 1077 (w), 1034 (m), 1020 (w), 874 (m), 846 (w), 841 (w), 805 (vs), 781 (w), 761 (w), 738 (s), 727 (m), 683 (w), 664 (w), 653 (w), 642 (w), 612 (s), 574 (w), 545 (w), 521 (m), 467 (m), 424 (m), 418 (m), 399 (w), 388 (w).

MS (ESI): m/z (%) = 2740 (2) $[M+H]^+$, 2739 (2) $[M]^+$. HRMS (ESI, $C_{191}H_{203}N_{15}O_2$): calcd 2738.6244, found 2738.6337.

Mp 308 °C.

2,6-Bis(5-(2,3,4,5,6-pentakis(3,6-di-*tert*-butyl-9*H*-carbazol-9-yl)phenyl)-1,3,4-oxadiazol-2-yl)pyridine (51)

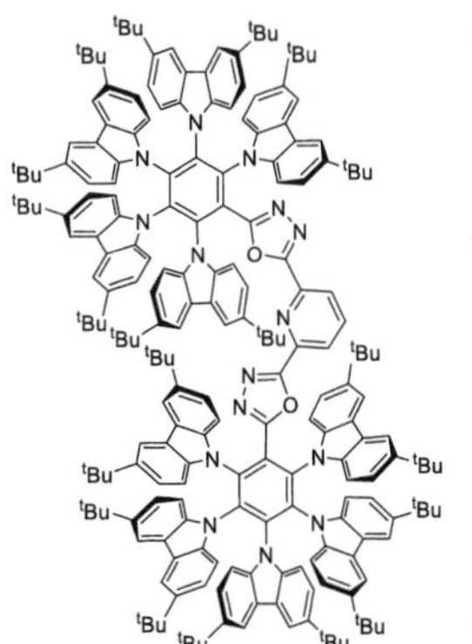

A sealable vial was charged with 9,9',9'',9''',9''''-((1r,2r,3r,4s,5r)-6-(2*H*-tetrazol-5-yl)benzene-1,2,3,4,5-pentayl)pentakis(3,6-di-tert-butyl-9*H*-carbazole) (**33**) (248 mg, 162 µmol, 2.10 equiv.) and pyridine-2,6-dicarbonyl dichloride (16.0 mg, 77.0 µmol, 1.00 equiv.). Chloroform (10 mL) was added, and the resulting mixture was heated at 100 °C for 16 h. After cooling to room temperature, the reaction mixture was poured into an excess of saturated aqueous sodium hydrogen carbonate solution (50 mL) and stirred for 15 min. Subsequently, the reaction mixture was extracted with dichloromethane (3 × 50 mL). The combined organic layers were washed with brine (50 mL), dried over sodium sulfate, and reduced in a vacuum. The crude product was purified by flash column chromatography over silica gel (*c*Hex/CH_2Cl_2, 1:1) to yield 38.6 mg of the title compound (12.3 µmol, 16%) as a brown solid.

R_f (*c*Hex/CH_2Cl_2, 1:3) = 0.65.

^1H NMR (500 MHz, $CDCl_3$, ppm) δ = 7.41 (d, *J* = 1.9 Hz, 8H, H_{ar}), 7.20 (d, *J* = 1.9 Hz, 12H, H_{ar}), 7.06 (d, *J* = 7.9 Hz, 1H, H_{ar}), 7.02 (d, *J* = 8.6 Hz, 4H, H_{ar}), 6.89 (d, *J* = 8.5 Hz, 8H, H_{ar}), 6.84 – 6.78 (m, 16H, H_{ar}), 6.60 (dd, *J* = 8.7, 2.0 Hz, 4H, H_{ar}), 6.55 (dd, *J* = 8.7, 1.9 Hz, 8H, H_{ar}), 6.35 (d, *J* = 8.1 Hz, 2H, H_{ar}), 1.21 (s, 72H, CH_3), 1.18 (s, 72H, CH_3), 1.16 (s, 36H, CH_3).

^{13}C NMR (126 MHz, $CDCl_3$, ppm) δ = 162.0, 159.2, 143.0, 142.9, 142.5, 141.3, 138.3, 138.0, 137.5, 137.4, 137.2, 124.4, 124.1, 124.0, 123.0, 122.3, 122.0, 115.6, 115.1, 114.9, 110.6, 110.1, 109.2, 34.5, 34.4, 32.2, 32.0, 31.9, 31.8, 29.9, 22.9.

IR (ATR, cm^{-1}) $\tilde{\nu}$ = 2953 (s), 2902 (w), 2864 (w), 1519 (m), 1487 (s), 1470 (vs), 1448 (vs), 1392 (w), 1363 (s), 1323 (m), 1310 (w), 1295 (vs), 1262 (vs), 1228 (s), 1201 (w), 1184 (w), 1152 (w), 1132 (w), 1102 (w), 1034 (w), 994 (s), 874 (m), 833 (m), 802 (vs), 773 (w), 761 (w), 739 (m), 671 (w), 654 (w), 609 (s), 594 (w), 558 (w), 552 (m), 492 (w), 469 (w), 421 (w).

MS (ESI): m/z (%) = 3140 (14) $[M+H]^+$, 3139 (11) $[M]^+$. HRMS (ESI, $C_{221}H_{243}N_{15}O_2$): calcd 3138.9374, found 3138.9438.

2,6-Bis(5-(3',5'-bis(3,6-di-*tert*-butyl-9*H*-carbazol-9-yl)-[1,1'-biphenyl]-4-yl)-1,3,4-oxadiazol-2-yl)pyridine (63a)

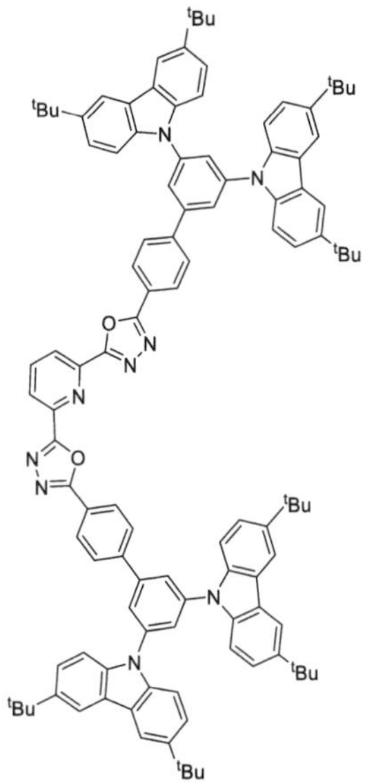

A sealable vial was charged with 9,9'-(4'-(2*H*-tetrazol-5-yl)-[1,1'-biphenyl]-3,5-diyl)bis(3,6-di-*tert*-butyl-9*H*-carbazole) (**61a**) (126 mg, 162 µmol, 2.10 equiv.) and pyridine-2,6-dicarbonyl dichloride (16.0 mg, 77.0 µmol, 1.00 equiv.). Chloroform (5 mL) was added, and the resulting mixture was heated at 100 °C for 16 h. After cooling to room temperature, the reaction mixture was poured into an excess of saturated aqueous sodium hydrogen carbonate solution (50 mL) and stirred for 15 min. Subsequently, the reaction mixture was extracted with dichloromethane (3 × 100 mL). The combined organic layers were washed with brine (50 mL), dried over sodium sulfate, and reduced in a vacuum. The crude product was purified by flash column chromatography over silica gel (*c*Hex/CH_2Cl_2, 3:1 to CH_2Cl_2/CH_3OH, 98:2) to yield 98.5 mg of the title compound (60.0 µmol, 77%) as a colorless solid.

R_f (CH_2Cl_2/CH_3OH, 99:1) = 0.24.

^{1}H NMR (500 MHz, $CDCl_3$, ppm) δ = 8.50 (d, *J* = 7.9 Hz, 2H, H_{ar}), 8.39 (d, *J* = 8.2 Hz, 4H, H_{ar}), 8.16 (d, *J* = 1.8 Hz, 9H, H_{ar}), 7.93 (d, *J* = 1.8 Hz, 4H, H_{ar}), 7.90 (d, *J* = 8.3 Hz, 4H, H_{ar}), 7.85 (s, 2H, H_{ar}), 7.56 – 7.47 (m, 16H, H_{ar}), 1.46 (s, 72H, CH_3).

^{13}C NMR (126 MHz, $CDCl_3$, ppm) δ = 165.8 (C_q), 163.5 (C_q), 144.5 (C_q), 143.6 (C_q), 143.4 (C_q), 143.1 (C_q), 140.6 (C_q), 139.0 (C_q), 138.8 (C_q), 128.4 (+, CH), 128.1 (+, CH), 124.1 (+, CH), 124.0 (+, CH), 123.9 (+, CH), 123.6 (+, CH), 123.3 (+, CH), 116.6 (+, CH), 109.3 (+, CH), 34.9 (C_q), 32.1 (+, CH_3).

IR (ATR, cm^{-1}) $\tilde{\nu}$ = 2951 (s), 2902 (w), 2866 (w), 1734 (vs), 1591 (s), 1574 (m), 1548 (w), 1487 (s), 1470 (vs), 1451 (vs), 1392 (m), 1363 (vs), 1343 (w), 1322 (s), 1295 (vs), 1261 (vs), 1230 (vs), 1200 (s), 1169 (vs), 1106 (m), 1082 (m), 1071 (m), 1034 (m), 1017 (m), 997 (w), 989 (w), 963 (w), 941 (w), 926 (w), 897 (w), 877 (s), 843 (s), 810 (vs), 788 (w), 779 (w), 748 (m), 739 (s), 720 (s), 694 (m), 683 (m), 653 (m), 613 (vs), 584 (w), 574 (w), 555 (w), 520 (w), 469 (w), 452 (w), 421 (m), 409 (w), 377 (w).

MS (ESI): m/z (%) = 1631 (22) $[M+3H]^+$, 1630 (59) $[M+2H]^+$, 1629 (100) $[M+H]^+$, 1628 (67) $[M]^+$. HRMS (ESI, $C_{113}H_{113}N_9O_2$): calcd 1627.9017, found 1627.9122.

2,6-Bis(5-(5-(3,5-bis(3,6-di-*tert*-butyl-9*H*-carbazol-9-yl)phenyl)pyridin-2-yl)-1,3,4-oxadiazol-2-yl)pyridine (63b)

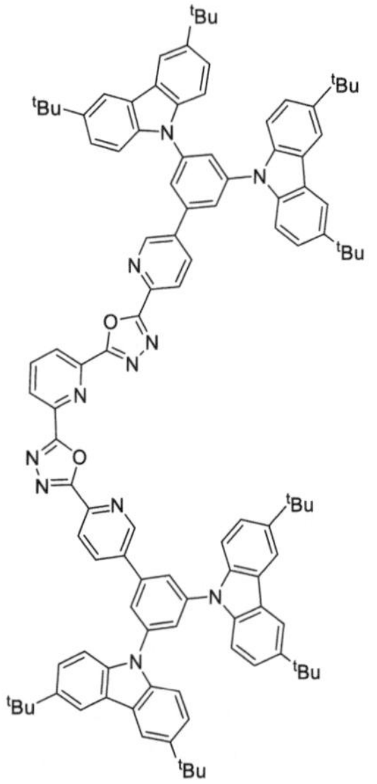

A sealable vial was charged with 9,9'-(5-(6-(2*H*-tetrazol-5-yl)pyridin-3-yl)-1,3-phenylene)bis(3,6-di-*tert*-butyl-9*H*-carbazole) (**61b**) (115 mg, 150 µmol, 2.10 equiv.) and pyridine-2,6-dicarbonyl dichloride (14.0 mg, 70.0 µmol, 1.00 equiv.). Chloroform (5 mL) was added, and the resulting mixture was heated at 100 °C for 16 h. After cooling to room temperature, the reaction mixture was poured into an excess of saturated aqueous sodium hydrogen carbonate solution (50 mL) and stirred for 15 min. Subsequently, the reaction mixture was extracted with dichloromethane (3 × 50 mL). The combined organic layers were washed with brine (50 mL), dried over sodium sulfate, and reduced in a vacuum. The crude product was purified by flash column chromatography over silica gel (toluene/EtOAc, 95:5 to 82:18) to yield 86.3 mg of the title compound (52.9 µmol, 77%) as a yellowish solid.

R_f (toluene/EtOAc, 9:1) = 0.31.

^{1}H NMR (500 MHz, $CDCl_3$, ppm) δ = 9.20 (d, *J* = 2.3 Hz, 2H, H_{ar}), 8.53 (dd, *J* = 12.0, 8.1 Hz, 4H, H_{ar}), 8.23 (dd, *J* = 8.2, 2.3 Hz, 2H, H_{ar}), 8.16 (d, *J* = 2.1 Hz, 9H, H_{ar}), 7.96 – 7.91 (m, 6H, H_{ar}), 7.56 – 7.48 (m, 16H, H_{ar}), 1.47 (s, 72H, CH_3).

^{13}C NMR (126 MHz, CD_2Cl_2, ppm) δ = 164.8 (C_q), 164.0 (C_q), 148.9 (+, CH), 144.2 (C_q), 143.7 (C_q), 142.5 (C_q), 140.9 (C_q), 139.8 (C_q), 138.7 (+, CH), 137.6 (C_q), 135.7 (+, CH), 125.8 (+, CH), 124.4 (+, CH), 124.0 (+, CH), 123.8 (+, CH), 123.3 (+, CH), 116.6 (+, CH), 109.1 (+, CH), 34.8 (C_q), 32.0 (+, CH_3).

MS (ESI): m/z (%) = 1631 (60) $[M+H]^+$, 1630 (100) $[M]^+$. HRMS (ESI, $C_{111}H_{111}N_{11}O_2$): calcd 1629.8922, found 1629.8998.

Mp 293 °C.

2,6-Bis(5-(4-(2,6-bis(3,6-di-*tert*-butyl-9*H*-carbazol-9-yl)pyridin-4-yl)phenyl)-1,3,4-oxadiazol-2-yl)pyridine (63c)

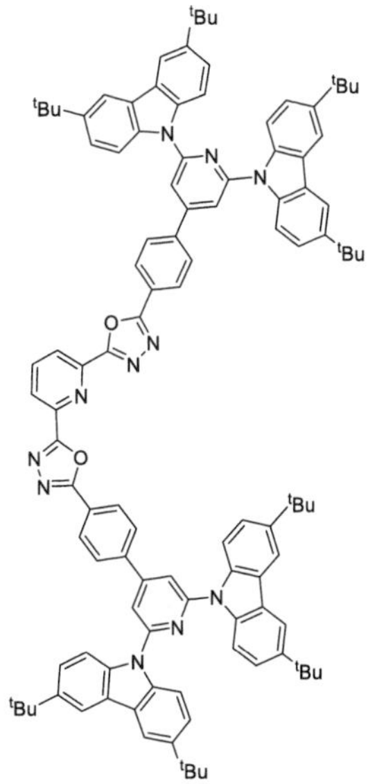

A sealable vial was charged with 9,9'-(4-(4-(2*H*-tetrazol-5-yl)phenyl)pyridine-2,6-diyl)bis(3,6-di-*tert*-butyl-9*H*-carbazole) (**61c**) (115 mg, 150 µmol, 2.10 equiv.) and pyridine-2,6-dicarbonyl dichloride (14.0 mg, 70.0 µmol, 1.00 equiv.). Chloroform (5 mL) was added, and the resulting mixture was heated at 100 °C for 16 h. After cooling to room temperature, the reaction mixture was poured into an excess of saturated aqueous sodium hydrogen carbonate solution (50 mL) and stirred for 15 min. Subsequently, the reaction mixture was extracted with dichloromethane (3 × 50 mL). The combined organic layers were washed with brine (50 mL), dried over sodium sulfate, and reduced in a vacuum. The crude product was purified by flash column chromatography over silica gel (*n*-pentane/CH_2Cl_2, 3:1 to CH_2Cl_2/CH_3OH, 99:1) to yield 85.4 mg of the title compound (52.4 µmol, 76%) as a yellow solid.

R_f (*n*-pentane/CH_2Cl_2, 1:1) = 0.13.

^1H NMR (500 MHz, $CDCl_3$, ppm) δ = 8.52 (d, *J* = 7.8 Hz, 2H, H_{ar}), 8.47 (d, *J* = 8.1 Hz, 4H, H_{ar}), 8.17 (t, *J* = 8.0 Hz, 1H, H_{ar}), 8.13 (d, *J* = 2.0 Hz, 8H, H_{ar}), 8.03 (d, *J* = 8.7 Hz, 8H, H_{ar}), 7.97 (d, *J* = 8.1 Hz, 4H, H_{ar}), 7.82 (s, 4H, H_{ar}), 7.47 (dd, *J* = 8.7, 2.0 Hz, 8H, H_{ar}), 1.45 (s, 72H, CH_3).

^{13}C NMR (126 MHz, $CDCl_3$, ppm) δ = 165.6 (C_q), 163.6 (C_q), 152.8 (C_q), 151.6 (C_q), 144.5 (C_q), 144.4 (C_q), 141.9 (C_q), 138.9 (C_q), 138.0 (C_q), 128.5 (+, CH), 128.2 (+, CH), 125.6 (+, CH), 124.9 (+, CH), 124.5 (+, CH), 124.2 (+, CH), 116.3 (+, CH), 111.9 (+, CH), 111.6 (+, CH), 34.9 (C_q), 32.0 (+, CH_3).

IR (ATR, cm^{-1}) $\tilde{\nu}$ = 2952 (m), 2902 (w), 2864 (w), 1598 (s), 1575 (w), 1555 (w), 1541 (m), 1490 (s), 1463 (vs), 1449 (vs), 1425 (vs), 1402 (vs), 1361 (s), 1295 (vs), 1261 (vs), 1228 (s), 1193 (s), 1167 (w), 1142 (w), 1125 (w), 1106 (w), 1085 (w), 1072 (w), 1034 (m), 1017 (w), 990 (w), 962 (w), 935 (w), 921 (w), 901 (w), 875 (m), 840 (m), 810 (vs), 781 (w), 764 (w), 747 (s), 738 (s), 714 (m), 691 (m), 683 (w), 670 (w), 653 (m), 642 (w), 612 (vs), 599 (m), 574 (w), 554 (w), 523 (w), 503 (w), 496 (w), 469 (m), 441 (w), 419 (m), 391 (w), 382 (w).

MS (ESI): m/z (%) = 1633 (20) $[M+3H]^+$, 1632 (61) $[M+2H]^+$, 1631 (100) $[M+H]^+$, 1630 (56) $[M]^+$. HRMS (ESI, $C_{111}H_{111}N_{11}O_2$): calcd 1629.8922, found 1629.8934.

Mp 293 °C.

2,6-Bis(5-(3,5-bis(3,6-di-*tert*-butyl-9*H*-carbazol-9-yl)phenyl)-1,3,4-oxadiazol-2-yl)pyridine (63d)

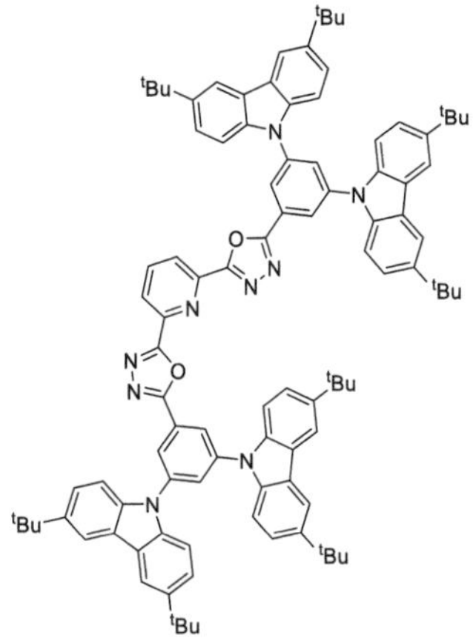

A sealable vial was charged with 9,9'-(5-(2*H*-tetrazol-5-yl)-1,3-phenylene)bis(3,6-di-*tert*-butyl-9*H*-carbazole) (**62**) (204 mg, 290 µmol, 2.10 equiv.) and pyridine-2,6-dicarbonyl dichloride (28.0 mg, 138 µmol, 1.00 equiv.). Chloroform (5 mL) was added, and the resulting mixture was heated at 100 °C for 16 h. After cooling to room temperature, the reaction mixture was poured into an excess of saturated aqueous sodium hydrogen carbonate solution (50 mL) and stirred for 15 min. Subsequently, the reaction mixture was extracted with dichloromethane (3 × 50 mL). The combined organic layers were washed with brine (50 mL), dried over sodium sulfate, and reduced in a vacuum. The crude product was purified by flash column chromatography over silica gel (*c*Hex/CH_2Cl_2, 3:1 to 0:1) to yield 37.4 mg of the title compound (25.3 µmol, 18%) as a colorless solid.

R_f (CH_2Cl_2) = 0.40.

^{1}H NMR (500 MHz, $CDCl_3$, ppm) δ = 8.58 (d, J = 8.0 Hz, 1H, H_{ar}), 8.43 (d, J = 1.9 Hz, 3H, H_{ar}), 8.16 (t, J = 7.9 Hz, 2H, H_{ar}), 8.12 (d, J = 1.5 Hz, 5H, H_{ar}), 7.96 (d, J = 2.2 Hz, 1H, H_{ar}), 7.93 (d, J = 1.8 Hz, 3H, H_{ar}), 7.81 (d, J = 7.7 Hz, 1H, H_{ar}), 7.46 (d, J = 2.4 Hz, 14H, H_{ar}), 7.32 – 7.28 (m, 3H, H_{ar}), 1.45 (s, 48H, CH_3), 1.38 (s, 24H, CH_3).

^{13}C NMR (126 MHz, $CDCl_3$, ppm) δ = 166.1 (C_q), 165.1 (C_q), 164.0 (C_q), 155.1 (C_q), 147.4 (C_q), 143.8 (C_q), 143.8 (C_q), 141.0 (C_q), 138.9 (+, CH), 138.8 (+, CH), 137.0 (C_q), 127.7 (+, CH), 127.0 (C_q), 126.9 (C_q), 125.3 (+, CH), 125.3 (+, CH), 124.8 (+, CH), 124.2 (+, CH), 124.0 (+, CH), 123.4 (+, CH), 116.6 (+, CH), 116.2 (+, CH), 115.7 (+, CH), 109.2 (+, CH), 34.9 (C_q), 32.1 (+, CH_3), 31.8 (+, CH_3).

IR (ATR, cm^{-1}) $\tilde{\nu}$ = 2952 (m), 2902 (w), 2866 (w), 1595 (m), 1543 (w), 1472 (vs), 1449 (s), 1392 (w), 1363 (s), 1317 (s), 1295 (vs), 1261 (vs), 1230 (s), 1201 (w), 1170 (w), 1035 (w), 877 (m), 807 (vs), 739 (s), 715 (w), 690 (m), 650 (w), 612 (vs), 469 (w), 418 (w).
MS (ESI): m/z (%) = 1478 (1) $[M+2H]^+$, 1477 (2) $[M+H]^+$, 1476 (1) $[M]^+$.

1,3-Bis(5-(3',5'-bis(3,6-di-*tert*-butyl-9*H*-carbazol-9-yl)-[1,1'-biphenyl]-4-yl)-1,3,4-oxadiazol-2-yl)benzene (64a)

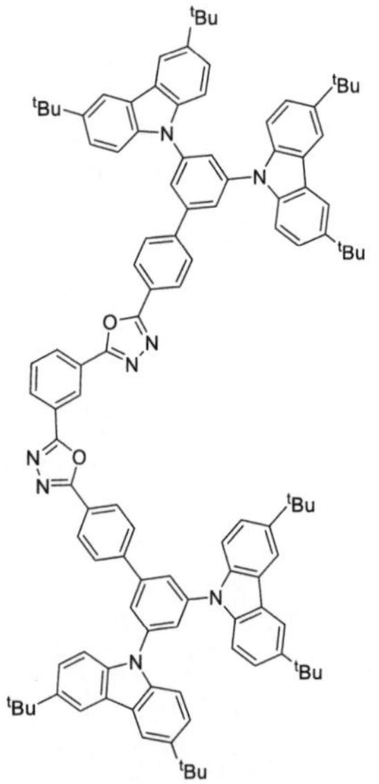

A sealable vial was charged with 9,9'-(4'-(2*H*-tetrazol-5-yl)-[1,1'-biphenyl]-3,5-diyl)bis(3,6-di-*tert*-butyl-9*H*-carbazole) (**61a**) (126 mg, 162 µmol, 2.10 equiv.) and isophthaloyl dichloride (16.0 mg, 77.0 µmol, 1.00 equiv.). Chloroform (5 mL) was added, and the resulting mixture was heated at 100 °C for 16 h. After cooling to room temperature, the reaction mixture was poured into an excess of saturated aqueous sodium hydrogen carbonate solution (50 mL) and stirred for 15 min. Subsequently, the reaction mixture was extracted with dichloromethane (3 × 100 mL). The combined organic layers were washed with brine (50 mL), dried over sodium sulfate, and reduced in a vacuum. The crude product was purified by flash column chromatography over silica gel (*c*Hex/CH_2Cl_2, 3:1 to CH_2Cl_2/CH_3OH, 97:3) to yield 71.2 mg of the title compound (44 µmol, 55%) as a yellow solid.
R_f (CH_2Cl_2/CH_3OH, 98:2) = 0.33.

^{1}H NMR (500 MHz, $CDCl_3$, ppm) δ = 8.98 – 8.95 (m, H_{ar}), 8.38 (dd, *J* = 7.8, 1.8 Hz, H_{ar}), 8.36 – 8.29 (m, H_{ar}), 8.17 (d, *J* = 1.8 Hz, H_{ar}), 8.00 – 7.84 (m, H_{ar}), 7.77 (t, *J* = 7.8 Hz, H_{ar}), 7.54 (h, *J* = 7.5, 6.9 Hz, H_{ar}), 1.48 (s, CH_3). The signals cannot be further analyzed due to the presence of rotamers.

^{13}C NMR (126 MHz, $CDCl_3$, ppm) δ = 143.6 (C_q), 143.1 (C_q), 140.6 (C_q), 139.0 (C_q), 128.1 (+, CH), 128.0 (+, CH), 124.1 (+, CH), 123.9 (+, CH), 123.6 (+, CH), 116.6 (+, CH), 109.3 (+, CH), 34.9 (C_q), 32.1 (+, CH_3).

IR (ATR, cm^{-1}) $\tilde{\nu}$ = 2952 (m), 1592 (s), 1575 (m), 1487 (s), 1472 (vs), 1452 (vs), 1392 (m), 1363 (s), 1322 (s), 1295 (vs), 1261 (vs), 1230 (s), 877 (s), 843 (m), 810 (vs), 742 (m), 720 (s), 694 (m), 684 (m), 613 (vs).

MS (ESI): m/z (%) = 1629 (5) $[M+2H]^+$, 1628 (8) $[M+H]^+$, 1627 (5) $[M]^+$. HRMS (ESI, $C_{114}H_{114}N_8O_2$): calcd 1626.9065, found 1626.9040.

Mp 312 °C.

1,2-Bis(5-(4-(2,6-bis(3,6-di-*tert*-butyl-9*H*-carbazol-9-yl)pyridin-4-yl)phenyl)-1,3,4-o-xadiazol-2-yl)benzene (64b)

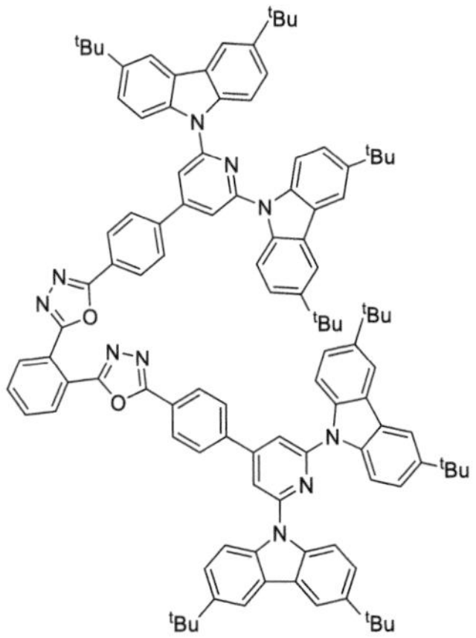

A sealable vial was charged with 9,9'-(4-(4-(2*H*-tetrazol-5-yl)phenyl)pyridine-2,6-diyl)bis(3,6-di-*tert*-butyl-9*H*-carbazole) (**61c**) (131 mg, 170 µmol, 2.10 equiv.) and phtaloyl dichloride (16.0 mg, 80.0 µmol, 1.00 equiv.). Chloroform (5 mL) was added, and the resulting mixture was heated at 100 °C for 16 h. After cooling to room temperature, the reaction mixture was poured into an excess of saturated aqueous sodium hydrogen carbonate solution (50 mL) and stirred for 15 min. Subsequently, the reaction mixture was extracted with dichloromethane (3 × 50 mL). The combined organic layers were washed with brine (50 mL), dried over sodium sulfate, and reduced in a vacuum. The crude product was purified by flash column chromatography over silica gel (*n*-pentane/CH_2Cl_2, 3:1 to CH_2Cl_2/CH_3OH, 95:5) to yield 111 mg of the title compound (68.2 µmol, 85%) as a yellow solid.

R_f (CH_2Cl_2/CH_3OH, 99:1) = 0.18.

^1H NMR (500 MHz, $CDCl_3$, ppm) δ = 8.21 (d, *J* = 8.1 Hz, 4H, H_{ar}), 8.17 (dd, *J* = 5.8, 3.3 Hz, 2H, H_{ar}), 8.13 (d, *J* = 1.9 Hz, 8H, H_{ar}), 8.00 (d, *J* = 8.7 Hz, 8H, H_{ar}), 7.88 (d, *J* = 8.3 Hz, 4H, H_{ar}), 7.82 (dd, *J* = 5.8, 3.4 Hz, 2H, H_{ar}), 7.76 (s, 4H, H_{ar}), 7.46 (dd, *J* = 8.7, 2.0 Hz, 8H, H_{ar}), 1.46 (s, 72H, CH_3).

^{13}C NMR (126 MHz, $CDCl_3$, ppm) δ = 164.8 (C_q), 163.6 (C_q), 152.7 (C_q), 151.4 (C_q), 144.5 (C_q), 141.4 (C_q), 138.0 (C_q), 132.1 (+, CH), 131.0 (+, CH), 128.1 (+, CH), 128.0 (+, CH), 124.9 (+, CH), 124.6 (+, CH), 124.2 (+, CH), 123.8 (+, CH), 116.3 (+, CH), 111.9 (+, CH), 111.5 (+, CH), 34.9 (C_q), 32.0 (+, CH_3).

IR (ATR, cm^{-1}) $\tilde{\nu}$ = 2952 (m), 2901 (w), 2864 (w), 1599 (s), 1574 (w), 1557 (w), 1541 (m), 1489 (m), 1465 (vs), 1451 (vs), 1425 (vs), 1402 (vs), 1361 (s), 1295 (vs), 1261 (vs), 1228 (s), 1191 (s), 1142 (w), 1106 (w), 1095 (w), 1077 (w), 1045 (w), 1034 (m), 1017 (m), 904 (m), 875 (m), 840 (s), 810 (vs), 762 (m), 745 (s), 730 (vs), 691 (m), 683 (w), 670 (w), 654 (m), 612 (vs), 601 (m), 575 (w), 561 (w), 551 (w), 523 (w), 469 (m), 450 (w), 443 (w), 421 (m), 398 (w) .

MS (ESI): m/z (%) = 1631 (65) $[M+2H]^+$,1630 (100) $[M+H]^+$, 1629 (73) $[M]^+$.

HRMS (ESI, $C_{112}H_{112}N_{10}O_2$): calcd 1628.8970, found 1628.9023.

1,3-Bis(5-(4-(2,6-bis(3,6-di-*tert*-butyl-9*H*-carbazol-9-yl)pyridin-4-yl)phenyl)-1,3,4-oxadiazol-2-yl)benzene (64c)

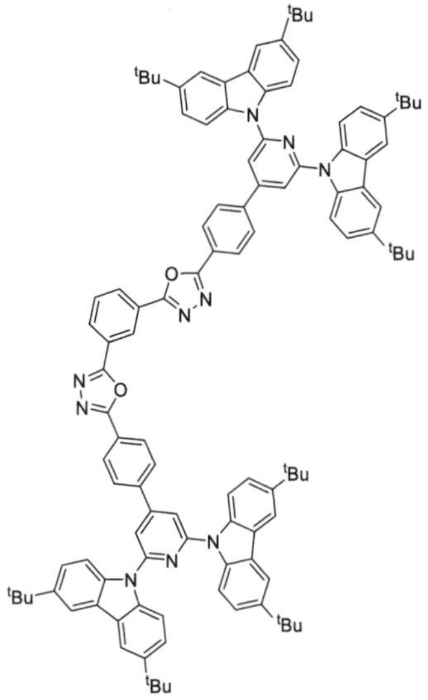

A sealable vial was charged with 9,9'-(4-(4-(2*H*-tetrazol-5-yl)phenyl)pyridine-2,6-diyl)bis(3,6-di-*tert*-butyl-9*H*-carbazole) (**61c**) (115 mg, 150 µmol, 2.10 equiv.) and isophthaloyl dichloride (14.0 mg, 70.0 µmol, 1.00 equiv.). Chloroform (5 mL) was added, and the resulting mixture was heated at 100 °C for 16 h. After cooling to room temperature, the reaction mixture was poured into an excess of saturated aqueous sodium hydrogen carbonate solution (50 mL) and stirred for 15 min. Subsequently, the reaction mixture was extracted with dichloromethane (3 × 50 mL). The combined organic layers were washed with brine (50 mL), dried over sodium sulfate, and reduced in a vacuum. The crude product was purified by flash column chromatography over silica gel (*c*Hex/CH_2Cl_2, 3:1 to CH_2Cl_2/CH_3OH, 9:1) to yield 42.0 mg of the title compound (25.8 µmol, 37%) as a yellow solid.

R_f (CH_2Cl_2/CH_3OH, 99:1) = 0.13.

^{1}H NMR (500 MHz, $CDCl_3$, ppm) δ = 8.40 (d, *J* = 7.9 Hz, 1H, H_{ar}), 8.20 (s, 4H, H_{ar}), 8.15 (s, 1H, H_{ar}), 8.04 (d, *J* = 8.5 Hz, 8H, H_{ar}), 7.93 – 7.87 (m, 8H, H_{ar}), 7.79 (d, *J* = 7.0 Hz, 4H, H_{ar}), 7.69 (s, 4H, H_{ar}), 7.48 (t, *J* = 8.1 Hz, 2H, H_{ar}), 7.33 (s, 8H, H_{ar}), 1.38 (s, 72H, CH_3).

^{13}C NMR (126 MHz, $CDCl_3$, ppm) δ = 158.2 (C_q), 152.7 (C_q), 151.6 (C_q), 144.6 (C_q), 144.5 (C_q), 140.8 (C_q), 138.0 (C_q), 128.3 (+, CH), 128.2 (+, CH), 125.8 (+, CH), 124.9 (+, CH), 124.8 (+, CH), 124.3 (+, CH), 116.3 (+, CH), 111.9 (+, CH), 111.7 (+, CH), 34.9 (C_q), 32.0 (+, CH_3).

IR (ATR, cm^{-1}) $\tilde{\nu}$ = 2956 (m), 2902 (w), 2864 (w), 1599 (s), 1575 (w), 1543 (m), 1489 (m), 1465 (vs), 1449 (vs), 1431 (vs), 1407 (vs), 1361 (s), 1295 (vs), 1261 (vs), 1228 (s), 1193 (m), 1160 (w), 1139 (w), 1095 (s), 1033 (s), 1018 (s), 901 (w), 875 (s), 840 (s), 809 (vs), 764 (s), 748 (m), 741 (m), 718 (w), 704 (w), 691 (m), 681 (m), 669 (w), 656 (m), 642 (w), 628 (w), 612 (s), 599 (m), 574 (w), 548 (w), 523 (m), 504 (w), 494 (w), 469 (m), 456 (m), 443 (w), 432 (w), 421 (s), 404 (m), 395 (m), 387 (m), 378 (m).

MS (ESI): m/z (%) = 1631 (21) $[M+2H]^+$, 1630 (34) $[M+H]^+$, 1629 (24) $[M]^+$. HRMS (ESI, $C_{112}H_{112}N_{10}O_2$): calcd 1628.8970, found 1628.8969.

1,4-Bis(5-(4-(2,6-bis(3,6-di-*tert*-butyl-9*H*-carbazol-9-yl)pyridin-4-yl)phenyl)-1,3,4-oxadiazol-2-yl)benzene (64d)

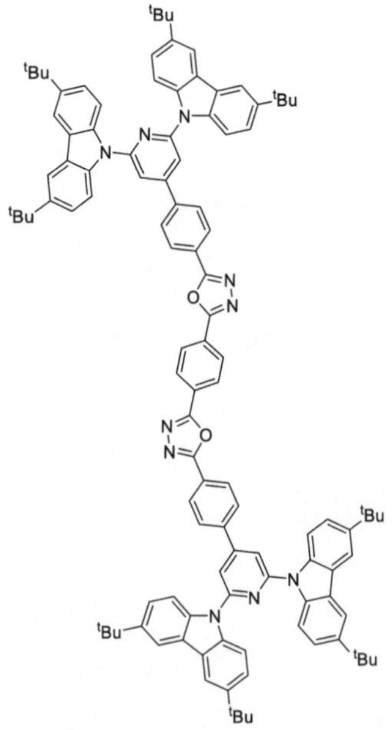

A sealable vial was charged with 9,9'-(4-(4-(2*H*-tetrazol-5-yl)phenyl)pyridine-2,6-diyl)bis(3,6-di-tert-butyl-9*H*-carbazole) (**61c**) (132 mg, 170 µmol, 2.10 equiv.) and terephthaloyl dichloride (16.0 mg, 80 µmol, 1.00 equiv.). Chloroform (5 mL) was added, and the resulting mixture was heated at 100 °C for 16 h. After cooling to room temperature, the reaction mixture was poured into an excess of saturated aqueous sodium hydrogen carbonate solution (50 mL) and stirred for 15 min. Subsequently, the reaction mixture was extracted with dichloromethane (3 × 50 mL). The combined organic layers were washed with brine

(50 mL), dried over sodium sulfate, and reduced in a vacuum. The crude product was purified by flash column chromatography over silica gel (toluene/EtOAc, 1:0 to 97:3) to yield 66.7 mg of the title compound (40.3 mmol, 51%) as a yellow solid.

R_f (CH_2Cl_2/CH_3OH, 99:1) = 0.38.

^{1}H NMR (500 MHz, $CDCl_3$, ppm) δ = 8.42 – 8.34 (m, 8H, H_{ar}), 8.16 (d, J = 2.0 Hz, 8H, H_{ar}), 8.05 (d, J = 8.7 Hz, 8H, H_{ar}), 8.00 – 7.95 (m, 4H, H_{ar}), 7.84 (s, 4H, H_{ar}), 7.50 (dd, J = 8.8, 2.0 Hz, 8H, H_{ar}), 1.48 (s, 72H, CH_3).

^{13}C NMR (126 MHz, $CDCl_3$, ppm) δ = 152.8 (C_q), 151.5 (C_q), 144.5 (C_q), 141.6 (C_q), 138.0 (C_q), 128.2 (+, CH), 128.1 (+, CH), 127.8 (+, CH), 124.9 (+, CH), 124.3 (+, CH), 116.3 (+, CH), 111.9 (+, CH), 111.6 (+, CH), 34.9 (C_q), 32.1 (+, CH_3).

IR (ATR, cm^{-1}) $\tilde{\nu}$ = 2953 (w), 2902 (w), 2867 (w), 1594 (s), 1572 (m), 1551 (m), 1537 (w), 1492 (s), 1476 (vs), 1466 (vs), 1452 (s), 1428 (vs), 1402 (vs), 1363 (m), 1324 (w), 1312 (m), 1296 (vs), 1264 (vs), 1228 (m), 1193 (m), 1108 (w), 1096 (w), 1068 (w), 1064 (w), 1034 (w), 1014 (w), 871 (m), 850 (w), 836 (s), 809 (vs), 789 (w), 764 (w), 742 (m), 711 (s), 693 (w), 684 (w), 657 (w), 615 (m), 599 (w), 513 (w), 497 (w), 470 (w), 418 (w).

MS (ESI): m/z (%) = 1632 (20) $[M+3H]^+$, 1631 (61) $[M+2H]^+$, 1630 (100) $[M+H]^+$, 1629 (49) $[M]^+$. HRMS (ESI, $C_{112}H_{112}N_{10}O_2$): calcd 1628.8970, found 1628.9021.

5,5'-Bis(3',5'-bis(3,6-di-*tert*-butyl-9*H*-carbazol-9-yl)-[1,1'-biphenyl]-4-yl)-2,2'-bi(1,3,4-oxadiazole) (65a)

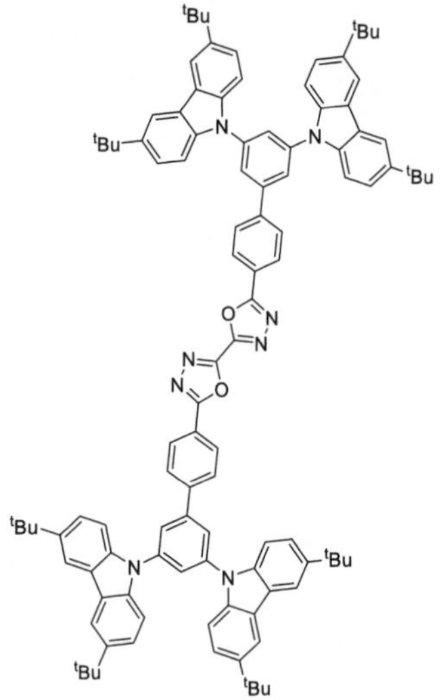

A sealable vial was charged with 9,9'-(4'-(2*H*-tetrazol-5-yl)-[1,1'-biphenyl]-3,5-diyl)bis(3,6-di-*tert*-butyl-9*H*-carbazole) (**61a**) (262 mg, 337 µmol, 2.14 equiv.) and oxalyl dichloride (20.0 mg, 158 µmol, 1.00 equiv.). Chloroform (10 mL) was added, and the resulting mixture was heated at 100 °C for 16 h. After cooling to room temperature, the reaction mixture was poured into an excess of saturated aqueous sodium hydrogen carbonate solution (50 mL) and stirred for 15 min. Subsequently, the reaction mixture was extracted with dichloromethane (3 × 50 mL). The combined organic layers were washed with brine (50 mL), dried over sodium sulfate, and reduced in a vacuum. The crude product was purified by flash column chromatography over silica gel (*c*Hex/CH_2Cl_2, 3:1 to 0:1) to yield 127 mg of the title compound (81.7 µmol, 52%) as a yellow solid.

R_f (CH_2Cl_2/CH_3OH, 98:2) = 0.39.

^1H NMR (500 MHz, $CDCl_3$, ppm) δ = 8.37 (d, *J* = 8.3 Hz, 4H, H_{ar}), 8.17 (d, *J* = 1.7 Hz, 8H, H_{ar}), 7.95 (d, *J* = 1.8 Hz, 4H, H_{ar}), 7.92 (d, *J* = 8.3 Hz, 4H, H_{ar}), 7.87 (s, 2H, H_{ar}), 7.57 – 7.50 (m, 16H, H_{ar}), 1.48 (s, 72H, CH_3).

^{13}C NMR (126 MHz, $CDCl_3$, ppm) δ = 143.7 (C_q), 140.9 (C_q), 139.0 (C_q), 128.6 (+, CH), 128.3 (+, CH), 124.1 (+, CH), 123.9 (+, CH), 116.7 (+, CH), 109.3 (+, CH), 34.9 (C_q), 32.2 (+, CH_3).

IR (ATR , cm^{-1}) $\tilde{\nu}$ = 2958 (m), 2904 (w), 2866 (w), 1592 (s), 1574 (m), 1489 (s), 1472 (vs), 1452 (vs), 1404 (w), 1394 (w), 1364 (m), 1322 (s), 1295 (vs), 1262 (s), 1228 (m), 1203 (w), 1170 (w), 1160 (w), 1084 (w), 1035 (w), 878 (m), 841 (m), 810 (vs), 745 (w), 742 (w), 720 (w), 696 (w), 683 (w), 654 (w), 615 (s), 470 (w), 421 (w).

MS (ESI): m/z (%) = 1554 (21) $[M+3H]^+$, 1553 (58) $[M+2H]^+$, 1552 (100) $[M+H]^+$, 1551 (75) $[M]^+$. HRMS (ESI, $C_{108}H_{110}N_8O_2$): calcd 1550.8752, found 1550.8810.

EA ($C_{108}H_{110}N_8O_2$) calc. C: 83.58, H: 7.14, N: 7.22; found C: 82.61, H: 7.00, N: 7.08.

Mp 347 °C.

5,5'-Bis(4-(2,6-bis(3,6-di-*tert*-butyl-9*H*-carbazol-9-yl)pyridin-4-yl)phenyl)-2,2'-bi(1,3,4-oxadiazole) (65b)

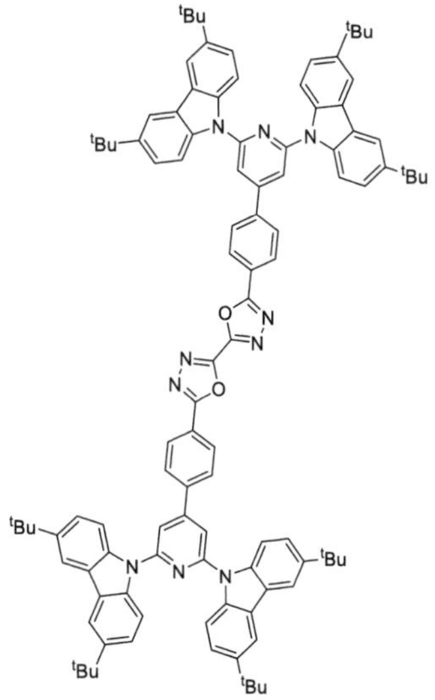

A sealable vial was charged with 9,9'-(4-(4-(2*H*-tetrazol-5-yl)phenyl)pyridine-2,6-diyl)bis(3,6-di-*tert*-butyl-9*H*-carbazole) (**61c**) (279 mg, 360 µmol, 2.10 equiv.) and oxalyl dichloride (22.0 mg, 170 µmol, 1.00 equiv.). Chloroform (10 mL) was added, and the resulting mixture was heated at 100 °C for 16 h. After cooling to room temperature, the reaction mixture was poured into an excess of saturated aqueous sodium hydrogen carbonate solution (50 mL) and stirred for 15 min. Subsequently, the reaction mixture was extracted with dichloromethane (3 × 50 mL). The combined organic layers were washed with brine (50 mL), dried over sodium sulfate, and reduced in a vacuum. The crude product was purified by flash column chromatography over silica gel (*n*-pentane/CH_2Cl_2, 3:1 to CH_2Cl_2/CH_3OH, 95:5) to yield 154 mg of the title compound (99.2 µmol, 57%) as a yellow solid.

R_f (CH_2Cl_2) = 0.21.

^{1}H NMR (500 MHz, $CDCl_3$, ppm) δ = 8.44 (d, J = 8.2 Hz, 4H, H_{ar}), 8.16 (d, J = 2.0 Hz, 8H, H_{ar}), 8.05 (d, J = 8.7 Hz, 8H, H_{ar}), 8.00 (d, J = 8.1 Hz, 4H, H_{ar}), 7.84 (s, 4H, H_{ar}), 7.50 (dd, J = 8.8, 2.0 Hz, 8H, H_{ar}), 1.48 (s, 72H, CH_3).

^{13}C NMR (126 MHz, $CDCl_3$, ppm) δ = 166.0 (C_q), 153.3 (C_q), 152.9 (C_q), 151.3 (C_q), 144.6 (C_q), 142.6 (C_q), 138.0 (C_q), 128.8 (+, CH), 128.4 (+, CH), 124.9 (+, CH), 124.3 (+, CH), 123.5 (+, CH), 116.3 (+, CH), 111.9 (+, CH), 111.6 (+, CH), 35.0 (C_q), 32.1 (+, CH_3).

IR (ATR, cm^{-1}) $\tilde{\nu}$ = 2952 (m), 2902 (w), 2866 (w), 1599 (s), 1572 (w), 1557 (w), 1543 (m), 1493 (s), 1465 (vs), 1451 (s), 1426 (vs), 1402 (vs), 1361 (s), 1323 (w), 1293 (vs), 1261 (vs), 1224 (m), 1191 (m), 1156 (w), 1143 (w), 1109 (w), 1098 (w), 1079 (w), 1034 (w), 1017 (w), 952 (w), 888 (w), 877 (w), 840 (m), 812 (vs), 802 (s), 765 (w), 747 (m), 711 (w), 691 (w), 680 (w), 656 (w), 613 (m), 602 (w), 470 (w), 419 (w).

MS (ESI): m/z (%) = 1555 (56) $[M+2H]^+$, 1554 (100) $[M+H]^+$, 1553 (46) $[M]^+$.

HRMS (ESI, $C_{106}H_{108}N_{10}O_2$): calcd 1552.8657, found 1552.8684.

EA ($C_{106}H_{108}N_{10}O_2$) calc. C: 81.92, H: 7.00, N: 9.01; found C: 81.20, H: 6.92, N: 8.87.

Mp > 400 °C.

1,2-Bis(5-(3',5'-bis(3,6-di-*tert*-butyl-9*H*-carbazol-9-yl)-[1,1'-biphenyl]-4-yl)-1,3,4-oxadiazol-2-yl)ethane (66a)

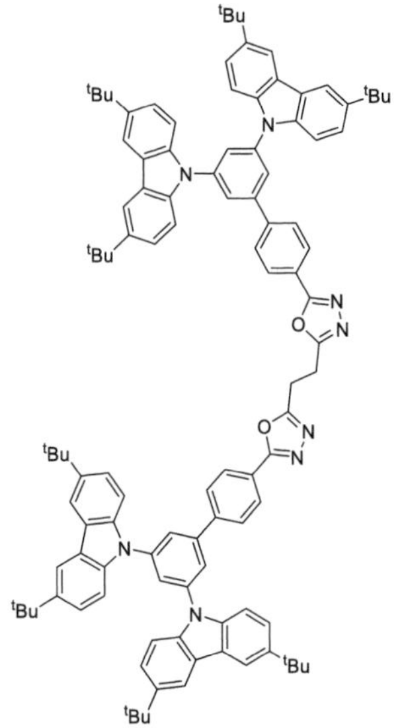

A sealable vial was charged with 9,9'-(4'-(2*H*-tetrazol-5-yl)-[1,1'-biphenyl]-3,5-diyl)bis(3,6-di-*tert*-butyl-9*H*-carbazole) (**61a**) (200 mg, 257 µmol, 2.10 equiv.) and succinyl dichloride (19.0 mg, 123 µmol, 1.00 equiv.). Chloroform (7 mL) was added, and the resulting mixture was heated at 100 °C for 16 h. After cooling to room temperature, the reaction mixture was poured into an excess of saturated aqueous sodium hydrogen carbonate solution (50 mL) and stirred for 15 min. Subsequently, the reaction mixture was extracted with dichloromethane (3 × 50 mL). The combined organic layers were washed with brine (50 mL), dried over sodium sulfate, and reduced in a vacuum. The crude product was purified by flash column chromatography over silica gel (CH_2Cl_2/CH_3OH, 1:0 to 98:2) to yield 76.5 mg of the title compound (48.4 µmol, 39%) as a colorless solid.

R_f (CH_2Cl_2) = 0.28.

^{1}H NMR (500 MHz, $CDCl_3$, ppm) δ = 8.29 – 8.25 (m, 4H, H_{ar}), 8.18 – 8.15 (m, 8H, H_{ar}), 7.92 (d, *J* = 1.8 Hz, 4H, H_{ar}), 7.85 (t, *J* = 1.9 Hz, 2H, H_{ar}), 7.83 – 7.79 (m, 4H, H_{ar}), 7.55 – 7.49 (m, 16H, H_{ar}), 2.93 (s, 4H, CH_2), 1.48 (s, 72H, CH_3).

^{13}C NMR (126 MHz, $CDCl_3$, ppm) δ = 170.4 (C_q), 158.2 (C_q), 144.4 (C_q), 143.6 (C_q), 142.8 (C_q), 140.6 (C_q), 139.0 (C_q), 132.1 (C_q), 130.5 (+, CH), 127.4 (+, CH), 124.2 (+, CH), 124.1 (+, CH), 123.9 (+, CH), 123.6 (+, CH), 116.6 (+, CH), 109.3 (+, CH), 34.9 (C_q), 32.1 (+, CH_3), 27.4 (-, CH_2).

IR (ATR, cm^{-1}) $\tilde{\nu}$ = 2958 (s), 2904 (w), 2866 (w), 1723 (vs), 1588 (vs), 1554 (w), 1487 (s), 1472 (vs), 1453 (vs), 1395 (m), 1363 (vs), 1323 (vs), 1295 (vs), 1261 (vs), 1231 (vs), 1181 (vs), 1174 (vs), 1139 (w), 1129 (w), 1106 (w), 1062 (m), 1035 (m), 999 (w), 941 (m), 898 (w), 878 (s), 844 (m), 810 (vs), 788 (w), 741 (m), 720 (w), 711 (w), 694 (w), 683 (m), 653 (s), 632 (w), 613 (vs), 594 (w), 470 (w), 442 (w), 421 (m), 398 (w), 390 (w).

MS (ESI): m/z (%) = 1578 (1) $[M]^{+}$.

(1s,3s,5r,7r)-1,3-Bis(5-(4-(2,6-bis(3,6-di-*tert*-butyl-9*H*-carbazol-9-yl)pyridin-4-yl)phenyl)-1,3,4-oxadiazol-2-yl)adamantane (66b)

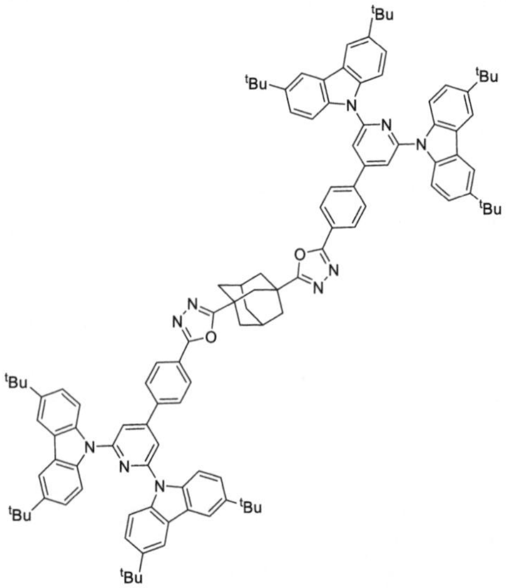

A sealable vial was charged with 9,9'-(4-(4-(2*H*-tetrazol-5-yl)phenyl)pyridine-2,6-diyl)bis(3,6-di-*tert*-butyl-9*H*-carbazole) (**61c**) (113 mg, 146 µmol, 2.10 equiv.) and (1s,3s,5r,7r)-adamantane-1,3-dicarbonyl dichloride (18.0 mg, 69.0 µmol, 1.00 equiv.). Chloroform (5 mL) was added, and the resulting mixture was heated at 100 °C for 16 h. After cooling to room temperature, the reaction mixture was poured into an excess of saturated aqueous sodium hydrogen carbonate solution (50 mL) and stirred for 15 min. Subsequently, the reaction mixture was extracted with dichloromethane (3 × 50 mL). The combined organic layers were washed with brine (50 mL), dried over sodium sulfate, and reduced in a vacuum. The crude product was purified by flash column chromatography over silica gel (*n*-pentane/CH_2Cl_2, 1:2 to CH_2Cl_2/CH_3OH, 99:0.5) to yield 75.2 mg of the title compound (44.5 µmol, 65%) as a yellow solid.

R_f (CH_2Cl_2/CH_3OH, 99:1) = 0.25.

^{1}H NMR (500 MHz, CD_2Cl_2, ppm) δ = 8.26 (d, *J* = 8.2 Hz, 4H, H_{ar}), 8.18 (d, *J* = 1.9 Hz, 8H, H_{ar}), 7.99 (dd, *J* = 13.2, 8.5 Hz, 12H, H_{ar}), 7.86 (s, 4H, H_{ar}), 7.50 (dd, *J* = 8.8, 2.0 Hz, 8H, H_{ar}), 2.58 (s, 2H, CH_2), 2.43 (s, 2H, CH_2), 2.27 (s, 8H, CH_2), 1.95 (s, 2H, CH_2), 1.46 (s, 72H, CH_3).

^{13}C NMR (126 MHz, CD_2Cl_2, ppm) δ = 172.4, 164.6 (C_q), 153.0 (C_q), 152.5 (C_q), 145.0 (C_q), 141.3 (C_q), 138.4 (C_q), 128.5 (+, CH), 128.2 (+, CH), 125.8 (+, CH), 125.1 (+, CH), 124.6 (+, CH), 116.8 (+, CH), 112.8 (+, CH), 112.0 (+, CH), 42.5 (-, CH_2), 39.5 (-, CH_2), 35.5 (C_q), 35.3 (C_q), 35.2 (C_q), 32.2 (CH_3, 24C), 28.3 (-, CH_2).

IR (ATR, cm^{-1}) $\tilde{\nu}$ = 2952 (s), 2904 (m), 2863 (w), 1599 (s), 1561 (m), 1541 (s), 1489 (s), 1465 (vs), 1451 (vs), 1425 (vs), 1401 (vs), 1361 (s), 1295 (vs), 1262 (vs), 1228 (s), 1191 (s), 1143 (w), 1106 (w), 1095 (w), 1086 (w), 1058 (w), 1034 (m), 1014 (m), 875 (m), 840 (s), 810 (vs), 764 (w), 748 (m), 741 (m), 718 (w), 703 (w), 691 (m), 677 (w), 670 (w), 654 (w), 643 (w), 612 (vs), 601 (m), 574 (w), 554 (w), 521 (w), 513 (w), 500 (w), 493 (w), 469 (m), 452 (w), 419 (m), 398 (w), 378 (w).

MS (ESI): m/z (%) = 1687 (5) $[M]^+$. HRMS (ESI, $C_{116}H_{122}N_{10}O_2$): calcd 1686.9752, found 1686.9757.

Mp 296 °C.

(1s,3s,5r,7r)-1,3-Bis(5-(3',5'-bis(3,6-di-*tert*-butyl-9*H*-carbazol-9-yl)-[1,1'-biphenyl]-4-yl)-1,3,4-oxadiazol-2-yl)adamantane (66c)

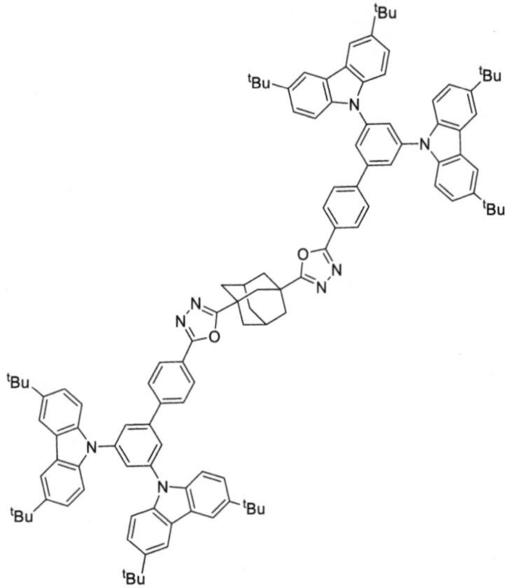

A sealable vial was charged with 9,9'-(4'-(2*H*-tetrazol-5-yl)-[1,1'-biphenyl]-3,5-diyl)bis(3,6-di-*tert*-butyl-9*H*-carbazole) (**61a**) (113 mg, 146 µmol, 2.10 equiv.) and (1s,3s,5r,7r)-adamantane-1,3-dicarbonyl dichloride (18.0 mg, 69.0 µmol, 1.00 equiv.). Chloroform (5 mL) was added, and the resulting mixture was heated at 100 °C for 16 h. After cooling to room temperature, the reaction mixture was poured into an excess of saturated aqueous sodium hydrogen carbonate solution (50 mL) and stirred for 15 min. Subsequently, the reaction mixture was extracted with dichloromethane (3 × 50 mL). The combined organic layers were washed with brine (50 mL), dried over sodium sulfate, and reduced in a vacuum. The crude product was purified by flash column chromatography over silica gel (*n*-pentane/CH_2Cl_2, 1:3 to CH_2Cl_2/CH_3OH, 98:2) to yield 74.0 mg of the title compound (43.9 µmol, 64%) as an off-white solid.

R_f (CH_2Cl_2/CH_3OH, 97:3) = 0.34.

^{1}H NMR (500 MHz, CD_2Cl_2, ppm) δ = 8.21 – 8.15 (m, 12H, H_{ar}), 7.96 (d, *J* = 1.9 Hz, 4H, H_{ar}), 7.91 – 7.87 (m, 4H, H_{ar}), 7.85 (t, *J* = 1.9 Hz, 2H, H_{ar}), 7.57 (d, *J* = 8.7 Hz, 8H, H_{ar}), 7.52 (dd, *J* = 8.6, 1.9 Hz, 8H, H_{ar}), 2.54 (s, 2H, CH_2), 2.40 (s, 2H, CH_2), 2.24 (s, 8H, CH_2), 1.92 (s, 2H, CH_2), 1.46 (s, 72H, CH_3).

^{13}C NMR (126 MHz, CD_2Cl_2, ppm) δ = 144.0 (C_q), 143.7 (C_q), 143.0 (C_q), 140.9 (C_q), 139.5 (C_q), 128.4 (+, CH), 128.0 (+, CH), 124.5 (+, CH), 124.2 (+, CH), 124.0 (+, CH),

117.0 (+, CH), 109.8 (+, CH), 39.5 (-, CH_2), 35.2 (C_q), 35.2 (C_q), 32.3 (CH_3, 24C), 28.3 (-, CH_2).

IR (ATR, cm^{-1}) $\tilde{\nu}$ = 2953 (s), 2904 (m), 2863 (w), 1592 (s), 1487 (s), 1472 (vs), 1453 (vs), 1394 (m), 1363 (s), 1322 (s), 1295 (vs), 1262 (vs), 1230 (vs), 1203 (m), 1170 (m), 1088 (w), 1061 (w), 1034 (m), 1014 (m), 877 (s), 841 (s), 810 (vs), 749 (m), 741 (s), 720 (m), 696 (s), 683 (m), 654 (m), 613 (vs), 574 (w), 517 (m), 500 (m), 493 (w), 484 (m), 470 (m), 452 (m), 421 (s), 402 (m), 388 (w), 384 (m), 377 (m).

MS (m/z, MALDI-TOF) = 1684 $[M]^{+}$.

Mp 352 °C.

12,43-Bis(5-(4-(2,6-bis(3,6-di-*tert*-butyl-9*H*-carbazol-9-yl)pyridin-4-yl)phenyl)-1,3,4-oxadiazol-2-yl)-1,4(1,4)-dibenzenacyclohexaphane (66d)

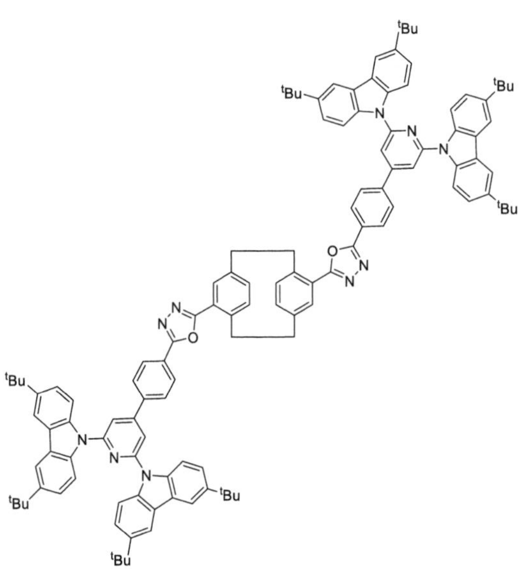

A sealable vial was charged with 9,9'-(4-(4-(2*H*-tetrazol-5-yl)-phenyl)pyridine-2,6-diyl)bis(3,6-di-tert-butyl-9*H*-carbazole) (**61c**) (245 mg, 315 µmol, 2.10 equiv.) and 1,4(1,4)-dibenzenacyclohexaphane-1^2,4^3-dicarbonyl dichloride (50.0 mg, 150 µmol, 1.00 equiv.). Chloroform (7 mL) was added, and the resulting mixture was heated at 100 °C for 16 h. After cooling to room temperature, the reaction mixture was poured into an excess of saturated aqueous sodium hydrogen carbonate solution (50 mL) and stirred for 15 min. Subsequently, the reaction mixture was extracted with dichloromethane (3 × 50 mL). The combined organic layers were washed with brine (50 mL), dried over sodium sulfate, and reduced in a vacuum. The crude product was purified by flash column chromatography over silica gel (*n*-pentane/EtOAc, 10:1 to 6:1) to yield 123 mg of the title compound (69.9 µmol, 47%) as a yellow solid.

R_f (*n*-pentane/EtOAc, 5:1) = 0.28.

^{1}H NMR (500 MHz, CD_2Cl_2, ppm) δ = 8.40 (d, *J* = 8.4 Hz, 4H, H_{ar}), 8.20 (d, *J* = 1.8 Hz, 8H, H_{ar}), 8.08 – 8.02 (m, 12H, H_{ar}), 7.91 (s, 4H, H_{ar}), 7.52 (dd, *J* = 8.8, 1.9 Hz, 8H, H_{ar}), 7.32 (s, 2H, H_{ar}), 6.71 (d, *J* = 7.9 Hz, 2H, H_{ar}), 6.68 (dd, *J* = 7.9, 1.5 Hz, 2H, H_{ar}), 4.40 – 4.29 (m, 2H, CH_2), 3.37 – 3.23 (m, 4H, CH_2), 3.21 – 3.14 (m, 2H, CH_2), 1.48 (s, 72H, CH_3).

^{13}C NMR (126 MHz, CD_2Cl_2, ppm) δ = 165.2 (C_q), 163.8 (C_q), 152.5 (C_q), 151.9 (C_q), 144.4 (C_q), 141.0 (C_q), 140.9 (C_q), 140.4 (C_q), 137.9 (C_q), 135.5 (+, CH), 133.5 (+, CH), 133.3 (+, CH), 128.2 (+, CH), 127.7 (+, CH), 125.1 (+, CH), 124.8 (+, CH), 124.6 (+, CH), 124.1 (+, CH), 116.3 (+, CH), 112.2 (+, CH), 111.4 (+, CH), 34.8 (+, CH_2), 34.7 (C_q), 34.0 (+, CH_2), 31.7 (+, CH_3).

IR (ATR, cm^{-1}) $\tilde{\nu}$ = 2953 (m), 2934 (m), 2905 (w), 2864 (w), 1595 (s), 1570 (w), 1553 (m), 1534 (m), 1492 (s), 1465 (vs), 1451 (vs), 1425 (vs), 1404 (vs), 1363 (s), 1326 (w), 1310 (m), 1295 (vs), 1261 (vs), 1227 (s), 1188 (s), 1147 (w), 1106 (w), 1096 (w), 1045 (w), 1034 (m), 1016 (w), 1009 (w), 965 (w), 902 (w), 890 (w), 871 (s), 843 (m), 832 (m), 809 (vs), 789 (w), 762 (w), 751 (m), 739 (m), 714 (w), 701 (w), 691 (w), 674 (w), 653 (w), 613 (s), 601 (m), 469 (w), 419 (w), 415 (w).

MS (m/z, MALDI-TOF) = 1758 $[M]^+$.

12,43-Bis(5-(3',5'-bis(3,6-di-*tert*-butyl-9*H*-carbazol-9-yl)-[1,1'-biphenyl]-4-yl)-1,3,4-oxadiazol-2-yl)-1,4(1,4)-dibenzenacyclohexaphane (66e)

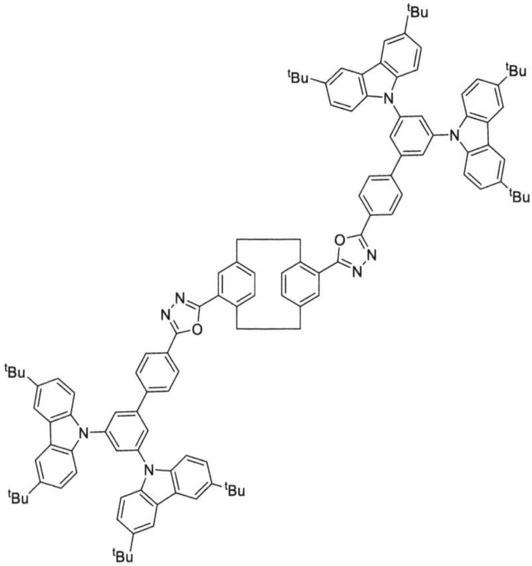

A sealable vial was charged with 9,9'-(4'-(2*H*-tetrazol-5-yl)-[1,1'-biphenyl]-3,5-diyl)bis(3,6-di-*tert*-butyl-9*H*-carbazole) (**61a**) (245 mg, 315 µmol, 2.10 equiv.) and 1,4(1,4)-dibenzenacyclohexaphane-12,43-dicarbonyl dichloride (50.0 mg, 150 µmol, 1.00 equiv.). Chloroform (7 mL) was added, and the resulting mixture was heated at 100 °C for 16 h. After cooling to room temperature, the reaction mixture was poured into an excess of saturated aqueous sodium hydrogen carbonate solution (50 mL) and stirred for 15 min. Subsequently, the reaction mixture was extracted with dichloromethane (3 × 50 mL). The combined organic layers were washed with brine (50 mL), dried over sodium sulfate, and reduced in a vacuum. The crude product was purified by flash column chromatography over silica gel (*n*-pentane/CH_2Cl_2, 1:2 to 0:1) to yield 166 mg of the title compound (94.3 µmol, 63%) as a beige solid.

R_f (CH_2Cl_2/CH_3OH, 99:1) = 0.26.

^{1}H NMR (500 MHz, CD_2Cl_2, ppm) δ = 8.32 (d, *J* = 8.4 Hz, 4H, H_{ar}), 8.20 (d, *J* = 1.4 Hz, 8H, H_{ar}), 8.01 (d, *J* = 1.7 Hz, 4H, H_{ar}), 7.97 (d, *J* = 8.2 Hz, 4H, H_{ar}), 7.89 (s, 2H, H_{ar}), 7.60 (d, *J* = 8.7 Hz, 8H, H_{ar}), 7.54 (dd, *J* = 8.7, 1.7 Hz, 8H, H_{ar}), 7.29 (s, 2H, H_{ar}), 6.67 (q, *J* = 9.3, 8.5 Hz, 4H, H_{ar}), 4.37 – 4.28 (m, 2H, CH_2), 3.32 – 3.20 (m, 4H, CH_2), 3.15 (td, *J* = 11.8, 10.6, 5.8 Hz, 2H, CH_2), 1.47 (s, 72H, CH_3).

^{13}C NMR (126 MHz, CD_2Cl_2, ppm) δ = 165.0 (C_q), 164.0 (C_q), 143.5 (C_q), 143.1 (C_q), 142.7 (C_q), 140.8 (C_q), 140.4 (C_q), 140.3 (C_q), 138.9 (C_q), 135.4 (+, CH), 133.5 (+, CH), 133.3 (+, CH), 128.0 (+, CH), 127.6 (+, CH), 124.8 (+, CH), 124.0 (+, CH), 123.8 (+, CH), 123.7 (+, CH), 123.6 (+, CH), 123.5 (+, CH), 116.5 (+, CH), 109.2 (+, CH), 34.8 (-, CH_2), 34.7 (C_q), 33.93 (-, CH_2), 31.7 (+, CH_3).

IR (ATR, cm^{-1}) $\tilde{\nu}$ = 2953 (m), 1592 (m), 1575 (m), 1487 (s), 1470 (vs), 1452 (vs), 1394 (m), 1363 (s), 1320 (s), 1295 (vs), 1261 (vs), 1230 (vs), 1203 (m), 1170 (m), 1051 (w), 1034 (m), 1017 (w), 898 (w), 877 (s), 840 (s), 810 (vs), 755 (m), 741 (s), 718 (m), 694 (m), 683 (m), 653 (m), 645 (m), 613 (vs), 574 (w), 555 (w), 521 (w), 483 (w), 469 (m), 458 (w), 450 (w), 419 (m), 401 (w), 375 (m).

MS (ESI): m/z (%) = 1758 (67) $[M+H]^+$, 1757 (100) $[M]^+$. HRMS (ESI, $C_{124}H_{124}N_8O_2$): calcd 1756.9847, found 1756.9866.

5.2.2.7. Precursors

(4-(3,6-Di-*tert*-butyl-9*H*-carbazol-9-yl)phenyl)boronic acid (67)

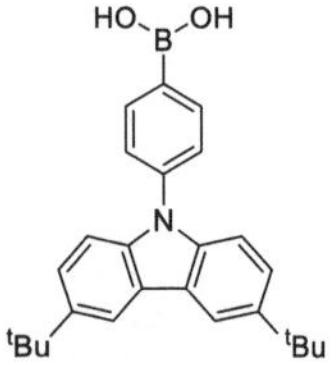

HO B OH N tBu tBu

A flame-dried Schlenk flask was charged with 9-(4-bromophenyl)-3,6-di-*tert*-butyl-9*H*-carbazole (955 mg, 2.20 mmol, 1.00 equiv.) dissolved in dry tetrahydrofuran (40 mL). *n*-Butyllithium (169 mg, 2.64 mmol, 1.20 equiv., 2.50 M solution in hexane) was added dropwise while stirring at -78 °C over a period of 30 min. Afterwards, trimethyl borate (343 mg, 3.30 mmol, 1.50 equiv.) was added smoothly, and the resulting mixture was stirred for 30 min at - 78 °C. The mixture was gradually warmed to room temperature, then stirred for 26 h at room temperature. Subsequently, the reaction mixture was quenched and extracted with dichloromethane (3 × 50 mL). The combined organic layers were washed with brine (50 mL), dried over sodium sulfate, and reduced in a vacuum. The crude product was purified by flash column chromatography over silica gel (*c*Hex/CH_2Cl_2, 3:1 to CH_2Cl_2/CH_3OH, 100:1) to yield 542 mg of the title compound (1.36 mmol, 62%) as an off-white solid.

R_f (CH_2Cl_2/CH_3OH, 100:1) = 0.32.

^1H NMR (500 MHz, DMSO-d_6, ppm) δ = 8.29 (dd, J = 2.0, 0.6 Hz, 2H, H_{ar}), 8.20 (s, 2H, OH), 8.05 (d, J = 8.4 Hz, 2H, H_{ar}), 7.58 (d, J = 8.3 Hz, 2H, H_{ar}), 7.48 (dd, J = 8.7, 2.0 Hz, 2H, H_{ar}), 7.35 (d, J = 8.5 Hz, 2H, H_{ar}), 1.42 (s, 18H, CH_3).

^{13}C NMR (126 MHz, DMSO-d_6, ppm) δ = 142.6 (C_q), 139.0 (C_q), 138.3 (C_q), 135.8 (+, CH), 124.9 (+, CH), 123.7 (+, CH), 122.9 (C_q), 116.7 (+, CH), 109.2 (+, CH), 34.5 (C_q), 31.8 (+, CH_3).

IR (ATR, cm^{-1}) $\tilde{\nu}$ = 2952 (m), 2902 (w), 1599 (s), 1489 (w), 1472 (m), 1421 (w), 1407 (w), 1341 (vs), 1322 (vs), 1293 (vs), 1259 (vs), 1231 (s), 1201 (m), 1174 (vs), 1152 (w), 1105 (m), 1082 (m), 1035 (w), 1016 (w), 945 (w), 877 (m), 841 (w), 810 (vs), 764 (w), 751 (m), 739 (m), 708 (s), 691 (s), 654 (m), 633 (m), 623 (w), 611 (vs), 596 (m), 511 (w), 506 (w), 470 (m), 412 (m), 402 (m), 391 (w).

MS (ESI): m/z (%) = 400 (23) $[M+H]^+$, 399 (100) $[M]^+$. HRMS (ESI, $C_{26}H_{30}BNO_2$): calcd 399.2370, found 399.2363.

1,4(1,4)-Dibenzenacyclohexaphane-12,43-dicarbonyl dichloride (68)

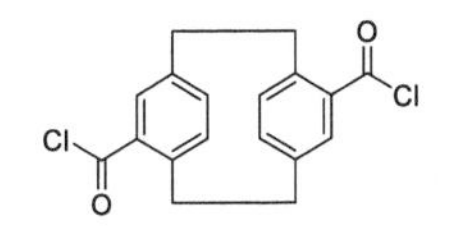

A flame-dried flask was charged with 1,4(1,4)-dibenzenacyclohexaphane-12,43-dicarboxylic acid (119 mg, 402 µmol, 1.00 equiv.), sulfurous dichloride (1.43 g, 12.1 mmol, 30.0 equiv.) and dimethylformamide (0.025 equiv., 3 drops). The resulting mixture was stirred at reflux 85 °C for 4 h. Afterwards, the mixture was concentrated and used in the next step without further purification.

5.3. Crystal Structures

Crystal structures were measured and solved by Dr. Olaf Fuhr at the Institute of Nanotechnology (INT) at Karlsruhe Institute of Technology.

Molecule name in this thesis	#	Code used by Dr. Fuhr
PXZ-MTT	**30a**	CEL-329
PXZ-MTT-Me	**30g**	CEL-293_P-1

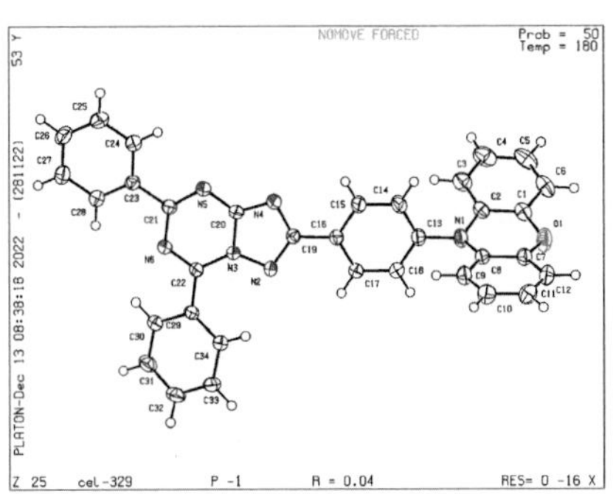

PXZ-MTT (30a), CEL-329

10-(4-(5,7-Diphenyl-[1,2,4]triazolo [1,5-a][1,3,5]triazin-2-yl)phenyl)-10*H*-phenoxazine

Identification code	CEL-329
Empirical formula	$C_{34}H_{22}N_6O$
Formula weight	530.57
Temperature/K	180
Crystal system	triclinic
Space group	P-1
a/Å	8.5335(13)
b/Å	10.3820(8)
c/Å	15.3021(18)
α/°	75.601(8)
β/°	77.885(11)
γ/°	82.328(10)
Volume/Å^3	1279.1(3)
Z	2
ρ_{calc}g/cm^3	1.378
μ/mm^{-1}	0.443
F(000)	552.0
Crystal size/mm^3	0.24 × 0.23 × 0.22
Radiation	GaKα (λ = 1.34143)
2Θ range for data collection/°	7.678 to 125.142
Index ranges	$-9 \le h \le 11$, $-13 \le k \le 13$, $-20 \le l \le 15$
Reflections collected	20529
Independent reflections	6068 [R_{int} = 0.0120, R_{sigma} = 0.0082]
Data/restraints/parameters	6068/0/370
Goodness-of-fit on F^2	1.064
Final R indexes [I>=2σ (I)]	R_1 = 0.0362, wR_2 = 0.0975
Final R indexes [all data]	R_1 = 0.0398, wR_2 = 0.1000
Largest diff. peak/hole / e Å^{-3}	0.20/-0.26

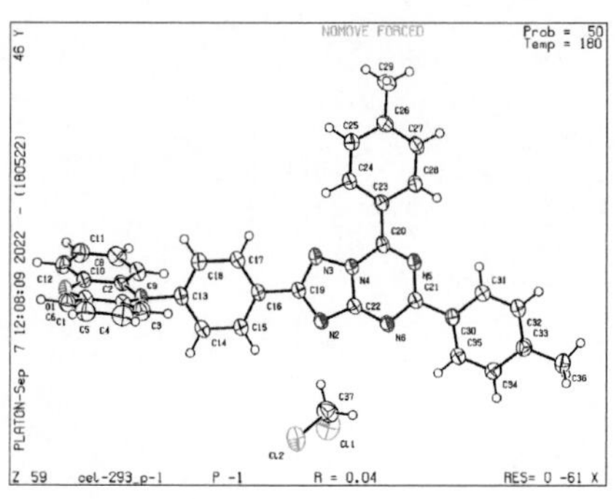

PXZ-MTT-Me (30g), CEL-293_P-1

10-(4-(5,7-Di-p-tolyl-[1,2,4] triazolo[1,5-a][1,3,5]triazin-2-yl)phenyl) - 10*H*-phenoxazine and one dichloromethane molecule

Identification code	CEL-293_P-1
Empirical formula	$C_{36}H_{26}N_6O + CH_2Cl_2$
Formula weight	643.55
Temperature/K	180.0
Crystal system	triclinic
Space group	P-1
a/Å	9.0545(6)
b/Å	11.4447(9)
c/Å	15.6954(17)
α/°	76.087(7)
β/°	89.942(7)
γ/°	80.416(6)
Volume/Å^3	1555.4(2)
Z	2
ρ_{calc}g/cm^3	1.374
μ/mm^{-1}	1.454
F(000)	668.0
Crystal size/mm^3	0.18 × 0.16 × 0.14
Radiation	GaKα (λ = 1.34143)
2Θ range for data collection/°	5.05 to 115.094
Index ranges	$-11 \le h \le 10, -14 \le k \le 13, -14 \le l \le 19$
Reflections collected	15562
Independent reflections	6305 [R_{int} = 0.0189, R_{sigma} = 0.0161]
Data/restraints/parameters	6305/0/417
Goodness-of-fit on F^2	1.075
Final R indexes [I>=2σ (I)]	R_1 = 0.0400, wR_2 = 0.1157
Final R indexes [all data]	R_1 = 0.0471, wR_2 = 0.1203
Largest diff. peak/hole / e Å^{-3}	0.35/-0.48

List of Abbreviations

ATR	Attenuated total reflectance IR spectroscopy
δ	Chemical shift (NMR)
°C	Celsius
Calc.	Calculated
CDCBs	Carbazolyl Dicyanobenzene Series
$CDCl_3$	Deuterated chloroform
*c*Hex	Cyclohexane
CH_2Cl_2	Dichloromethane
^1CT	Singlet Charge Transfer state
^3CT	Triplet Charge Transfer state
CT	Charge Transfer
CV	Cyclic Voltammetry
CzSi	9-(4-*Tert*-butylphenyl)-3,6-bis(triphenylsilyl)-9*H*-carbazole
d	Doublet (NMR)
DCM	Dichloromethane
DFT	Density Functional Theory
DMAC	9,9-Dimethyl-9,10-dihydroacridine
DMF	Dimethylformamide
DMSO	Dimethyl sulfoxide
DPAC	9,9-Diphenyl-9,10-dihydroacridine
DPEPO	Bis[2-(diphenylphosphino)phenyl] ether oxide
DTT	Ditriazolotriazine
EA	Elemental Analysis
EML	Emission Layer
equiv.	Equivalent
EQE	External Quantum Efficiency
ESI	Electrospray Ionization
Et	Ethyl
et. al.	And others

ETL	Electron Transport Layer
EtOAc	Ethyl Acetate
FAB	Fast Atom Bombardment
FRET	Fluorescence Resonance Energy Transfer
FWHM	Full Width at Half Maximum
HOMO	Highest Occupied Molecular Orbital
H_{SO}	Spin-orbit constant
HTL	Hole Transport Layer
IC	Internal Conversion
IQE	Internal Quantum Efficiency
IR	Infrared spectroscopy
ISC	Intersystem Crossing
ITO	Indium Tin Oxide
J	Exchange Energy Integral or Coupling Constant (NMR)
k_B	Boltzmann constant
k_F	Rate of Fluorescence
k_{ISC}	Rate of Intersystem Crossing
k_{NR}	Rate of non-radiative processes
k_{rISC}	Rate of Reverse Intersystem Crossing
^1LE_D	Locally excited singlet state on the Donor
^3LE	Locally excited triplet States
^3LE_A	Locally excited triplet state on the Acceptor
^3LE_D	Locally excited triplet state on the Donor
LED	Light Emitting Diode
LUMO	Lowest Unoccupied Molecular Orbital
λ	Wavelength
λ_{max}	Emission maxima
m	Middle (IR)
mCP	N,N′-dicarbazolyl-3,5-benzene
mL	Milliliter
mmol	Millimol

MALDI-ToF-MS	Matrix-Assisted Laser Desorption Ionisation Time of Flight Mass Spectrometry
Me	Methyl
Me-Cy-Hex	Methylcyclohexane
Mp	Melting point
ms	Mass spectrometry
MTT	Monotriazolotriazine
NMR	Nuclear Magnetic Resonance
ns	Nanoseconds
OLED	Organic Light Emitting Diode
ph	Phenyl
PLQY/Φ_{PL}	Photoluminescence Quantum Yield
PMMA	Poly (Methyl Methacrylate)
ppm	Parts per million
PXZ	10*H*-Phenoxazine
q	Quartett (NMR)
rt	Room temperature
rISC	Reverse Intersystem Crossing
s	Singlet (NMR), Strong (IR)
S_0	Ground state
S_1	First excited singlet state
S_NAr	Nucleophilic Aromatic Substitution
SOC	Spin-Orbit Coupling
t	Triplet (NMR)
T	Temperature
T_1	First excited triplet state
TADF	Thermally Activated Delayed Fluorescence
tBu	*Tert*-butyl
tCz	3,6-Di-*tert*-butyl-9*H*-carbazole
TD-DFT	Time Dependent Density Functional Theory
TGA	Thermogravimetric Analysis

THF	Tetrahydrofuran
TLC	Thin Layer Chromatography
TTA	Triplet-Triplet-Annihilation
TTT	Tristriazolotriazine
TV	Television
τ_{DF}	Lifetime Delayed Fluorescence
τ_{PF}	Lifetime Prompt Fluorescence
UV	Ultraviolet
w	Weak (IR)
wt%	Weight percent
XRD	Single crystal X-ray diffraction
ΔE_{ST}	Energy gap between S_1 and T_1
Φ	HOMO Wavefunction
Ψ	LUMO Wavefunction

Bibliography

[1] H. Uoyama, K. Goushi, K. Shizu, H. Nomura, C. Adachi, *Nature* **2012,** *492* (7428), 234-8.

[2] D. Zhang, M. Cai, Y. Zhang, D. Zhang, L. Duan, *Mater. Horiz.* **2016,** *3* (2), 145-151.

[3] "Efficient lighting for developing and emerging countries" can be found under http://www.enlighten-initiative.org/ visited 18.10.2022.

[4] "Anzahl der Smartphone-Nutzer weltweit von 2016 bis 2020 und Prognose bis 2024" can be found under https://de.statista.com/statistik/daten/studie/309656/umfrage/prognose-zur-anzahl-der-smartphone-nutzer-weltweit/ published 24.01.2022 visited 19.10.2022.

[5] Ein Patent dass die Welt veränderte" can be found under http://www.focus.de/wissen/mensch/geschichte/erfindungen/technikgeschichte-ein-patent-das-die-welt-veraenderte¬_aid_474269.html visited 18.10.2022.

[6] "Wie hoch ist der Wirkungsgrad und Effizienz bei LED Lampen?" can be found under https://www.gluehbirne.de/led-ratgeber-wirkungsgrad-effizienz-led-lampen#:~:text=Der%20Wirkungsgrad%20einer%20Gl%C3%BChlampe%20liegt,10%20bis%2015%20Lumen%2FWatt visited 18.10.2022.

[7] V. Balzani, G. Bergamini, P. Ceroni, *Angew. Chem. Int. Ed.* **2015,** *54* (39), 11320-37.

[8] P. Von Dollen, S. Pimputkar, J. S. Speck, *Angew. Chem. Int. Ed.* **2014,** *53* (51), 13978-80.

[9] K. O. Udovychenko, *Core.ac.uk* **2019**.

[10] Y. Takahashi, Y. Furuki, S. Yoshida, T. Otani, M. Muto, Y. Suga, Y. Ito, *SID Symposium Digest of Technical Papers* **2014,** *45* (1), 381-384.

[11] A. Arjona-Esteban, D. Volz, *Highly Efficient OLEDs: Materials Based on Termally Activated Delayed Fluorescence - Status and Next Steps of TADF Technology: An Industrial Perspective,* Wiley-VCH Verlag GmbH&Co. KG, **2018**, p 543-572.

[12] C. Huang-Jen, C. Shih-Jen, *IEEE Transactions on Industrial Electronics* **2007,** *54* (5), 2751-2760.

[13] V. C. Coffey, *Optics and Photonics News* **2017,** *28*, 34-41.

[14] B. Valeur, M. N. Berberan-Santos, *J. Chem. Educ.* **2011,** *88* (6), 731-738.

[15] S. E. Braslavsky, *Pure Appl. Chem.* **2007,** *79* (3), 293-465.

[16] J. R. Partington, *Annals of Science* **2006,** *11* (1), 1-26.

[17] M. Muyskens, *J. Chem. Educ.* **2006,** *83.*

[18] A. U. Acuña, F. Amat-Guerri, *Fluorescence of Supermolecules, Polymers, and Nanosystems,* Springer, **2007**.

[19] A. U. Acuña, F. Amat-Guerri, P. Morcillo, M. Liras, B. Rodriguez, *Org. Lett.* **2009,** *11,* 3020–3023.

[20] G. G. Stokes, *Philos. Trans.* **1852,** *142,* 463–562.

[21] G. G. Stokes, *Philos. Trans.* **1853,** *143,* 385–396.

[22] E. Becquerel, *Ann. Chim. Phys.* **1842,** *9,* 257–322.

[23] E. Becquerel, *La Lumiere. Ses Causes et ses Effets,* Firmin Didot, Paris, **1867**, Vol. 1.

[24] J. Zimmermann, A. Zeug, B. Röder, *Phys. Chem. Chem. Phys.* **2003,** *5* (14), 2964-2969.

[25] G. Feng, G. Q. Zhang, D. Ding, *Chem. Soc. Rev.* **2020,** *49* (22), 8179-8234.

[26] J. R. Lakowicz, *Principles of Fluorescence Spectroscopy Third Edition,* Springer, **2006**.

[27] "Phosphorescence" can be found under https://www.britannica.com/science/phosphorescence visited 09.10.2022.

[28] "What are Fluorescence and Phosphorescence?" can be found under https://www.chemistryviews.org/details/education/10468955/What_are_Fluorescence_and_Phosphorescence/ visited 19.10.2022.

[29] P. W. Atkins, *Physical Chemistry,* Oxford University Press, **1994**.

[30] P. W. Atkins, J. d. Paula, *Kurzlehrbuch Physikalische Chemie* Wiley-VCH, Weinheim, **2008**.

[31] G. N. Lewis, M. Kasha, *J. Am. Chem. Soc.* **2002,** *66* (12), 2100-2116.

[32] H. Yersin, *Highly Efficient OLEDs - Photophysics of Thermally Activated Delayed Fluorescence,* Wiley-VCH, Weinheim, **2019**.

[33] C. A. Parker, C. G. Hatchard, *Trans. Faraday Soc.* **1961,** *57,* 1894-1904.

[34] M. N. Berberan-Santos, J. M. M. Garcia, *J. Am. Chem. Soc.* **1996,** *118,* 9391-9394.

[35] A. Endo, M. Ogasawara, A. Takahashi, D. Yokoyama, Y. Kato, C. Adachi, *Adv. Mater.* **2009,** *21* (47), 4802-6.

[36] K. Goushi, C. Adachi, *Appl. Phys. Lett.* **2012,** *101* (2), 023306.

[37] K. Goushi, K. Yoshida, K. Sato, C. Adachi, *Nature Photonics* **2012,** *6* (4), 253-258.

[38] F. B. Dias, K. N. Bourdakos, V. Jankus, K. C. Moss, K. T. Kamtekar, V. Bhalla, J. Santos, M. R. Bryce, A. P. Monkman, *Adv. Mater.* **2013,** *25* (27), 3707-14.

[39] M. Pope, H. P. Kallmann, P. Magnante, *J. Chem. Phys.* **1963,** *38* (8), 2042-2043.

[40] W. Helfrich, W. G. Schneider, *Phys. Rev. Lett.* **1965,** *14* (7), 229-231.

[41] B. Minaev, G. Baryshnikov, H. Agren, *Phys. Chem. Chem. Phys.* **2014,** *16* (5), 1719-58.

[42] C. W. Tang, S. A. VanSlyke, *Appl. Phys. Lett.* **1987,** *51* (12), 913-915.

[43] A. Kohler, H. Bassler, *Electronic Processes in Organic Semiconductors,* Wiley - VCH, **2015**.

[44] W. Brutting, C. Adachi, *Physics of Organic Semiconductors,* Wiley - VCH., **2012**.

[45] B. Lüssem, M. Riede, K. Leo, *Phys. Status Solidi A* **2013,** *210* (1), 9-43.

[46] H. Sasabe, J. Kido, *Eur. J. Org. Chem.* **2013,** *2013* (34), 7653-7663.

[47] M. A. Baldo, D. F. O'Brien, Y. You, A. Shoustikov, S. Sibley, M. E. Thompson, S. R. Forrest, *Nature* **1998,** *395,* 151-154.

[48] A. P. Monkman, C. Rothe, S. M. King, *Proceedings of the IEEE* **2009,** *97* (9), 1597-1605.

[49] D. Y. Kondakov, T. D. Pawlik, T. K. Hatwar, J. P. Spindler, *J. Appl. Phys.* **2009,** *106* (12), 124510.

[50] S.-J. Su, T. Chiba, T. Takeda, J. Kido, *Adv. Mater.* **2008,** *20* (11), 2125-2130.

[51] D. Chen, S.-J. Su, Y. Cao, *J. Mater. Chem. C* **2014,** *2* (45), 9565-9578.

[52] S.-J. Su, H. Sasabe, T. Takeda, J. Kido, *Chem. Mater.* **2008,** *20,* 1691–1693.

[53] M. Mońka, I. E. Serdiuk, K. Kozakiewicz, E. Hoffman, J. Szumilas, A. Kubicki, S. Y. Park, P. Bojarski, *J. Mater. Chem. C* **2022,** *10* (20), 7925-7934.

[54] K. H. Kim, C. K. Moon, J. H. Lee, S. Y. Kim, J. J. Kim, *Adv. Mater.* **2014,** *26* (23), 3844-7.

[55] C. W. Lee, J. Y. Lee, *Adv. Mater.* **2013,** *25* (38), 5450-4.

[56] C. Adachi, M. A. Baldo, M. E. Thompson, S. R. Forrest, *Journal of Applied Physics* **2001,** *90* (10), 5048-5051.

[57] C. Adachi, M. A. Baldo, S. R. Forrest, M. E. Thompson, *Appl. Phys. Lett.* **2000,** *77* (6), 904-906.

[58] P. T. Furuta, L. Deng, S. Garon, M. E. Thompson, J. M. J. Fréchet, *J. Am. Chem. Soc.* **2004,** *126*, 15388–15389.

[59] C. M. Che, C. C. Kwok, S. W. Lai, A. F. Rausch, W. J. Finkenzeller, N. Zhu, H. Yersin, *Chem. Eur. J.* **2010,** *16* (1), 233-47.

[60] P. T. Chou, Y. Chi, *Eur. J. Inorg. Chem.* **2006,** *2006* (17), 3319-3332.

[61] F.-I. Wu, P.-I. Shih, Y.-H. Tseng, G.-Y. Chen, C.-H. Chien, C.-F. Shu, Y.-L. Tung, Y. Chi, A. K.-Y. Jen, *J. Phys. Chem.* **2005,** *109*, 14000–14005.

[62] J. Lu, Y. Tao, Y. Chi, Y. Tung, *Synthetic Metals* **2005,** *155* (1), 56-62.

[63] P. T. Chou, Y. Chi, *Chem. Eur. J.* **2007,** *13* (2), 380-95.

[64] M. Baldo, S. Forrest, *Phys. Rev. B: Condens. Matter Mater. Phys.* **2000,** *62*, 10958.

[65] T. T. Bui, F. Goubard, M. Ibrahim-Ouali, D. Gigmes, F. Dumur, *Beilstein J. Org. Chem.* **2018,** *14*, 282-308.

[66] X. K. Chen, D. Kim, J. L. Bredas, *Acc. Chem. Res.* **2018,** *51* (9), 2215-2224.

[67] R. Mertens "Is TADF the future of efficient OLED emitters?" can be found under https://www.oled-info.com/tadf-future-efficient-oled-emitters-premium-article visited 18.10.2022.

[68] M. Mimuro, S. Akimoto, T. Tomo, M. Yokono, H. Miyashita, T. Tsuchiya, *Biochimica et Biophysica Acta* **2007,** *1767* (4), 327-34.

[69] M. Pope, C. E. S. ed., *Electronic Processes in Organic Crystals and Polymers.*, Oxford University Press., **1999**.

[70] C. Murawski, K. Leo, M. C. Gather, *Adv. Mater.* **2013,** *25* (47), 6801-27.

[71] S. Wehrmeister, L. Jäger, T. Wehlus, A. F. Rausch, T. C. G. Reusch, T. D. Schmidt, W. Brütting, *PHYS. REV. APPLIED* **2015,** *3* (2).

[72] Y. Zhang, S. R. Forrest, *Phys. Rev. Lett.* **2012,** *108* (26), 267404.

[73] H. van Eersel, P. A. Bobbert, R. A. J. Janssen, R. Coehoorn, *Appl. Phys. Lett.* **2014,** *105* (14), 143303.

[74] R. Coehoorn, H. van Eersel, P. Bobbert, R. Janssen, *Adv. Funct. Mater.* **2015,** *25* (13), 2024-2037.

[75] M. J. Leitl, V. A. Krylova, P. I. Djurovich, M. E. Thompson, H. Yersin, *J. Am. Chem. Soc.* **2014,** *136* (45), 16032-8.

[76] T. Chen, L. Zheng, J. Yuan, Z. An, R. Chen, Y. Tao, H. Li, X. Xie, W. Huang, *Sci. Rep.* **2015,** *5*, 10923.

[77] L. Yao, B. Yang, Y. Ma, *Sci China Chem* **2014,** *57* (3), 335-345.

[78] F. B. Dias, T. J. Penfold, A. P. Monkman, *Methods Appl. Fluoresc.* **2017,** *5* (1), 012001.

[79] C. Baleizao, M. N. Berberan-Santos, *J. Chem. Phys.* **2007,** *126* (20), 204510.

[80] G. Baryshnikov, B. Minaev, H. Agren, *Chem. Rev.* **2017,** *117* (9), 6500-6537.

[81] M. Montalti, *Handbook of Photochemistry,* Taylor & Francis Group Third Edition, **2006**.

[82] B. T. Lim, S. Okajima, A. K. Chandra, E. C. Lim, *Chem. Phys. Lett.* **1981,** *79*, 22–7.

[83] T. J. Penfold, E. Gindensperger, C. Daniel, C. M. Marian, *Chem. Rev.* **2018,** *118* (15), 6975-7025.

[84] C. Baleizao, S. Nagl, S. M. Borisov, M. Schaferling, O. S. Wolfbeis, M. N. Berberan-Santos, *Chem. Eur. J.* **2007,** *13* (13), 3643-51.

[85] Y. Olivier, B. Yurash, L. Muccioli, G. D'Avino, O. Mikhnenko, J. C. Sancho-García, C. Adachi, T. Q. Nguyen, D. Beljonne, *Phys. Rev. Materials* **2017,** *1* (7).

[86] T. Hosokai, H. Nakanotani, S. Santou, H. Noda, Y. Nakayama, C. Adachi, *Synthetic Metals* **2019,** *252*, 62-68.

[87] M. K. Etherington, J. Gibson, H. F. Higginbotham, T. J. Penfold, A. P. Monkman, *Nat. Commun.* **2016,** *7*, 13680.

[88] T. Saragi, T. Spehr, A. Siebert, T. Fuhrmann - Lieker, J. Salbeck, *Chem. Rev.* **2007,** *107*, 1011.

[89] Y. Shi, H. Ma, Z. Sun, W. Zhao, G. Sun, Q. Peng, *Angew. Chem. Int. Ed.* **2022,** e202213463.

[90] W. Huang, M. Einzinger, T. Zhu, H. S. Chae, S. Jeon, S.-G. Ihn, M. Sim, S. Kim, M. Su, G. Teverovskiy, T. Wu, T. Van Voorhis, T. M. Swager, M. A. Baldo, S. L. Buchwald, *Chem. Mater.* **2018,** *30* (5), 1462-1466.

[91] S.-J. Woo, Y.-H. Kim, J.-J. Kim, *Chem. Mater.* **2021,** *33* (14), 5618-5630.

[92] J. W. Sun, J. H. Lee, C. K. Moon, K. H. Kim, H. Shin, J. J. Kim, *Adv. Mater.* **2014,** *26* (32), 5684-8.

[93] X. Liang, Z. L. Tu, Y. X. Zheng, *Chem. Eur. J.* **2019,** *25* (22), 5623-5642.

[94] Y. K. Wang, C. C. Huang, H. Ye, C. Zhong, A. Khan, S. Y. Yang, M. K. Fung, Z. Q. Jiang, C. Adachi, L. S. Liao, *Adv. Optical Mater.* **2019,** *8* (2), 1901150.

[95] E. Spuling, N. Sharma, I. D. W. Samuel, E. Zysman-Colman, S. Brase, *Chem. Commun.* **2018,** *54* (67), 9278-9281.

[96] S. Kumar, L. G. Franca, K. Stavrou, E. Crovini, D. B. Cordes, A. M. Z. Slawin, A. P. Monkman, E. Zysman-Colman, *J. Phys. Chem. Lett.* **2021,** *12* (11), 2820-2830.

[97] T. Hatakeyama, K. Shiren, K. Nakajima, S. Nomura, S. Nakatsuka, K. Kinoshita, J. Ni, Y. Ono, T. Ikuta, *Adv. Mater.* **2016,** *28* (14), 2777-2781.

[98] J. Han, Z. Huang, X. Lv, J. Miao, Y. Qiu, X. Cao, C. Yang, *Adv. Optical Mater.* **2021,** *10* (4), 2102092.

[99] Y. Liu, X. Xiao, Y. Ran, Z. Bin, J. You, *Chem. Sci.* **2021,** *12* (27), 9408-9412.

[100] D. Hall, S. M. Suresh, P. L. dos Santos, E. Duda, S. Bagnich, A. Pershin, P. Rajamalli, D. B. Cordes, A. M. Z. Slawin, D. Beljonne, A. Köhler, I. D. W. Samuel, Y. Olivier, E. Zysman - Colman, *Adv. Optical Mater.* **2019,** *8* (2), 1901627.

[101] D. Sun, S. M. Suresh, D. Hall, M. Zhang, C. Si, D. B. Cordes, A. M. Z. Slawin, Y. Olivier, X. Zhang, E. Zysman-Colman, *Mater. Chem. Front.* **2020,** *4* (7), 2018-2022.

[102] "TADF: What is Thermally Activated Delayed Fluorescence?" can be found under https://www.edinst.com/blog/tadf-thermally-activated-delayed-fluorescence/ visited on 19.10.2022.

[103] T. A. Lin, T. Chatterjee, W. L. Tsai, W. K. Lee, M. J. Wu, M. Jiao, K. C. Pan, C. L. Yi, C. L. Chung, K. T. Wong, C. C. Wu, *Adv. Mater.* **2016,** *28* (32), 6976-83.

[104] D. R. Lee, B. S. Kim, C. W. Lee, Y. Im, K. S. Yook, S. H. Hwang, J. Y. Lee, *ACS Appl. Mater. Interfaces* **2015,** *7* (18), 9625-9.

[105] Q. Zhang, H. Kuwabara, W. J. Potscavage, Jr., S. Huang, Y. Hatae, T. Shibata, C. Adachi, *J. Am. Chem. Soc.* **2014,** *136* (52), 18070-81.

[106] F. Hundemer, E. Crovini, Y. Wada, H. Kaji, S. Bräse, E. Zysman-Colman, *Mater. Adv.* **2020,** *1* (8), 2862-2871.

[107] S. K. Pathak, Y. Xiang, M. Huang, T. Huang, X. Cao, H. Liu, G. Xie, C. Yang, *RSC Adv.* **2020,** *10* (26), 15523-15529.

[108] S. Wang, X. Wang, K. H. Lee, S. Liu, J. Y. Lee, W. Zhu, Y. Wang, *Dyes Pigments* **2020,** *182*, 108589.

[109] R. Huisgen, H. J. Sturm, M. Seidel, *Chemische Berichte* **2006,** *94* (6), 1555-1562.

[110] T. Rieth, N. Roder, M. Lehmann, H. Detert, *Chem. Eur. J.* **2018,** *24* (1), 93-96.

[111] R. Cristiano, J. Eccher, I. H. Bechtold, C. N. Tironi, A. A. Vieira, F. Molin, H. Gallardo, *Langmuir* **2012,** *28* (31), 11590-8.

[112] T. Rieth, S. Glang, D. Borchmann, H. Detert, *Mol. Cryst. Liq. Cryst.* **2015,** *610* (1), 89-99.

[113] R. Su, Y. Zhao, F. Yang, L. Duan, J. Lan, Z. Bin, J. You, *Science Bulletin* **2021,** *66* (5), 441-448.

[114] H. Cheng, Y. Su, Y. Hu, X. Zhang, Z. Cai, *Polymers* **2018,** *10* (7).

[115] X. Y. Liu, F. Liang, L. S. Cui, X. D. Yuan, Z. Q. Jiang, L. S. Liao, *Chem. Asian J.* **2015,** *10* (6), 1402-9.

[116] M. W. Wong, R. Leung-Toung, C. Wentrup, *J. Am. Chem. Soc.* **2002,** *115* (6), 2465-2472.

[117] V. G. Kiselev, P. B. Cheblakov, N. P. Gritsan, *J. Phys. Chem. A* **2011,** *115* (9), 1743-53.

[118] C. Leonhardt, *Unpublished M. Sc. Thesis: Synthetic Strategies towards TADF Emitters for OLED Applications with an advanced Acceptor Motive: Tris[1,2,4]-triazolo[1,3,5]-triazine,* Karlsruhe Institute of Technology, **2020.**

[119] P. Guerret, R. Jacquier, G. Maury, *J. Heterocyclic Chem.* **1971,** *8*, 643-650.

[120] F. L. Rose, G. J. Stacey, P. J. Taylor, T. W. Thompson, *Chem. Commun.* **1970,** 1524.

[121] Mamedov V. A, Zhukova N. A, K. M. S., *Chem. Heterocycl. Compd.* **2021,** *57* (4), 342-368.

[122] S. Ye, Y. Liu, C. Di, H. Xi, W. Wu, Y. Wen, K. Lu, C. Du, Y. Liu, G. Yu, *Chem. Mater.* **2009,** *21*, 1333–1342.

[123] Q. Wang, J. Ding, D. Ma, Y. Cheng, L. Wang, F. Wang, *Adv. Mater.* **2009,** *21* (23), 2397-2401.

[124] T. Serevicius, R. Skaisgiris, D. Gudeika, K. Kazlauskas, S. Jursenas, *Phys. Chem. Chem. Phys.* **2021,** *24* (1), 313-320.

[125] D. Berenis, G. Kreiza, S. Juršėnas, E. Kamarauskas, V. Ruibys, O. Bobrovas, P. Adomėnas, K. Kazlauskas, *Dyes and Pigments* **2020,** *182*, 108579.

[126] H. Li, J. Li, D. Liu, T. Huang, D. Li, *Chem. Eur. J.* **2020,** *26* (30), 6899-6909.

[127] J. S. Price, N. C. Giebink, *SID Symposium Digest of Technical Papers* **2017,** *48* (1), 565-565.

[128] Y. Deng, C. Keum, S. Hillebrandt, C. Murawski, M. C. Gather, *Adv. Optical Mater.* **2020,** *9* (14), 2001642.

[129] D. Zhang, X. Song, A. J. Gillett, B. H. Drummond, S. T. E. Jones, G. Li, H. He, M. Cai, D. Credgington, L. Duan, *Adv. Mater.* **2020,** *32* (19), e1908355.

[130] C. Y. Chan, L. S. Cui, J. U. Kim, H. Nakanotani, C. Adachi, *Adv. Funct. Mater.* **2018,** *28* (11), 1706023.

[131] T. Kitazaki, T. Ichikawa, A. Tasaka, H. Hosono, Y. Matsushita, R. Hayashi, K. Okonogi, K. Itoh, *Chem. Pharm. Bull.* **2000,** *48*, 1935-1946.

[132] D. Zhang, X. Cao, Q. Wu, M. Zhang, N. Sun, X. Zhang, Y. Tao, *J. Mater. Chem. C* **2018,** *6* (14), 3675-3682.

[133] R. Huisgen, J. Sauer, H. J. Sturm, J. H. Markgraf, *Chemische Berichte* **2006,** *93* (9), 2106-2124.

[134] C. Leonhardt, *Unpublished B. Sc. Thesis: Synthese und Charakterisierung Akzeptor-modifizierter TADF-Emitter* Karlsruhe Institute of Technology **2017**.

[135] J. Liu, Z. Li, T. Hu, T. Gao, Y. Yi, P. Wang, Y. Wang, *Adv. Optical Mater.* **2022,** *10* (8), 2102558.

[136] E. Cho, M. Hong, Y. S. Yang, Y. J. Cho, V. Coropceanu, J.-L. Brédas, *J. Mater. Chem. C* **2022,** *10* (12), 4629-4636.

[137] A. Monkman, *ACS Appl. Mater. Interfaces* **2022,** *14* (18), 20463-20467.

[138] C.-Y. Chan, M. Tanaka, Y.-T. Lee, Y.-W. Wong, H. Nakanotani, T. Hatakeyama, C. Adachi, *Nature Photonics* **2021,** *15* (3), 203-207.

[139] L. E. de Sousa, L. dos Santos Born, P. H. de Oliveira Neto, P. de Silva, *J. Mater. Chem. C* **2022,** *10* (12), 4914-4922.

[140] G. Hong, X. Gan, C. Leonhardt, Z. Zhang, J. Seibert, J. M. Busch, S. Brase, *Adv. Mater.* **2021,** *33* (9), e2005630.

Appendix

Curriculum Vitae

Personal Information

Céline Leonhardt
18.06.1995 in Rastatt
Schützenstraße 40, 76137 Karlsruhe

Education

Jul 2020 – Jan 2023 **KARLSRUHE INSTITUTE OF TECHNOLOGY (KIT)**, *KARLSRUHE (GERMANY)*
Doctorate in organic chemistry in the research group of Prof. Dr. S. Bräse
- Member of KSOP – Karlsruhe School of Optics & Photonics

May 2018 – May 2020 **KARLSRUHE INSTITUTE OF TECHNOLOGY (KIT)**, *KARLSRUHE (GERMANY)*
MSc in Chemistry focused on organic chemistry in the Bräse Group
- Thesis: Synthetic Strategies towards TADF Emitters for OLED Applications with an advanced Acceptor Motive: Tris[1,2,4]-triazolo[1,3,5]-triazine (grade 1.0)
- Modules included: Advanced Physical Chemistry, Radiochemistry

Oct 2014 – May 2018 **KARLSRUHE INSTITUTE OF TECHNOLOGY (KIT)**, *KARLSRUHE (GERMANY)*
BSc in Chemistry focused on organic chemistry in the Bräse Group
- Thesis: Modular Modification of highly efficient TADF Emitters (grade 1.0)
- Modules included: Advanced Inorganic Chemistry, Advanced Organic Chemistry, Physical Chemistry, Experimental Physics, Applied Chemistry, Toxicology

Sep 2011 – Mar 2014 **HELENE-LANGE-SCHULE**, *MANNHEIM (GERMANY)*
- German Abitur: Chemistry, German, Mathematics, Biology

Sep 2005 – Jul 2011 **KARL-FRIEDRICH GYMNASIUM**, *MANNHEIM (GERMANY)*

International Experience

Oct 2018 – Feb 2019 **MONASH UNIVERSITY**, *MELBOURNE (AUSTRALIA)*
Visiting Researcher in the Glen B. Deacon Group
- C-F-activation through fluorine-metal interactions of organic-lanthanide complexes with acyl amides as a novel ligand class in lanthanide chemistry

Teaching Experience

May 2019– Aug 2019	Teaching assistant of organic chemistry lectures, INSTITUTE OF ORGANIC CHEMISTRY, KIT
May 2018– Aug 2018	Teaching assistant of organic chemistry lectures, INSTITUTE OF ORGANIC CHEMISTRY, KIT

Practical Experience

Jun 2022 – Sep 2022	Hosting an internship through the course of the DAAD Rise Germany program for a student from the University of Dublin (Ireland)
Sep 2021 – Sep 2022	Supervision and Instruction of Laboratory Chemist Trainee (second year mentorship)
Jul 2021– Sep 2021	Hosting an internship through the course of the DAAD Rise Germany program for a student from the University of Birmingham (United Kingdom)
Sep 2020 – Sep 2021	Supervision and Instruction of Laboratory Chemist Trainee (first year mentorship)
Jul 2018 – Oct 2018	Research Assistant in the Bräse Group – Development of efficient light emitting materials for application in OLED devices, INSTITUTE OF ORGANIC CHEMISTRY, KIT

Awards and Scholarships

Jun 2021 – Jun 2022	Scholarship of the Karlsruhe School of Optics and Photonics
Mar 2014	Graduate Award of the German Agricultural Society

Extracurricular Activities

Jul 2020 – Jan 2023	**SAFETY ADVISOR**, INSTITUTE OF ORGANIC CHEMISTRY, KIT
Jul 2020 – Jan 2023	**FIRST RESPONDER**, INSTITUTE OF ORGANIC CHEMISTRY, KIT

Skills

MBA Fundamentals	Project Management Operational Research Entrepreneurship Marketing
Language Skills	German (native) English (fluent) Latin (Latin proficiency certificate) French (basic)

IT Skills	MS Office, ChemOffice, MestReNova, Origin, Turbomole, Adobe Lightroom & Photoshop

List of Publications

Articles published and in preparation

1) C. Leonhardt, E. Crovini, M. Nieger, E. Zysman-Colman, S. Bräse, *in preparation*
Mono- and Ditriazolotriazine as a novel acceptor for blue Thermally Activated Delayed Fluorescence

2) C. Leonhardt, I. Garin Fernandez, A. Mauri, M. Kozlowska, W. Wenzel, U. Lemmer, S. Bräse, *in preparation*
Nitrile Derivatization of the 5TCzBN TADF System

3) G. Hong, X. Gan, C. Leonhardt, Z. Zhang, J. Seibert, J. M. Busch, S. Bräse, *Adv. Mater.* **2021**, 33, 2005630.
A Brief History of OLEDs – Emitter Development and Industry Milestones

4) Z. Guo, R. Huo, Y. Q. Tan, N. T. Flosbach, N. Wang, C. Leonhardt, A. Urbatsch, G. B. Deacon, P. C. Junk, E. I. Izgorodina, V. L. Blair, *Journal of Coordination Chemistry* **2021**, 74, 2947–2958.
A new twist on an old molecule: a rotameric isomer of bis(pentafluorophenyl)mercury

5) F. Hundemer, L. Graf von Reventlow, C. Leonhardt, M. Polamo, M. Nieger, S. M. Seifermann, A. Colsmann, S. Bräse, *Chemistry Open* **2019**, 8, 1413–1420.
Acceptor Derivatization of the 4CzIPN TADF System: Color Tuning and Introduction of Functional Groups

Conference Posters

1) C. Leonhardt, *A novel Acceptor Core for blue TADF Emitters*, MRS Spring Meeting **2022**, Honolulu, Hawaii.

2) C. Leonhardt, A. Jung, J. Seibert, L. Holzhauer, C. Adam, S. Sarwar, C. Bednarek, S. Bräse, *Design and Synthesis of TADF Emitters*, Kick-off Meeting, **2022**, London, United Kingdom.

3) C. Leonhardt, *Modular Modification of Organic TADF Emitters – Design, Synthesis and Application in OLEDs*, KDOP, **2021**, Karlsruhe, Germany.

4) C. Leonhardt, J. Moon, *C–F Activation in Rare-Earth Complexes*, raci Inorganic Symposium Day, **2018**, Melbourne, Australia.

Acknowledgements

First and foremost, I want to express my gratitude to my supervisor Prof. Dr. Stefan Bräse. Thank you for letting me work on this amazing topic, with all the freedom one researcher can imagine. Thank you for your help, guidance and listening through my time here. Thank you for the opportunity to attend a graduate school and conferences. All I've learned here shaped me and I will always be grateful that you gave me the opportunity to work in your group.

I want to acknowledge all my collaboration partners that worked with me through the course of this thesis. Thanks to Ettore Crovini from the University of St. Andrews for doing DFT calculations and photophysical measurements for project N°1. Furthermore, a big thank you to Anna Mauri and Mariana Kozlowska from KIT for conducting the calculations for project N°2. And at last, a very much thank you to Idoia Garin and Ian Howard from KIT for taking care of the photophysical measurements of project N°2. Without all your work and meetings this thesis would not be as successful.

I want to thank Dr. Andreas Rapp, Angelika Mösle, Lara Hirsch and Carolin Kohnle for the analytical service here at IOC, KIT. Thank you, Dr. Martin Nieger from the University of Helsinki and Dr. Olaf Fuhr from KIT for crystal structure analyses. Thank you, Richard von Budberg (and Kiwi) for doing all the special glassware.

Thank you, Janine Bolz and Christiane Lampert for being always helpful with bureaucracy and organizational tasks. Thanks also to Dr. Christin Bednarek, for having a sympathetic ear especially in tough times and for helping to get through.

Thanks for being part of the graduate school KSOP, who pushed me going beyond research to learn other skills and furthermore for their financial support especially taking care of travel costs to being able to attend conferences. I'm very grateful to was able to be part of the MRS Spring Meeting in Hawaii, 2022.

I'm so happy to have chosen the Bräse research group for my PhD. The work atmosphere is like nowhere else. Thanks for letting me be part of this awesome group and for all the support may it be intellectual or social.
Thank you so much for proofreading my work and giving helpful comments, Clara and Julian.

I'm very grateful for the friendships that were shaped during working here. I really do appreciate all of you. Lisa-Lou for being a sunshine and entertainingly clumsy. Clara for having her own funny words, and for our trip to the other end of the world. Hannes for being super funny and super kind. Thank you, Sarah for your kindness and for introducing me to your horse Nils.

Thank you, Simon, Lisa, Julian and Hannes for our weekly swim sessions.

A huge shout out to all my students Felix S., Lucy, Lisa, Felix B., Patrick, Milada, David, Clioana and Martina who worked so hard either through their theses, laboratory courses or apprenticeships. Especially, thank you so much Felix S. for being the most hard working and most enthusiastic of them all. And of course, because of the constant snack supply. At last, a big thanks to Lucy for attending the DAAD Rise Program and coming last minute to Germany. I'm very grateful for our friendship ever since.

Finally, I want to speak out my gratefulness for my parents. You supported me all my life unconditionally. I can't say how much that means to be. Thank you for always being the constant in my life. Thanks to my brother, Marius, for being the bright light everyone needs in their life. You spread happiness where you go.
And to save the best for last, Yvo. Thank you for being my partner in crime, travel partner, surf companion, shoulder to cry on, fun and crazy one. You are the funniest, most loving, and respectful person I know. Thank you for putting up with me and my madness. Having you in my life makes it complete.